Introduction to Instrumentation, Sensors, and Process Control

For a listing of related titles from *Artech House*,
turn to the back of this book

Introduction to Instrumentation, Sensors, and Process Control

William C. Dunn

ARTECH HOUSE
BOSTON | LONDON
artechhouse.com

Library of Congress Cataloging-in-Publication Data
Dunn, William C.
   Introduction to instrumentation, sensors, and process control/William C. Dunn.
      p. cm. —(Artech House Sensors library)

   ISBN 1-58053-011-7 (alk. paper)
   1. Process control. 2. Detectors. I. Title. II. Series.

TS156.8.D86 2005
670.42'7—dc22   2005050832

British Library Cataloguing in Publication Data
Dunn, William C.
   Introduction to instrumentation, sensors, and process control. —(Artech House sensors library)
   1. Engineering instruments 2. Electronic instruments 3. Process control
   I. Title
   681.2

   ISBN-10:   1-58053-011-7

Cover design by Cameron Inc.

© 2006 ARTECH HOUSE, INC.
685 Canton Street
Norwood, MA 02062

All rights reserved. Printed and bound in the United States of America. No part of this book may be reproduced or utilized in any form or by any means, electronic or mechanical, including photocopying, recording, or by any information storage and retrieval system, without permission in writing from the publisher.
   All terms mentioned in this book that are known to be trademarks or service marks have been appropriately capitalized. Artech House cannot attest to the accuracy of this information. Use of a term in this book should not be regarded as affecting the validity of any trademark or service mark.

International Standard Book Number: 1-58053-011-7

10 9 8 7 6 5 4 3 2 1

Contents

Preface	xv
Acknowledgment	xvi

CHAPTER 1
Introduction to Process Control — 1

1.1	Introduction	1
1.2	Process Control	1
	1.2.1 Sequential Process Control	2
	1.2.2 Continuous Process Control	2
1.3	Definition of the Elements in a Control Loop	4
1.4	Instrumentation and Sensors	5
	1.4.1 Instrument Parameters	5
1.5	Control System Evaluation	9
	1.5.1 Stability	9
	1.5.2 Regulation	9
	1.5.3 Transient Response	9
1.6	Analog and Digital Data	10
	1.6.1 Analog Data	10
	1.6.2 Digital Data	10
	1.6.3 Pneumatic Data	10
	1.6.4 Smart Sensors	11
1.7	Process Facility Considerations	11
1.8	Summary	12
	Definitions	12
	References	14

CHAPTER 2
Units and Standards — 15

2.1	Introduction	15
	2.1.1 Units and Standards	15
2.2	Basic Units	16
2.3	Units Derived from Base Units	16
	2.3.1 Units Common to Both the English and SI Systems	16
	2.3.2 English Units Derived from Base Units	16
	2.3.3 SI Units Derived from Base Units	18
	2.3.4 Conversion Between English and SI Units	18

		2.3.5	Metric Units not Normally Used in the SI System	20
2.4	Standard Prefixes			21
2.5	Standards			22
	2.5.1	Physical Constants		22
	2.5.2	Standards Institutions		22
2.6	Summary			23
	References			23

CHAPTER 3

Basic Electrical Components   25

3.1	Introduction		25
3.2	Circuits with R, L, and C		25
	3.2.1	Voltage Step Input	25
	3.2.2	Time Constants	27
	3.2.3	Sine Wave Inputs	28
3.3	RC Filters		32
3.4	Bridge Circuits		34
	3.4.1	Voltage Dividers	34
	3.4.2	dc Bridge Circuits	34
	3.4.3	ac Bridge Circuits	38
3.5	Summary		39
	References		40

CHAPTER 4

Analog Electronics   41

4.1	Introduction		41
4.2	Analog Circuits		41
	4.2.1	Operational Amplifier Introduction	41
	4.2.2	Basic Op-Amp	42
	4.2.3	Op-Amp Characteristics	42
4.3	Types of Amplifiers		45
	4.3.1	Voltage Amplifiers	45
	4.3.2	Converters	50
	4.3.3	Current Amplifiers	52
	4.3.4	Integrating and Differentiating Amplifiers	53
	4.3.5	Nonlinear Amplifiers	54
	4.3.6	Instrument Amplifiers	55
	4.3.7	Input Protection	57
4.4	Amplifier Applications		57
4.5	Summary		58
	References		58

CHAPTER 5

Digital Electronics   59

5.1	Introduction	59
5.2	Digital Building Blocks	59
5.3	Converters	61

		5.3.1 Comparators	62
		5.3.2 Digital to Analog Converters	64
		5.3.3 Analog to Digital Converters	68
		5.3.4 Sample and Hold	72
		5.3.5 Voltage to Frequency Converters	72
5.4	Data Acquisition Devices		74
		5.4.1 Analog Multiplexers	74
		5.4.2 Digital Multiplexers	74
		5.4.3 Programmable Logic Arrays	75
		5.4.4 Other Interface Devices	75
5.5	Basic Processor		75
5.6	Summary		76
	References		77

CHAPTER 6
Microelectromechanical Devices and Smart Sensors — 79

6.1	Introduction	79
6.2	Basic Sensors	80
	6.2.1 Temperature Sensing	80
	6.2.2 Light Intensity	80
	6.2.3 Strain Gauges	81
	6.2.4 Magnetic Field Sensors	82
6.3	Piezoelectric Devices	84
	6.3.1 Time Measurements	86
	6.3.2 Piezoelectric Sensors	87
	6.3.3 PZT Actuators	88
6.4	Microelectromechanical Devices	88
	6.4.1 Bulk Micromachining	89
	6.4.2 Surface Micromachining	91
6.5	Smart Sensors Introduction	94
	6.5.1 Distributed System	95
	6.5.2 Smart Sensors	96
6.6	Summary	96
	References	97

CHAPTER 7
Pressure — 99

7.1	Introduction	99
7.2	Pressure Measurement	99
	7.2.1 Hydrostatic Pressure	99
	7.2.2 Specific Gravity	100
	7.2.3 Units of Measurement	101
	7.2.4 Buoyancy	103
7.3	Measuring Instruments	105
	7.3.1 Manometers	105
	7.3.2 Diaphragms, Capsules, and Bellows	106
	7.3.3 Bourdon Tubes	108

	7.3.4 Other Pressure Sensors	109
	7.3.5 Vacuum Instruments	110
7.4	Application Considerations	111
	7.4.1 Selection	111
	7.4.2 Installation	112
	7.4.3 Calibration	112
7.5	Summary	113
	Definitions	113
	References	114

CHAPTER 8
Level — 115

8.1	Introduction	115
8.2	Level Measurement	115
	8.2.1 Direct Level Sensing	115
	8.2.2 Indirect Level Sensing	118
	8.2.3 Single Point Sensing	124
	8.2.4 Level Sensing of Free-Flowing Solids	125
8.3	Application Considerations	126
8.4	Summary	128
	References	128

CHAPTER 9
Flow — 129

9.1	Introduction	129
9.2	Fluid Flow	129
	9.2.1 Flow Patterns	129
	9.2.2 Continuity Equation	131
	9.2.3 Bernoulli Equation	132
	9.2.4 Flow Losses	134
9.3	Flow Measuring Instruments	136
	9.3.1 Flow Rate	136
	9.3.2 Total Flow	142
	9.3.3 Mass Flow	144
	9.3.4 Dry Particulate Flow Rate	144
	9.3.5 Open Channel Flow	145
9.4	Application Considerations	145
	9.4.1 Selection	145
	9.4.2 Installation	147
	9.4.3 Calibration	147
9.5	Summary	147
	Definitions	148
	References	148

CHAPTER 10
Temperature and Heat — 149

- 10.1 Introduction — 149
- 10.2 Temperature and Heat — 149
  - 10.2.1 Temperature Units — 149
  - 10.2.2 Heat Energy — 151
  - 10.2.3 Heat Transfer — 153
  - 10.2.4 Thermal Expansion — 155
- 10.3 Temperature Measuring Devices — 157
  - 10.3.1 Expansion Thermometers — 157
  - 10.3.2 Resistance Temperature Devices — 160
  - 10.3.3 Thermistors — 161
  - 10.3.4 Thermocouples — 162
  - 10.3.5 Pyrometers — 164
  - 10.3.6 Semiconductor Devices — 165
- 10.4 Application Considerations — 166
  - 10.4.1 Selection — 166
  - 10.4.2 Range and Accuracy — 166
  - 10.4.3 Thermal Time Constant — 167
  - 10.4.4 Installation — 168
  - 10.4.5 Calibration — 168
  - 10.4.6 Protection — 168
- 10.5 Summary — 169
- Definitions — 169
- References — 170

CHAPTER 11
Position, Force, and Light — 171

- 11.1 Introduction — 171
- 11.2 Position and Motion Sensing — 171
  - 11.2.1 Position and Motion Measuring Devices — 171
  - 11.2.2 Position Application Considerations — 176
- 11.3 Force, Torque, and Load Cells — 177
  - 11.3.1 Force and Torque Introduction — 178
  - 11.3.2 Stress and Strain — 178
  - 11.3.3 Force and Torque Measuring Devices — 181
  - 11.3.4 Strain Gauge Sensors — 183
  - 11.3.5 Force and Torque Application Considerations — 186
- 11.4 Light — 186
  - 11.4.1 Light Introduction — 186
  - 11.4.2 EM Radiation — 186
  - 11.4.3 Light Measuring Devices — 188
  - 11.4.4 Light Sources — 188
  - 11.4.5 Light Application Considerations — 189
- 11.5 Summary — 190
- Definitions — 190
- References — 191

CHAPTER 12

Humidity and Other Sensors — 193

- 12.1 Humidity — 193
  - 12.1.1 Humidity Introduction — 193
  - 12.1.2 Humidity Measuring Devices — 194
  - 12.1.3 Humidity Application Considerations — 197
- 12.2 Density and Specific Gravity — 198
  - 12.2.1 Density and Specific Gravity Introduction — 198
  - 12.2.2 Density Measuring Devices — 199
  - 12.2.3 Density Application Considerations — 202
- 12.3 Viscosity — 202
  - 12.3.1 Viscosity Introduction — 202
  - 12.3.2 Viscosity Measuring Instruments — 203
- 12.4 Sound — 204
  - 12.4.1 Sound Measurements — 204
  - 12.4.2 Sound Measuring Devices — 205
  - 12.4.3 Sound Application Considerations — 206
- 12.5 pH Measurements — 206
  - 12.5.1 pH Introduction — 206
  - 12.5.2 pH Measuring Devices — 207
  - 12.5.3 pH Application Considerations — 207
- 12.6 Smoke and Chemical Sensors — 208
  - 12.6.1 Smoke and Chemical Measuring Devices — 208
  - 12.6.2 Smoke and Chemical Application Consideration — 208
- 12.7 Summary — 209
- Definitions — 209
- References — 210

CHAPTER 13

Regulators, Valves, and Motors — 211

- 13.1 Introduction — 211
- 13.2 Pressure Controllers — 211
  - 13.2.1 Pressure Regulators — 211
  - 13.2.2 Safety Valves — 213
  - 13.2.3 Level Regulators — 214
- 13.3 Flow Control Valves — 215
  - 13.3.1 Globe Valve — 215
  - 13.3.2 Butterfly Valve — 217
  - 13.3.3 Other Valve Types — 218
  - 13.3.4 Valve Characteristics — 219
  - 13.3.5 Valve Fail Safe — 219
  - 13.3.6 Actuators — 220
- 13.4 Power Control — 221
  - 13.4.1 Electronic Devices — 222
  - 13.4.2 Magnetic Control Devices — 227
- 13.5 Motors — 227
  - 13.5.1 Servo Motors — 228

		13.5.2	Stepper Motors	228
		13.5.3	Synchronous Motors	229
	13.6	Application Considerations		230
		13.6.1	Valves	230
		13.6.2	Power Devices	231
	13.7	Summary		231
		References		232

CHAPTER 14
Programmable Logic Controllers — 233

	14.1	Introduction		233
	14.2	Programmable Controller System		233
	14.3	Controller Operation		235
	14.4	Input/Output Modules		236
		14.4.1	Discrete Input Modules	236
		14.4.2	Analog Input Modules	238
		14.4.3	Special Function Input Modules	238
		14.4.4	Discrete Output Modules	239
		14.4.5	Analog Output Modules	240
		14.4.6	Smart Input/Output Modules	240
	14.5	Ladder Diagrams		243
		14.5.1	Switch Symbols	243
		14.5.2	Relay and Timing Symbols	244
		14.5.3	Output Device Symbols	244
		14.5.4	Ladder Logic	245
		14.5.5	Ladder Gate Equivalent	245
		14.5.6	Ladder Diagram Example	246
	14.6	Summary		249
		References		249

CHAPTER 15
Signal Conditioning and Transmission — 251

	15.1	Introduction		251
	15.2	General Sensor Conditioning		251
		15.2.1	Conditioning for Offset and Span	252
		15.2.2	Linearization in Analog Circuits	253
		15.2.3	Temperature Correction	253
		15.2.4	Noise and Correction Time	255
	15.3	Conditioning Considerations for Specific Types of Devices		255
		15.3.1	Direct Reading Sensors	255
		15.3.2	Capacitive Sensors	255
		15.3.3	Magnetic Sensors	256
		15.3.4	Resistance Temperature Devices	257
		15.3.5	Thermocouple Sensors	259
		15.3.6	LVDTs	259
		15.3.7	Semiconductor Devices	260
	15.4	Digital Conditioning		260

		15.4.1 Conditioning in Digital Circuits	260
15.5	Pneumatic Transmission		261
	15.5.1	Signal Conversion	261
15.6	Analog Transmission		262
	15.6.1	Noise Considerations	262
	15.6.2	Voltage Signals	262
	15.6.3	Current Signals	264
15.7	Digital Transmission		264
	15.7.1	Transmission Standards	264
	15.7.2	Foundation Fieldbus and Profibus	265
15.8	Wireless Transmission		267
	15.8.1	Short Range Protocols	267
	15.8.2	Telemetry Introduction	267
	15.8.3	Width Modulation	268
	15.8.4	Frequency Modulation	268
15.9	Summary		269
	Definitions		269
	References		270

CHAPTER 16
Process Control — 271

16.1	Introduction		271
16.2	Sequential Control		271
16.3	Discontinuous Control		273
	16.3.1	Discontinuous On/Off Action	273
	16.3.2	Differential Closed Loop Action	273
	16.3.3	On/Off Action Controller	274
	16.3.4	Electronic On/Off Controller	275
16.4	Continuous Control		275
	16.4.1	Proportional Action	276
	16.4.2	Derivative Action	278
	16.4.3	Integral Action	280
	16.4.4	PID Action	281
	16.4.5	Stability	284
16.5	Process Control Tuning		285
	16.5.1	Automatic Tuning	286
	16.5.2	Manual Tuning	286
16.6	Implementation of Control Loops		287
	16.6.1	On/Off Action Pneumatic Controller	287
	16.6.2	Pneumatic Linear Controller	288
	16.6.3	Pneumatic Proportional Mode Controller	289
	16.6.4	PID Action Pneumatic Controller	289
	16.6.5	PID Action Control Circuits	290
	16.6.6	PID Electronic Controller	293
16.7	Summary		294
	Definitions		295
	References		296

CHAPTER 17

Documentation and P&ID — 297

- 17.1 Introduction — 297
- 17.2 Alarm and Trip Systems — 297
    - 17.2.1 Safety Instrumented Systems — 297
    - 17.2.2 Safe Failure of Alarm and Trip — 298
    - 17.2.3 Alarm and Trip Documentation — 299
- 17.3 PLC Documentation — 300
- 17.4 Pipe and Instrumentation Symbols — 300
    - 17.4.1 Interconnect Symbols — 301
    - 17.4.2 Instrument Symbols — 302
    - 17.4.3 Functional Identification — 302
    - 17.4.4 Functional Symbols — 304
- 17.5 P&ID Drawings — 308
- 17.6 Summary — 309
- References — 311

Glossary — 313

About the Author — 321

Index — 323

Preface

Industrial process control was originally performed manually by operators using their senses of sight and feel, making the control totally operator-dependent. Industrial process control has gone through several revolutions and has evolved into the complex modern-day microprocessor-controlled system. Today's technology revolution has made it possible to measure parameters deemed impossible to measure only a few years ago, and has made improvements in accuracy, control, and waste reduction.

This reference manual was written to provide the reader with a clear, concise, and up-to-date text for understanding today's sensor technology, instrumentation, and process control. It gives the details in a logical order for everyday use, making every effort to provide only the essential facts. The book is directed towards industrial control engineers, specialists in physical parameter measurement and control, and technical personnel, such as project managers, process engineers, electronic engineers, and mechanical engineers. If more specific and detailed information is required, it can be obtained from vendor specifications, application notes, and references given at the end of each chapter.

A wide range of technologies and sciences are used in instrumentation and process control, and all manufacturing sequences use industrial control and instrumentation. This reference manual is designed to cover the aspects of industrial instrumentation, sensors, and process control for the manufacturing of a cost-effective, high quality, and uniform end product.

Chapter 1 provides an introduction to industrial instrumentation, and Chapter 2 introduces units and standards covering both English and SI units. Electronics and microelectromechanical systems (MEMS) are extensively used in sensors and process control, and are covered in Chapters 3 through 6. The various types of sensors used in the measurement of a wide variety of physical variables, such as level, pressure, flow, temperature, humidity, and mechanical measurements, are discussed in Chapters 7 through 12. Regulators and actuators, which are used for controlling pressure, flow, and other input variables to a process, are discussed in Chapter 13. Industrial processing is computer controlled, and Chapter 14 introduces the programmable logic controller. Sensors are temperature-sensitive and nonlinear, and have to be conditioned. These sensors, along with signal transmission, are discussed in Chapter 15. Chapter 16 discusses different types of process control action, and the use of pneumatic and electronic controllers for sensor signal amplification and control. Finally, Chapter 17 introduces documentation as applied to instrumentation and control, together with standard symbols recommended by the Instrument Society of America for use in instrumentation control diagrams.

Every effort has been made to ensure that the text is accurate, easily readable, and understandable.

Both engineering and scientific units are discussed in the text. Each chapter contains examples for clarification, definitions, and references. A glossary is given at the end of the text.

Acknowledgment

I would like to thank my wife Nadine for her patience, understanding, and many helpful suggestions during the writing of this text.

CHAPTER 1
Introduction to Process Control

1.1 Introduction

The technology of controlling a series of events to transform a material into a desired end product is called process control. For instance, the making of fire could be considered a primitive form of process control. Industrial process control was originally performed manually by operators. Their sensors were their sense of sight, feel, and sound, making the process totally operator-dependent. To maintain a process within broadly set limits, the operator would adjust a simple control device. Instrumentation and control slowly evolved over the years, as industry found a need for better, more accurate, and more consistent measurements for tighter process control. The first real push to develop new instruments and control systems came with the Industrial Revolution, and World Wars I and II added further to the impetus of process control. Feedback control first appeared in 1774 with the development of the fly-ball governor for steam engine control, and the concept of proportional, derivative, and integral control during World War I. World War II saw the start of the revolution in the electronics industry, which has just about revolutionized everything else. Industrial process control is now highly refined with computerized controls, automation, and accurate semiconductor sensors [1].

1.2 Process Control

Process control can take two forms: (1) sequential control, which is an event-based process in which one event follows another until a process sequence is complete; or (2) continuous control, which requires continuous monitoring and adjustment of the process variables. However, continuous process control comes in many forms, such as domestic water heaters and heating, ventilation, and air conditioning (HVAC), where the variable temperature is not required to be measured with great precision, and complex industrial process control applications, such as in the petroleum or chemical industry, where many variables have to be measured simultaneously with great precision. These variables can vary from temperature, flow, level, and pressure, to time and distance, all of which can be interdependent variables in a single process requiring complex microprocessor systems for total control. Due to the rapid advances in technology, instruments in use today may be obsolete tomorrow. New and more efficient measurement techniques are constantly being introduced. These changes are being driven by the need for higher accuracy, quality, precision, and performance. Techniques that were thought to be impossible a few years ago have been developed to measure parameters.

1.2.1 Sequential Process Control

Control systems can be sequential in nature, or can use continuous measurement; both systems normally use a form of feedback for control. Sequential control is an event-based process, in which the completion of one event follows the completion of another, until a process is complete, as by the sensing devices. Figure 1.1 shows an example of a process using a sequencer for mixing liquids in a set ratio [2]. The sequence of events is as follows:

1. Open valve A to fill tank A.
2. When tank A is full, a feedback signal from the level sensor tells the sequencer to turn valve A Off.
3. Open valve B to fill tank B.
4. When tank B is full, a feedback signal from the level sensor tells the sequencer to turn valve B Off.
5. When valves A and B are closed, valves C and D are opened to let measured quantities of liquids A and B into mixing tank C.
6. When tanks A and B are empty, valves C and D are turned Off.
7. After C and D are closed, start mixing motor, run for set period.
8. Turn Off mixing motor.
9. Open valve F to use mixture.
10. The sequence can then be repeated after tank C is empty and Valve F is turned Off.

1.2.2 Continuous Process Control

Continuous process control falls into two categories: (1) elementary On/Off action, and (2) continuous control action.

On/Off action is used in applications where the system has high inertia, which prevents the system from rapid cycling. This type of control only has only two states, On and Off; hence, its name. This type of control has been in use for many decades,

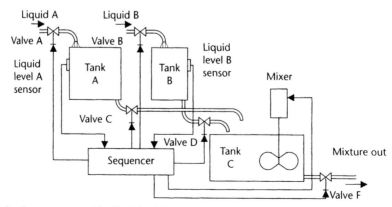

Figure 1.1 Sequencer used for liquid mixing.

long before the introduction of the computer. HVAC is a prime example of this type of application. Such applications do not require accurate instrumentation. In HVAC, the temperature (measured variable) is continuously monitored, typically using a bimetallic strip in older systems and semiconductor elements in newer systems, as the sensor turns the power (manipulated variable) On and Off at preset temperature levels to the heating/cooling section.

Continuous process action is used to continuously control a physical output parameter of a material. The parameter is measured with the instrumentation or sensor, and compared to a set value. Any deviation between the two causes an error signal to be generated, which is used to adjust an input parameter to the process to correct for the output change. An example of an unsophisticated automated control process is shown in Figure 1.2. A float in a swimming pool is used to continuously monitor the level of the water, and to bring the water level up to a set reference point when the water level is low. The float senses the level, and feedback to the control valve is via the float arm and pivot. The valve then controls the flow of water (manipulated variable) into the swimming pool, as the float moves up and down.

A more complex continuous process control system is shown in Figure 1.3, where a mixture of two liquids is required. The flow rate of liquid A is measured with a differential pressure (DP) sensor, and the amplitude of the signal from the DP measuring the flow rate of the liquid is used by the controller as a reference signal (set point) to control the flow rate of liquid B. The controller uses a DP to measure the flow rate of liquid B, and compares its amplitude to the signal from the DP monitoring the flow of liquid A. The difference between the two signals (error signal) is used to control the valve, so that the flow rate of liquid B (manipulated variable) is directly proportional to that of liquid A, and then the two liquids are combined [3].

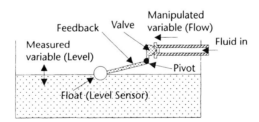

Figure 1.2 Automated control system.

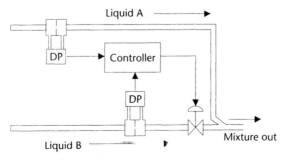

Figure 1.3 Continuous control for liquid mixing.

1.3 Definition of the Elements in a Control Loop

In any process, there are a number of inputs (i.e., from chemicals to solid goods). These are manipulated in the process, and a new chemical or component emerges at the output. To get a more comprehensive look at a typical process control system, it will be broken down into its various elements. Figure 1.4 is a block diagram of the elements in a continuous control process with a feedback loop.

Process is a sequence of events designed to control the flow of materials through a number of steps in a plant to produce a final utilitarian product or material. The process can be a simple process with few steps, or a complex sequence of events with a large number of interrelated variables. The examples shown are single steps that may occur in a process.

Measurement is the determination of the physical amplitude of a parameter of a material; the measurement value must be consistent and repeatable. Sensors are typically used for the measurement of physical parameters. A sensor is a device that can convert the physical parameter repeatedly and reliably into a form that can be used or understood. Examples include converting temperature, pressure, force, or flow into an electrical signal, measurable motion, or a gauge reading. In Figure 1.3, the sensor for measuring flow rates is a DP cell.

Error Detection is the determination of the difference between the amplitude of the measured variable and a desired set reference point. Any difference between the two is an error signal, which is amplified and conditioned to drive a control element. The controller sometimes performs the detection, while the reference point is normally stored in the memory of the controller.

Controller is a microprocessor-based system that can determine the next step to be taken in a sequential process, or evaluate the error signal in continuous process control to determine what action is to be taken. The controller can normally condition the signal, such as correcting the signal for temperature effects or nonlinearity in the sensor. The controller also has the parameters of the process input control element, and conditions the error sign to drive the final element. The controller can monitor several input signals that are sometimes interrelated, and can drive several control elements simultaneously. The controllers are normally referred to as programmable logic controllers (PLC). These devices use ladder networks for programming the control functions.

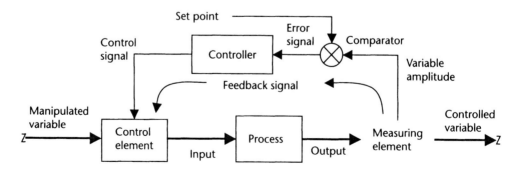

Figure 1.4 Block diagram of the elements that make up the feedback path in a process control loop.

Control Element is the device that controls the incoming material to the process (e.g., the valve in Figure 1.3). The element is typically a flow control element, and can have an On/Off characteristic or can provide liner control with drive. The control element is used to adjust the input to the process, bringing the output variable to the value of the set point.

The control and measuring elements in the diagram in Figure 1.4 are oversimplified, and are broken down in Figure 1.5. The measuring element consists of a sensor to measure the physical property of a variable, a transducer to convert the sensor signal into an electrical signal, and a transmitter to amplify the electrical signal, so that it can be transmitted without loss. The control element has an actuator, which changes the electrical signal from the controller into a signal to operate the valve, and a control valve. In the feedback loop, the controller has memory and a summing circuit to compare the set point to the sensed signal, so that it can generate an error signal. The controller then uses the error signal to generate a correction signal to control the valve via the actuator and the input variable. The function and operation of the blocks in different types of applications will be discussed in a later chapter. The definitions of the terms used are given at the end of the chapter.

1.4 Instrumentation and Sensors

The operator's control function has been replaced by instruments and sensors that give very accurate measurements and indications, making the control function totally operator-independent. The processes can be fully automated. Instrumentation and sensors are an integral part of process control, and the quality of process control is only as good as its measurement system. The subtle difference between an instrument and a sensor is that an instrument is a device that measures and displays the magnitude of a physical variable, whereas a sensor is a device that measures the amplitude of a physical variable, but does not give a direct indication of the value. The same physical parameters normally can be applied to both devices [4].

1.4.1 Instrument Parameters

The choice of a measurement device is difficult without a good understanding of the process. All of the possible devices should be carefully considered. It is also important to understand instrument terminology. ANSI/ISA-51.1-R1979 (R1993)

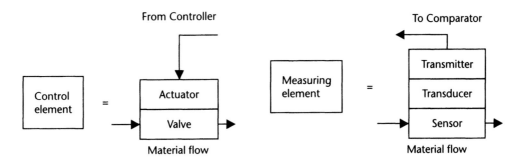

Figure 1.5 Breakdown of measuring and control elements.

Process Instrumentation Terminology gives the definitions of the terms used in instrumentation in the process control sector. Some of the more common terms are discussed below.

Accuracy of an instrument or device is the error or the difference between the indicated value and the actual value. Accuracy is determined by comparing an indicated reading to that of a known standard. Standards can be calibrated devices, and may be obtained from the National Institute of Standards and Technology (NIST). The NIST is a government agency that is responsible for setting and maintaining standards, and developing new standards as new technology requires it. Accuracy depends on linearity, hysteresis, offset, drift, and sensitivity. The resulting discrepancy is stated as a plus-or-minus deviation from true, and is normally specified as a percentage of reading, span, or of full-scale reading or deflection (% FSD), and can be expressed as an absolute value. In a system where more than one deviation is involved, the total accuracy of the system is statistically the root mean square (rms) of the accuracy of each element.

Example 1.1

A pressure sensor has a span of 25 to 150 psi. Specify the error when measuring 107 psi, if the accuracy of the gauge is (a) ±1.5% of span, (b) ±2% FSD, and (c) ±1.3% of reading.

a. Error = ±0.015 (150 − 25) psi = ±1.88 psi.
b. Error = ±0.02 × 150 psi = ±3 psi.
c. Error = ± 0.013 × 103 psi = ±1.34 psi.

Example 1.2

A pressure sensor has an accuracy of ±2.2% of reading, and a transfer function of 27 mV/kPa. If the output of the sensor is 231 mV, then what is the range of pressures that could give this reading?

The pressure range = 231/27 kPa ± 2.2% = 8.5 kPa ± 2.2% = 8.313 to 8.687 kPa

Example 1.3

In a temperature measuring system, the transfer function is 3.2 mV/k ± 2.1%, and the accuracy of the transmitter is ±1.7%. What is the system accuracy?

System accuracy = $\pm[(0.021)^2 + (0.017)^2]^{1/2}$ = ±2.7%

Linearity is a measure of the proportionality between the actual value of a variable being measured and the output of the instrument over its operating range. The deviation from true for an instrument may be caused by one or several of the above factors affecting accuracy, and can determine the choice of instrument for a particular application. Figure 1.6 shows a linearity curve for a flow sensor, which is the output from the sensor versus the actual flow rate. The curve is compared to a best-fit straight line. The deviation from the ideal is 4 cm/min., which gives a linearity of ±4% of FSD.

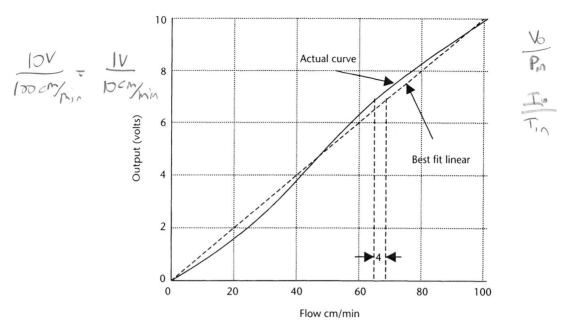

Figure 1.6 Linearity curve or a comparison of the sensor output versus flow rate, and the best-fit straight line.

Sensitivity is a measure of the change in the output of an instrument for a change in the measured variable, and is known as a transfer function. For example, when the output of a flow transducer changes by 4.7 mV for a change in flow of 1.3 cm/s, the sensitivity is 3.6 mV/cm/s. High sensitivity in an instrument is desired, since this gives a higher output, but has to be weighed against linearity, range, and accuracy.

Reproducibility is the inability of an instrument to consistently reproduce the same reading of a fixed value over time under identical conditions, creating an uncertainty in the reading.

Resolution is the smallest change in a variable to which the instrument will respond. A good example is in digital instruments, where the resolution is the value of the least significant bit.

Example 1.4

A digital meter has 10-bit accuracy. What is the resolution on the 16V range?

Decade equivalent of 10 bits = 2^{10} = 1,024

Resolution = 16/1,024 = 0.0156V = 15.6 mV

Hysteresis is the difference in readings obtained when an instrument approaches a signal from opposite directions. For example, if an instrument reads a midscale value beginning at zero, it can give a different reading than if it read the value after making a full-scale reading. This is due to stresses induced into the material of the instrument by changing its shape in going from zero to full-scale deflection. A hysteresis curve for a flow sensor is shown in Figure 1.7, where the output

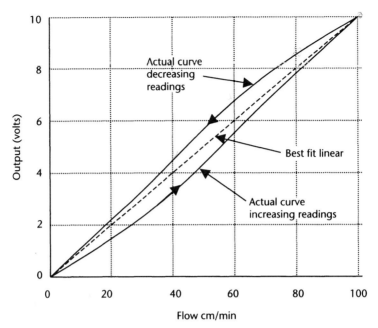

Figure 1.7 Hysteresis curve showing the difference in readings when starting from zero, and when starting from full scale.

initiating from a zero reading and initiating from a maximum reading are different. For instance, the output from zero for a 50 cm/min is 4.2V, compared to 5.6V when reading the same flow rate after a maximum reading.

Time constant of a sensor to a sudden change in a measured parameter falls into two categories, termed first-order and second-order responses. The first-order response is the time the sensor takes to reach its final output after a transient change. For example, a temperature measuring device will not change immediately following a change in temperature, due to the thermal mass of the sensor and the thermal conductivity of the interface between the hot medium and the sensing element. The response time to a step change in temperature is an exponential given by:

$$A(t) = A_0 + (A_f - A_0)(1 - e^{-t/\tau}) \tag{1.1}$$

where A(t) is the amplitude at time t, A_0 is the initial amplitude, A_f is the final amplitude, and τ is the time constant of the sensor.

The second-order response occurs when the effect of a transient on the monitoring unit is to cause oscillations in the output signal before settling down. The response can be described by a second-order equation.

Other parameters used in instrumentation are Range, Span, Precision, Offset, Drift, and Repeatability. The definitions of these parameters are given at the end of the chapter.

Example 1.5

A linear pressure sensor has a time constant of 3.1 seconds, and a transfer function of 29 mV/kPa. What is the output after 1.3 seconds, if the pressure changes from 17 to 39 kPa? What is the pressure error at this time?

Initial output voltage $A_0 = 17 \times 29$ mV $= 493$ mV

Final output voltage $A_f = 29 \times 39$ mV $= 1{,}131$ mV

$$A(1.3) = 493 + (1131 - 493)(1 - e^{-1.3/3.1})$$

$$A(1.3) = 493 + 638 \times 0.66 = 914.1 \text{ mV}$$

Pressure after 1.3 sec $= 914.1/29$ kPa $= 31.52$ kPa

Error $= 39 - 31.52 = 7.48$ kPa

1.5 Control System Evaluation

A general criterion for evaluating the performance of a process control system is difficult to establish. In order to obtain the quality of the performance of the controller, the following have to be answered:

1. Is the system stable?
2. How good is the steady state regulation?
3. How good is the transient regulation?
4. What is the error between the set point and the variable?

1.5.1 Stability

In a system that uses feedback, there is always the potential for stability. This is due to delays in the system and feedback loop, which causes the correction signal to be in-phase with the error signal change instead of out-of-phase. The error and correction signal then become additive, causing instability. This problem is normally corrected by careful tuning of the system and damping, but this unfortunately comes at the expense of a reduction in the response time of the system.

1.5.2 Regulation

The regulation of a variable is the deviation of the variable from the set point or the error signal. The regulation should be as tight as possible, and is expressed as a percentage of the set point. A small error is always present, since this is the signal that is amplified to drive the actuator to control the input variable, and hence controls the measured variable. The smaller the error, the higher the systems gain, which normally leads to system instability. As an example, the set point may be 120 psi, but the regulation may be 120 ± 2.5 psi, allowing the pressure to vary from 117.5 to 122.5 psi.

1.5.3 Transient Response

The transient response is the system's reaction time to a sudden change in a parameter, such as a sudden increase in material demand, causing a change in the measured variable or in the set point. The reaction can be specified as a dampened response or as a limited degree of overshoot of the measured variable, depending on the process, in order to return the measured variable to the set point in a specified time. The topic is covered in more detail in Chapter 16.

1.6 Analog and Digital Data

Variables are analog in nature, and before digital processing evolved, sensor signals were processed using analog circuits and techniques, which still exist in many processing facilities. Most modern systems now use digital techniques for signal processing [5].

1.6.1 Analog Data

Signal amplitudes are represented by voltage or current amplitudes in analog systems. Analog processing means that the data, such as signal linearization, from the sensor is conditioned, and corrections that are made for temperature variations are all performed using analog circuits. Analog processing also controls the actuators and feedback loops. The most common current transmission range is 4 to 20 mA, where 0 mA is a fault indication.

Example 1.6

The pressure in a system has a range from 0 to 75 kPa. What is the current equivalent of 27 kPa, if the transducer output range is from 4 to 20 mA?

Equivalent range of 75 kPa = 16 mA

Hence, 27 kPa = (4 + 16 × 27/75) mA = 9.76 mA

1.6.2 Digital Data

Signal amplitudes are represented by binary numbers in digital systems. Since variables are analog in nature, and the output from the sensor needs to be in a digital format, an analog to digital converter (ADC) must be used, or the sensor's output must be directly converted into a digital signal using switching techniques. Once digitized, the signal will be processed using digital techniques, which have many advantages over analog techniques, and few, if any, disadvantages. Some of the advantages of digital signals are: data storage, transmission of signals without loss of integrity, reduced power requirements, storage of set points, control of multiple variables, and the flexibility and ease of program changes. The output of a digital system may have to be converted back into an analog format for actuator control, using either a digital to analog converter (DAC) or width modulation techniques.

1.6.3 Pneumatic Data

Pressure was used for data transmission before the use of electrical signals, and is still used in conditions where high electrical noise could affect electrical signals, or in hazardous conditions where an electrical spark could cause an explosion or fire hazard. The most common range for pneumatic data transmission is 3 to 15 psi (20 to 100 kPa in SI units), where 0 psi is a fault condition.

1.6.4 Smart Sensors

The digital revolution also has brought about large changes in the methodology used in process control. The ability to cost-effectively integrate all the controller functions, along with ADCs and DACs, have produced a family of Smart Sensors that combine the sensor and control function into a single housing. This device reduces the load on the central processor and communicates to the central processor via a single serial bus (Fieldbus), reducing facility wiring requirements and making the concept of plug-and-play a reality when adding new sensors.

1.7 Process Facility Considerations

The process facility has a number of basic requirements, including well-regulated and reliable electrical, water, and air supplies, and safety precautions.

An electrical supply is required for all control systems, and must meet all standards in force at the plant. The integrity of the electrical supply is most important. Many facilities have backup systems to provide an uninterruptible power supply (UPS) to take over in case of the loss of external power. Power failure can mean plant shutdown and the loss of complete production runs. Isolating transformer should be used in the power supply lines to prevent electromagnetic interference (EMI) generated by devices, such as motors, from traveling through the power lines and affecting sensitive electronic control instruments.

Grounding is a very important consideration in a facility for safety reasons. Any variations in the ground potential between electronic equipment can cause large errors in signal levels. Each piece of equipment should be connected to a heavy copper bus that is properly grounded. Ground loops also should be avoided by grounding cable screens and signal return lines at only one end. In some cases, it may be necessary to use signal isolators to alleviate grounding problems in electronic devices and equipment.

An air supply is required to drive pneumatic actuators in most facilities. Instrument air in pneumatic equipment must meet quality standards. The air must be free of dirt, oil, contamination, and moisture. Contaminants, such as frozen moisture or dirt, can block or partially block restrictions and nozzles, giving false readings or causing complete equipment failure. Air compressors are fitted with air dryers and filters, and have a reservoir tank with a capacity large enough for several minutes of supply in case of system failure. Dry, clean air is supplied at a pressure of 90 psig (630 kPa-g), and with a dew point of 20°F (10°C) below the minimum winter operating temperature at atmospheric pressure. Additional information on the quality of instrument air can be found in ANSI/ISA – 7.0.01 – 1996 Standard for Instrument Air.

A water supply is required in many cleaning and cooling operations and for steam generation. A domestic water supply contains large quantities of particulates and impurities, and while it may be satisfactory for cooling, it is not suitable for most cleaning operations. Filtering and other operations can remove some of contaminants, making the water suitable for some cleaning operations, but if ultrapure water is required, then a reverse osmosis system may be required.

Installation and maintenance must be considered when locating devices, such as instruments and valves. Each device must be easily accessible for maintenance and inspection. It also may be necessary to install hand-operated valves, so that equipment can be replaced or serviced without complete plant shutdown. It may be necessary to contract out maintenance of certain equipment, or have the vendor install equipment, if the necessary skills are not available in-house.

Safety is a top priority in a facility. The correct materials must be used in container construction, plumbing, seals, and gaskets, to prevent corrosion and failure, leading to leakage and spills of hazardous materials. All electrical equipment must be properly installed to Code, with breakers. Electrical systems must have the correct fire retardant. More information can be found in ANSI/ISA – 12.01.01 – 1999, — "Definitions and Information Pertaining to Electrical Apparatus in Hazardous Locations."

1.8 Summary

This chapter introduced the concept of process control, and the differences between sequential, continuous control and the use of feedback loops in process control. The building blocks in a process control system, the elements in the building blocks, and the terminology used, were defined.

The use of instrumentation and sensors in process parameter measurements was discussed, together with instrument characteristics, and the problems encountered, such as nonlinearity, hysteresis, repeatability, and stability. The quality of a process control loop was introduced, together with the types of problems encountered, such as stability, transient response, and accuracy.

The various methods of data transmission used are analog data, digital data, and pneumatic data; and the concept of the smart sensor as a plug-and-play device was given.

Considerations of the basic requirements in a process facility, such as the need for an uninterruptible power supply, a clean supply of pressurized air, clean and pure water, and the need to meet safety regulations, were covered.

Definitions

Absolute Accuracy of an instrument is the deviation from true expressed as a number.

Accuracy of an instrument or device is the difference between the indicated value and the actual value.

Actuators are devices that control an input variable in response to a signal from a controller.

Automation is a system where most of the production process, movement, and inspection of materials are performed automatically by specialized testing equipment, without operator intervention.

Controlled or Measured Variable is the monitored output variable from a process, where the value of the monitored output parameter is normally held within tight given limits.

Controllers are devices that monitor signals from transducers and keep the process within specified limits by activating and controlling the necessary actuators, according to a predefined program.

Converters are devices that change the format of a signal without changing the energy form (e.g., from a voltage to a current signal).

Correction Signal is the signal that controls power to the actuator to set the level of the input variable.

Drift is the change in the reading of an instrument of a fixed variable with time.

Error Signal is the difference between the set point and the amplitude of the measured variable.

Feedback Loop is the signal path from the output back to the input, which is used to correct for any variation between the output level and the set level.

Hysteresis is the difference in readings obtained when an instrument approaches a signal from opposite directions.

Instrument is the name of any various device types for indicating or measuring physical quantities or conditions, performance, position, direction, and so forth.

Linearity is a measure of the proportionality between the actual value of a variable being measured and the output of the instrument over its operating range.

Manipulated Variable is the input variable or parameter to a process that is varied by a control signal from the processor to an actuator.

Offset is the reading of the instrument with zero input.

Precision is the limit within which a signal can be read, and may be somewhat subjective.

Range of an instrument is the lowest and highest readings that it can measure.

Reading Accuracy is the deviation from true at the point the reading is being taken, and is expressed as a percentage.

Repeatability is a measure of the closeness of agreement between a number of readings taken consecutively of a variable.

Reproducibility is the ability of an instrument to repeatedly read the same signal over time, and give the same output under the same conditions.

Resolution is the smallest change in a variable to which the instrument will respond.

Sensitivity is a measure of the change in the output of an instrument for a change in the measured variable.

Sensors are devices that can detect physical variables.

Set Point   is the desired value of the output parameter or variable being monitored by a sensor; any deviation from this value will generate an error signal.

Span   of an instrument is its range from the minimum to maximum scale value.

Transducers   are devices that can change one form of energy into another.

Transmitters   are devices that amplify and format signals, so that they are suitable for transmission over long distances with zero or minimal loss of information.

References

[1] Battikha, N. E., *The Condensed Handbook of Measurement and Control*, 2nd ed., ISA, 2004, pp. 1–8.

[2] Humphries J. T., and L. P. Sheets, *Industrial Electronics*, 4th ed., Delmar, 1993, pp. 548–550.

[3] Sutko, A., and J. D. Faulk, *Industrial Instrumentation*, 1st ed., Delmar Publishers, 1996, pp. 3–14.

[4] Johnson, C. D., *Process Control Instrumentation Technology*, 7th ed., Prentice Hall, 2003, pp. 6–43.

[5] Johnson, R. N., "Signal Conditioning for Digital Systems," Proceedings Sensors Expo, October 1993, pp. 53–62.

CHAPTER 2
Units and Standards

2.1 Introduction

The measurement and control of physical properties require the use of well-defined units. Units commonly used today are defined in either the English system or the Systéme International d'Unités (SI) system [1]. The advent of the Industrial Revolution, developing first in England in the eighteenth century, showed how necessary it was to have a standardized system of measurements. Consequently, a system of measurement units was developed. Although not ideal, the English system (and U.S. variants; see gallon and ton) of measurements became the accepted standard for many years. This system of measurements has slowly been eroded by the development of more acceptable scientific units developed in the SI system. However, it should be understood that the base unit dimensions in the English or SI system are artificial quantities. For example, the units of distance (e.g., feet, meter), time, and mass, and the use of water to define volume, were chosen by the scientific community solely as reference points for standardization.

2.1.1 Units and Standards

As with all disciplines' sets of units and standards have evolved over the years to ensure consistency and avoid confusion. The units of measurement fall into two distinct systems: the English system and the SI system [2].

The SI units are sometimes referred to as the centimeter-gram-second (CGS) units and are based on the metric system but it should be noted that not all of the metric units are used. The SI system of units is maintained by the Conférence Genérale des Poids et Measures. Because both systems are in common use it is necessary to understand both system of units and to understand the relationship between them. A large number of units (electrical) in use are common to both systems. Older measurement systems are calibrated in English units, where as newer systems are normally calibrated in SI units

The English system has been the standard used in the United States, but the SI system is slowly making inroads, so that students need to be aware of both systems of units and be able to convert units from one system to the other. Confusion can arise over the use of the pound (lb) as it can be used for both mass and weight and also its SI equivalent being. The pound mass is the Slug (no longer in common use as a scientific unit) The slug is the equivalent of the kg in the SI system of units, where as the pound weight is a force similar to the Newton, which is the unit of force in the SI system. The practical unit in everyday use in the English system of units is the lb weight, where as, in the SI system the unit of mass or kg is used. The conversion factor of 1 lb = 0.454 kg which is used to convert mass (weight) between the two systems, is in effect equating 1 lb force to 0.454 kg mass this being the mass that will produce a force of 4.448 N under the influence of gravity which is a force of 1 lb. Care must be taken not to mix units from the two systems. For consistency some units may have to be converted before they can be used in an Equation. The Instrument Society of America (ISA) has developed a complete list of symbols for instruments, instrument identification, and process control drawings, which will be discussed in Chapter 17. Other standards used in process control have been developed in other disciplines.

2.2 Basic Units

Table 2.1 gives a list of the base units used in instrumentation and measurement in the English and SI systems. Note that the angle units are supplementary geometric units.

2.3 Units Derived from Base Units

All other units are derived from the base units. The derived units have been broken down into units used in both systems (e.g., electrical units), the units used in the English system, and the units used in the SI system.

2.3.1 Units Common to Both the English and SI Systems

The units used in both systems are given in Table 2.2.

2.3.2 English Units Derived from Base Units

Table 2.3 lists some commonly used units in the English system. The correct unit for mass is the slug, which is now not normally used. The English system uses weight to infer mass, which can lead to confusion. The units for the pound in energy and horsepower are mass, whereas the units for the pound in pressure is a force. Note that the lb force = lb mass (m) × g = lb (m) ft s^{-2} [3].

Table 2.1 Basic Units

Quantity	English Units	English Symbol	SI Units	SI Symbol
Length	foot	ft	meter	m
Mass	pound (slug)	lb	kilogram	kg
Time	second	s	second	s
Temperature	rankine	°R	Kelvin	K
Electric current	Ampere	A	ampere	A
Amount of substance			mole	mol
Luminous intensity	candle	c	lumen	lm
Angle	degree	°	radian	rad
Solid angle			steradian	sr

2.3 Units Derived from Base Units

Table 2.2 Electrical Units Common to the English and SI Systems

Quantity	Name	Symbol	Units
Frequency	hertz	Hz	s^{-1}
Wavelength	meter	λ	m
Resistance	ohm	Ω	$kg\ m^2\ s^{-3}\ A^{-2}$
Conductance	siemens	S	A/V, or $m^{-2}\ kg^{-1}\ s^3\ A^2$
Electromotive force	volt	V	A Ω, or $m^2\ kg\ s^{-3}\ A^{-1}$
Electronic quantity	coulomb	C	A s
Capacitance	farad	F	$s^4\ A^2\ kg^{-1}\ m^{-2}$
Energy density	joule per cubic meter	J/m^3	$kg\ m^{-1}\ s^{-2}$
Electric field strength	volts per meter	V/m	$V\ m^{-1}$
Electric charge density	coulombs per cubic meter	C/m^3	$C\ m^{-3}$
Surface flux density	coulombs per square meter	C/m^2	$C\ m^{-2}$
Current density	amperes per square meter	A/m^2	$A\ m^{-2}$
Magnetic field strength	amperes per meter	A/m	$A\ m^{-1}$
Permittivity	farads per meter	F/m	$A^2\ s^4\ m^{-3}\ kg^{-1}$
Inductance	henry	H	$kg\ m^2\ s^{-2}\ A^{-2}$
Permeability	henrys per meter	H/m	$m\ kg\ s^{-2}\ A^{-2}$
Magnetic flux density	tesla	T	Wb/m^2, or $kg\ s^{-2}\ A^{-1}$
Magnetic flux	weber	Wb	V s, or $m^2\ kg\ s^{-2}\ A^{-1}$

Table 2.3 English Units Derived from Base Units

Quantity	Name	Symbol	Units
Frequency	revolutions per minute	r/min	s^{-1}
Speed		ft/s	$ft\ s^{-1}$
—Linear	feet per second		
—Angular	degrees per second	degree/s	$degree\ s^{-1}$
Acceleration		ft/s^2	$ft\ s^{-2}$
—Linear	feet per second squared		
—Angular	degrees per second squared	degree/s^2	$degree\ s^{-2}$
Energy	foot-pound	ft-lb	$lb\ (m)\ ft^2\ s^{-2}$
Force	pound	lb	$lb\ (m)\ ft\ s^{-2}$
Pressure	pounds per square in	psi	$lb\ (m)\ ft^{-1}\ s^{-2}$
Power	horsepower	hp	$lb\ (m)\ ft^2\ s^{-3}$
Density	pound (slug) per cubic foot	lb (slug)/ft^3	$lb\ (m)\ ft^{-3}$
Specific weight	pound per cubic foot	lb/ft^3	$lb\ (m)\ ft^{-2}\ s^{-2}$
Surface tension	pound per foot	lb/ft	$lb\ (m)\ s^{-2}$
Quantity of heat	British thermal unit	Btu	$lb\ (m)\ ft^2\ s^{-2}$
Specific heat		Btu/lb (m) °F	$ft^2\ s^{-2}\ °F^{-1}$
Thermal conductivity		Btu/ft h °F	$lb\ (m)\ ft\ s^{-3}\ °F^{-1}$
Thermal convection		Btu/h ft^2 °F	$lb\ (m)\ s^{-3}\ °F^{-1}$
Thermal radiation		Btu/h ft^2 °R^4	$lb\ (m)\ s^{-3}\ °R^{-4}$
Stress		σ	$lb\ (m)\ ft^{-1}\ s^{-2}$
Strain		ε	dimensionless
Gauge factor		G	dimensionless
Young's modulus		lb/ft^2	$lb\ (m)ft^{-1}\ s^{-2}$
Viscosity dynamic	poise	P	$lb\ (m)\ ft^{-1}\ s^{-1}$
Viscosity kinematic	stoke	St	$ft^2\ s^{-1}$
Torque (moment of force)		lb ft	$lb\ (m)\ ft^2\ s^{-2}$

Conversion between English units is given in Table 2.4. This table gives the conversion between units of mass, length, and capacity in the English system. Note the difference in U.S. and English gallon and ton.

2.3.3 SI Units Derived from Base Units

The SI system of units is based on the CGS or metric system, but not all of the units in the metric system are used. Table 2.5 lists the metric units used in the SI system. It should be noted that many of the units have a special name [4].

Conversion between SI units is given in Table 2.6. This table gives the conversion between mass, length, and capacity in the SI system.

2.3.4 Conversion Between English and SI Units

Table 2.7 gives the factors for converting units between the English and SI systems [5].

Example 2.1

How many meters are there in 2.5 miles?

2.5 miles = 2.5 × 5,280 × 0.305m = 4,026m = 4.026 km

Example 2.2

What is the weight of 3.7-lb mass in newtons?

3.7 lb mass = 3.7 × 32.2 lb weight = 119.1 lb

119.1 lb = 4.448 × 119.1N = 530N

Example 2.3

What is the pressure equivalent of 423 Pa in lb/ft^2?

423 Pa = 0.423/6.897 psi = 0.061 psi

0.061 psi = 0.061 × 12 × 12 psf = 8.83 psf

Table 2.4 Conversion Between Mass, Length, and Capacity in the English System

Quantity	Name	Symbol	Conversion
Length	mile	1 mi	5,280 ft
Capacity to volume	gallon (U.S.)	1 gal	0.1337 ft^3
	imperial gallon	1 imp gal	0.1605 ft^3
Capacity to weight (water)	1 gal (U.S.)		8.35 lb
	1 imp gal		10 lb
Weight	ton (U.S.)	ton short	2,000 lb
	imperial ton	ton long	2,240 lb

2.3 Units Derived from Base Units

Table 2.5 SI units Derived from Base Units

Quantity	Name	Symbol	Other Units	Base Units
Frequency	hertz	Hz	s^{-1}	s^{-1}
Speed — Linear	meters per second		m/s	$m\ s^{-1}$
— Angular	radians per second		rad/s	$rad\ s^{-1}$
Acceleration — Linear	meters per second squared		m/s^2	$m\ s^{-2}$
— Angular	radians per second squared		rad/s^2	$rad\ s^{-2}$
Wave number	per meter		m^{-1}	m^{-1}
Density	kilograms per cubic meter		kg/m^3	$kg\ m^{-3}$
Specific weight	weight per cubic meter		kN/m^3	$kg\ m^{-2}\ s^{-2}$
Concentration of amount of substance	mole per cubic meter		mol/m^3	$mol\ m^{-3}$
Specific volume	cubic meters per kilogram		m^3/kg	$kg^{-1}m^3$
Energy	joule	J	N m	$kg\ m^2\ s^{-2}$
Force	newton	N	$m\ kg/s^2$	$kg\ m\ s^{-2}$
Pressure	pascal	Pa	N/m^2	$kg\ m^{-1}\ s^{-2}$
Power	watt	W	J/s	$kg\ m^2\ s^{-3}$
Luminance	lux	lx	lm/m^2	$m^{-2}\ cd\ sr$
Luminous flux	lumen	lm	cd sr	cd sr
Quantity of heat	joule	J	N m	$kg\ m^2\ s^{-2}$
Heat flux density irradiance	watts per square meter		W/m^2	$kg\ s^{-3}$
Heat capacity entropy	joules per kelvin		J/K	$kg\ m^2\ s^{-2}\ K^{-1}$
Specific heat entropy			J/kg K	$m^2\ s^{-2}\ K^{-1}$
Specific energy	joules per kilogram		J/kg	$m^2\ s^{-2}$
Thermal conductivity			W/m K	$kg\ m\ s^{-3}\ K^{-1}$
Thermal convection			$W/m^2\ K$	$kg\ s^{-3}\ K^{-1}$
Thermal radiation				$kg\ s^{-3}\ K^{-4}$
Stress		σ	Pa	$kg\ m^{-1}\ s^{-2}$
Strain		ε	$\delta m/m$	Dimensionless
Gauge Factor		G	$\delta R/R$ per ε	Dimensionless
Young's modulus			N/m^2	$kg\ m^{-1}\ s^{-2}$
Viscosity dynamic	Poiseuille	Po	kg/m s	$kg\ m^{-1}\ s^{-1}$
Viscosity kinematic	Stokes	St	cm^2/s	$m^2\ s^{-1}$
Surface tension	newtons per meter		N/m	$kg\ s^{-2}$
Torque (moment)	newton meter		N m	$kg\ m^2\ s^{-2}$
Molar energy	joules per mole		J/mol	$kg\ m^2\ s^{-2}\ mol^{-1}$
Molar entropy, heat capacity	joules per mole kelvin		J/(mol K)	$kg\ m^2\ s^{-2}\ K^{-1}\ mol^{-1}$
Radioactivity	Becquerel	Bq	per sec	s^{-1}
Absorbed radiation	Gray	Gy	J/kg	$m^2\ s^{-2}$

Table 2.6 Conversion Between Mass, Length, and Capacity and Other Units in the SI System

Quantity	Name	Symbol	Conversion
Capacity	liter	L	$1L = 1\ dm^3$ ($1{,}000L = 1\ m^3$)
Weight	liter	L	1L water = 1 kg
Area	hectare	ha	1 ha = 10,000 m^2
Charge	electron volt	eV	1 eV = 1.602×10^{-19} J
Mass	unified atomic mass unit	μ	1.66044×10^{-27} kg

Table 2.7 Conversion Between English and SI Units

Quantity	English Units	SI Units
Length	1 ft	0.305 m
Speed	1 mi/h	1.61 km/h
Acceleration	1 ft/s^2	0.305 m/s^2
Mass	1 lb (m)	14.59 kg
Weight	1 lb	0.454 kg
Capacity	1 gal (U.S.)	3.78 L
Force	1 lb	4.448 N
Angle	1 degree	2π/360 rad
Temperature	1°F	5/9°C
Temperature	1°R	5/9 K
Energy	1 ft lb	1.356 J
Pressure	1 psi	6.897 kPa
Power	1 hp	746 W
Quantity of heat	1 Btu	252 cal or 1,055 J
Thermal conduction	1 Btu/hr ft °F	1.73 W/m K
Specific heat	1 Btu/lb (m) °F	J/kg K
Thermal convection	Btu/h ft^2 °F	W/m^2 K
Thermal radiation	Btu/h ft^2 °R^4	W/m^2 K^4
Expansion	1 α/°F	1.8 α/°C
Specific weight	1 lb/ft^3	0.157 kN/m^3
Density	1 lb (m)/ft^3	0.516 kg/m^3
Dynamic viscosity	1 lb s/ft^2	49.7 Pa s (4.97 P)
Kinematic viscosity	1 ft^2/s	9.29 × 10−2 m^2/s (929 St)
Torque	1 lb ft	1.357 N m
Stress	1 psi	6.897 kPa
Young's modulus	1 psi	6.897 kPa

Example 2.4

A steam boiler generates 7.4 kBtu/h. The steam is used to drive a 47% efficient steam engine. What is the horsepower of the engine?

7.4 kBtu/h = 7,400 × 1,055/60 W = 130 kW

130 kW @ 47% = 130,000 × 0.47/746 = ~~81.97 hp~~

Example 2.5

A 110V electric motor uses 5.8A. If the motor is 87% efficient, then how many horsepower will the motor generate?

Watts = 110 × 5.8 × 0.87 W = 555.1 W

hp = 555.1/746 = 0.74 hp

2.3.5 Metric Units not Normally Used in the SI System

There are a large number of units in the metric system, but all of these units are not required in the SI system of units because of duplication. A list of some of the units not used is given in Table 2.8.

2.4 Standard Prefixes

Table 2.8 Metric Units not Normally Used in the SI System

Quantity	Name	Symbol	Equivalent
Length	Angstrom	Å	1Å = 0.1 nm
	Fermi	fm	1 fm = 1 femtometer
	X unit		1 X unit = 100.2 fm
Volume	Stere	st	1 st = 1 m^3
	Lambda	λ	1 mm^3
Mass	metric carat		1 metric carat = 200 mg
	Gamma	γ	1 γ = 1 μg
Force	Dyne	dyn	1 dyn = 10 μN
Pressure	Torr	torr	1 torr = 133 Pa
	Bar	bar	1 bar = 100 kPa = 1.013 atm
Energy	Calorie	cal	1 cal = 4.1868J
	Erg	erg	1 erg = 0.1 μJ
Viscosity dynamic	Poise	P	1 P = 0.1 Pa s
kinematic	Stoke	St	1 St = 1 cm^2/s
Conductance	mho	mho	1 mho = 1 S
Magnetic field strength	Oersted	Oe	1 Oe = (1,000/4π) A/m
Magnetic flux	Maxwell	Mx	1 Mx = 0.01 μWb
Magnetic flux density	Gauss	Gs (G)	1 QsG = 0.1 mT
Magnetic Induction	Gamma	γ	1 g = 1 nT
Radioactivity	Curie	Ci	1 Ci = 37 GBq
Absorbed Rradiation	rad	rad	1 rad = 10 mGy

2.4 Standard Prefixes

Standard prefixes are commonly used for multiple and sub-multiple quantities, in order to cover the wide range of values used in measurement units. These are given in Table 2.9.

Digital Standard Prefixes are now common practice in the digital domain. The International Electrotechnical Commission (IEC), an international organization for standardization in electrotechnology, approved in December 1998 the following standards for binary numbers, as given in Table 2.10. The Institute of Electrical and Electronic Engineers (IEEE) also has adopted this convention.

These definitions allow the SI prefixes to be used for their original values; for example, k, M, and G represent 1,000, 10^6, and 10^9, respectively. As an example:

1 kilobit = 1 kb = 10^3 bits = 1,000 bits

Table 2.9 Standard Prefixes

Multiple	Prefix	Symbol	Multiple	Prefix	Symbol
10^{18}	exa	E	10^{-1}	deci	d
10^{15}	peta	P	10^{-2}	centi	c
10^{12}	tera	T	10^{-3}	milli	m
10^9	giga	G	10^{-6}	micro	μ
10^6	mega	M	10^{-9}	nano	n
10^3	kilo	k	10^{-12}	pico	p
10^2	hecto	h	10^{-15}	femto	f
10	deka	da	10^{-10}	atto	a

Table 2.10 Binary Prefixes and Numbers

Name	Prefix	Symbol	Factor
Kilobinary	kibi	Ki	2^{10}
Megabinary	mebi	Mi	$(2^{10})^2 = 2^{20}$
Gigabinary	gibi	Gi	$(2^{10})^3 = 2^{30}$
Terabinary	tebi	Ti	$(2^{10})^4 = 2^{40}$
Petabinary	pepi	Pi	$(2^{10})^5 = 2^{50}$
Exabinary	exbi	Ei	$(2^{10})^6 = 2^{60}$

1 kilobinarybit = 1 kibibit = 1 KiB = 2^{10} bits = 1,024 bits

1 megabyte = 1 MB = 10^6 Bytes = 1,000,000 Bytes

1 megabinarybyte = 1 megabyte = 1 MiB = 2^{20} Bytes = 1,048,576 Bytes

2.5 Standards

There are two types of standards: the accepted physical constants, and the standards developed by various institutions for uniformity of measurement and conformity between systems.

2.5.1 Physical Constants

A number of commonly encountered physical constants are given in Table 2.11.

2.5.2 Standards Institutions

Instrumentation and process control use the disciplines from several technical fields, and therefore, use the industrial and technical standards that have evolved in these various disciplines. A list of these technical institutions and their Web sites is given in Table 2.12.

Each of the institutions has developed a large number of accepted standards for consistency and uniformity of measurement and control. A list of these standards, along with further information for each institution, can be obtained from their Web

Table 2.11 Physical Constants

Quantity	English Units	SI Units	Comments
Gravitational acceleration	32.2 ft/s^2	9.8 m/s^2	
Atmospheric pressure	14.7 psi	101.3 kPa	sea level
Absolute temperature	−459.6°F	−273.15°C	
Sound intensity Reference level		12^{-16} W/cm^2	@ 1 kHz
Sound pressure Reference level		20 μN/m^2	@ 1 kHz
Specific weight of water	62.43 lb/ft^3	9.8 kN/m^3	@ 4°C
E/M velocity	0.98 Gft/s 185.7 kmi/h	0.299 Gm/s	vacuum

Table 2.12 Web Addresses of Technical Institutions

1. Institute of Electrical and Electronic Engineers	www.ieee.org
2. Instrumentation, Systems, and Automation Society	www.isa.org
3. National Institute of Standards and Technology	www.nist.gov
4. American National Standards Institute	www.ansi.org
5. National Electrical Manufactures Association	www.nema.org
6. Industrial Control and Plant Automation	www.xnet.com
7. Society of Automotive Engineers	www.sae.org/servlets/index
8. The American Institute of Physics	www.aip.org
9. American Chemical Society	www.acs.org
10. International Electrotechnical Commission	www.iec.ch
11. American Institute of Chemical Engineers	www.aiche.org
12. American Association of Mechanical Engineers	www.asme.org
13. American Society for Testing and Materials	www.astm.org
14. Occupational Safety and Health Administration	www.osha.gov
15. Environmental Protection Agency	www.epa.gov

sites. The Web addresses of OSHA and EPA are included, since many of their rules and regulations affect plant operation and safety.

2.6 Summary

This chapter discussed the need for well-defined units for physical measurements. The English system originally was the most widely used, but is being replaced by the more scientifically acceptable SI system. SI units are based on centigrade-gram-second units from the metric system. Measurement units were given in both systems, along with their relation to the base units, and conversion factors between the two systems. Other commonly used metric units not required because of duplication were given as they may be encountered. Standard prefixes are given to cover the wide range of measurements that require the use of multiple and submultiple units.

The digital domain also requires the use of prefixes that have been defined for the base 2, to distinguish between binary and digital numbers. Some of the more common physical constants were given, and the Web addresses of institutions that set industrial standards were given, so that the reader can obtain more specific information.

References

[1] Taylor, B. N., (ed.), The International System of Units (SI), National Institute of Standards Special Publications 330, Government Printing Office, Washington, DC, 1991.

[2] Eccles, L. H., "The Presentation of Physical Units in IEEE 1451.2," Sensors Magazine, Vol. 16, No. 4, April 1999.

[3] Johnson, C. D., *Process Control Instrumentation Technology*, 2nd ed., Prentice Hall, 2003, pp. 597–600.

[4] Battikha, N. E., *The Condensed Handbook of Measurement and Control*, 2nd ed., ISA, 2004, pp. 275–283.

[5] www.efunda.com/units/index.

Basic Electrical Components

3.1 Introduction

Resistors, capacitors, and inductors—these are the three basic passive elements used in electrical circuits, either as individual devices or in combination. These elements are used as loads, delays, and current limiting devices. Capacitors are used as dc blocking devices, in level shifting, integrating, differentiating, filters, frequency determination, selection, and delay circuits. Inductive devices can be extended to cover analog meter movements, relays, audio to electrical conversion, electrical to audio conversion, and electromagnetic devices. They are also the basis for transformers and motors.

3.2 Circuits with R, L, and C

Passive components are extensively used in ac circuits for frequency selection, noise suppression, and so forth, and are always present as parasitic components, limiting signal response and introducing unwanted delays. These components also cause phase shift between voltages and currents, which has to be taken into account when evaluating the performance of ac circuits.

3.2.1 Voltage Step Input

When a dc voltage is applied to the series resistor-capacitor circuit shown in Figure 3.1(a) a current flows in the elements, charging the capacitor. Figure 3.1(b) shows the input voltage step, the resulting current flowing, and the voltages across the resistor and the capacitor. Initially, all the voltage is dropped across the resistor. Although current is flowing into the capacitor, there is no voltage drop across the capacitor. As the capacitor charges, the voltage across the capacitor builds up exponentially, and the voltage across the resistor starts to decline, until eventually the capacitor is fully charged and current ceases to flow. The voltage across the capacitor is then equal to the supply voltage, and the voltage across the resistor is zero.

It should be noted that the current flowing through the resistor and into the capacitor is the same for both components, but the voltages across each component are different. That is, when the current flowing in the resistor is a maximum, the voltage across the resistor is a maximum, given by $E = IR$, and the voltage is said to be in phase with the current. In the case of the capacitor, the voltage is zero when the current flowing is a maximum, and the voltage is a maximum when the current is

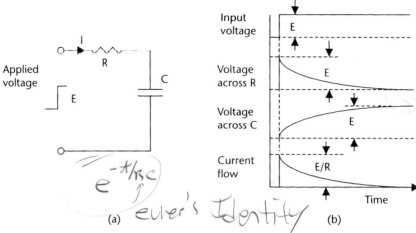

Figure 3.1 (a) Input voltage transient to a circuit with resistance and capacitance, and (b) associated waveforms.

zero. In this case, the voltage lags the current, or there is a phase shift between the voltage and the current of 90°. The voltage across the capacitor builds up exponentially, at a rate determined by the values of the resistor and the capacitor.

When a dc voltage is applied to a series inductance circuit resistance, as shown in Figure 3.2(a) a current will build up, Figure 3.2(b) shows the input voltage step, the resulting current buildup, and the voltages across the resistor and the inductor the inductance will initially appear as a high impedance preventing current from flowing. The current will be zero, the supply voltage will appear across the inductance, and there will be zero voltage across the resistor. After the initial turn-on, current will start to flow and build up, the voltage across the resistor will increase, and then start to decrease across the inductance. This allows the current to build up exponentially, until the current flow is limited by the resistance at its maximum value and the voltage across the inductance will be zero. The effects are similar to those that occur when a dc voltage is applied to a series R.C. network. The voltage and current in the resistor are in phase, but in the inductor are out of phase. That is,

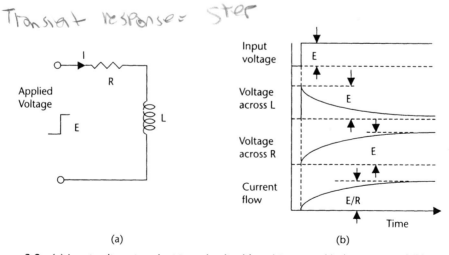

Figure 3.2 (a) Input voltage transient to a circuit with resistance and inductance, and (b) associated waveforms.

3.2 Circuits with R, L, and C

in this case, the voltage appears across the inductance before the current starts to flow, and goes to zero when the current is at its maximum, so that the voltage leads the current, and there is a phase shift between the voltage and the current of 90°. The voltage across the resistor increases exponentially, at a rate determined by the value of the inductance and resistance.

3.2.2 Time Constants

In an RC network when a step voltage is applied, as shown in Figure 3.1(a), the voltage across the capacitor is given by the equation [1]:

$$E_C = E\left(1 - e^{-t/RC}\right) \quad (3.1)$$

where E_C is the voltage across the capacitor at any instant of time, E is the source voltage, t is the time (seconds) after the step is applied, R is in ohms and C is in farads.

Conversely, after the capacitor is fully charged, and the step input voltage is returned to zero, C discharges and the voltage across the capacitor is given by the equation:

$$E_C = Ee^{-t/RC} \quad (3.2)$$

Similar equations apply to the rise and fall of currents in the inductive circuit shown in Figure 3.2(a).

Example 3.1

What is the voltage across the capacitor in Figure 3.1(a) 35 μs after the input voltage steps from 0V to 54V, if the value of the resistor is 47 kΩ and the capacitor is 1.5 μF?

RC = 70.5 ms

$E_c = 54(1 - e^{-35 \times 1000000/1000 \times 47 \times 1000 \times 1.5})V = 54(1 - 1/e^{0.5})V$

$E_c = 54(1 - 0.606) = 21.3V$ = 26.3 mV

The time constant of the voltage in a capacitive circuit from (3.1) and (3.2) is defined as:

$$t = CR \quad (3.3)$$

where t is the time (seconds) it takes for the voltage to reach 63.2% of its final or aiming voltage after the application of an input voltage step (charging or discharging). For example, by the end of the first time constant, the voltage across the capacitor will reach 9.48V when a 15V step is applied. During the second time constant, the voltage across the capacitor will rise another 63.2% of the remaining voltage step; that is, (15 − 9.48)V × 63.2% = 3.49V. At the end of two time constant periods, the voltage across the capacitor will be 12.97V, and at the end of three periods, the voltage will be 14.25V, and so forth. The voltage across the capacitor reaches 99% of its aiming value in 5 CR.

Example 3.2

What is the time constant for the circuit shown in Figure 3.1(a), if the resistor has a value of 220 kΩ, and the capacitor is 2.2 μF?

$$t = 2.2 \times 10^{-6} \times 220 \times 10^{3} \text{ sec} = 484 \times 10^{-3} \text{ sec} = 0.484 \text{ sec}$$

The RC time constant is often used as the basis for time delays. A comparator circuit is set to detect when a voltage across a capacitor in a CR network reaches 63.2% of the input step voltage. The time delay generated is then 1 CR.

In the case of an inductive circuit, the time constant for the current is given by;

$$t = L/R \tag{3.4}$$

where L is the inductance in Henrys, and t gives the time for the current to increase to 63.2% of its final current through the inductor.

Time constants apply not only to the rate of change of currents and voltages in electrical circuits when a step voltage is applied, but also to sensor outputs when there is a change in the measured variable. The output signal from the sensor changes exponentially, so that there is a delay before the sensor output reaches its final value. In the case of a temperature sensor, the time constant of the sensor is determined by factors such as its thermal mass.

3.2.3 Sine Wave Inputs

When an ac sine wave is applied to C, L, and R circuits, as shown in Figure 3.3(a) the same phase shift between voltage and current occurs as when a step voltage is applied. Figure 3.3(b) shows the relationship between the input voltage, the current flowing and the voltage across C, L, and R as can be seen. In resistive elements, the current and voltage are in phase; in capacitive circuits, the current leads the voltage

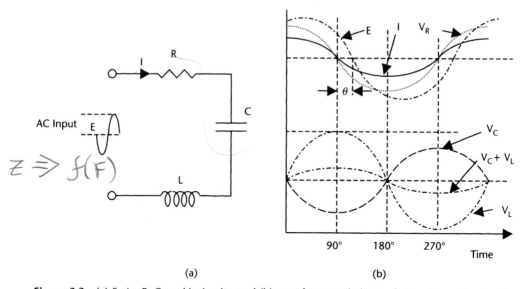

Figure 3.3 (a) Series R, C, and L circuits, and (b) waveforms and phase relations in a series circuit.

3.2 Circuits with R, L, and C

by 90° (Figure 3.1); and in inductive circuits, the current lags the voltage by 90° (Figure 3.2).

Since the voltages and the currents are not in phase in capacitive and inductive ac circuits, these devices have impedance and not resistance. Impedance and resistance cannot be directly added. If a resistor, capacitor, and inductor are connected in series as shown in Figure 3.3(a), then the same current will flow through all three devices. However, the voltages in the capacitor and inductor are 180° out of phase and 90° out of phase with the voltage in the resistor, respectively, as shown in Figure 3.3(b). However, they can be combined using vectors to give:

$$E^2 = V_R^2 + (V_L - V_C)^2 \quad (3.5)$$

where E is the supply voltage, V_R is the voltage across the resistor, V_L is the voltage across the inductor, and V_C is the voltage across the capacitor.

The vector addition of the voltages is shown in Figure 3.4. In Figure 3.4(a), the relations between V_R, V_L, and V_C are given. V_L and V_C lie on the x-axis, with one positive and the other negative, since they are 180° out of phase. Since they have opposite signs, they can be subtracted to give the resulting $V_C - V_L$ vector. V_R lies at right angles (90°) on the y-axis. In Figure 3.4(b), the $V_C - V_L$ and V_R vectors are shown with the resulting E vector, which, from the trigonometry function, gives (3.5).

The impedance (Z) of the ac circuit, as seen by the input is given by:

$$Z = \sqrt{(R^2 + [X_L - X_C]^2)} \quad (3.6)$$

where $X_C = 1/2\pi fC$, which is the impedance to an ac frequency of f hertz by a capacitor C farads, and $X_L = 2\pi fL$, which is the impedance to an ac frequency of f hertz by an inductance of L henries.

The current flowing in the circuit can be obtained from Ohm's Law, as follows:

$$I = E/Z \quad (3.7)$$

Example 3.3

What is the current flowing in the circuit shown in Figure 3.3 (a), if R = 27 kΩ, C = 2.2 nF, L = 33 mH, E = 20 V and the input frequency = 35 kHz?

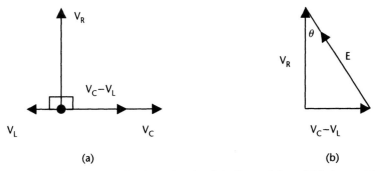

Figure 3.4 (a) The voltage vectors for the series circuit in Figure 3.5, and (b) the resulting voltage E vector.

$$X_L = 2\pi fL = 2 \times 3.142 \times 35 \times 10^3 \times 33 \times 10^{-3} = 7.25\,\Omega$$

$$X_C = \frac{1}{2\pi fC} = \frac{1}{2 \times 3.142 \times 35 \times 10^3 \times 2.2 \times 10^{-9}}\,\Omega = \frac{10^6}{483.56}\,\Omega = 2.1k\Omega$$

$$Z = \sqrt{\left(R^2 + [X_L - X_C]^2\right)} = \sqrt{\left\{(27 \times 10^3)^2 + [7.25 \times 10^3 - 2.1 \times 10^3]^2\right\}}$$

$$Z = \sqrt{\{729 \times 10^6 + 26.5 \times 10^6\}}$$

$$Z = \sqrt{(755.5 \times 10^6)} = 27.5 \times 10^3\,\Omega = 27.5k\Omega$$

$$I = E/Z = 20/27.5 \times 10^3 = 0.73mA$$

X_L and X_C are frequency-dependent, since as the frequency increases, X_L increases and X_C decreases. A frequency can be reached where X_L and X_C are equal, and the voltages across these components are equal, opposite, and cancel. At this frequency, $Z = R$, $E = IR$, and the current is a maximum. This frequency is called the *resonant frequency* of the circuit. At resonance:

$$2\pi fL = \frac{1}{2\pi fC} \tag{3.8}$$

which can be rewritten for frequency as

$$f = \frac{1}{2\pi\sqrt{LC}}\,Hz \tag{3.9}$$

When the input frequency is below the resonant frequency X_C is larger than X_L, and the circuit is capacitive; above the resonant frequency, X_L is larger than X_C, and the circuit is inductive. Plotting the input current against the input frequency shows a peak in the input current at the resonant frequency, as shown in Figure 3.5(a).

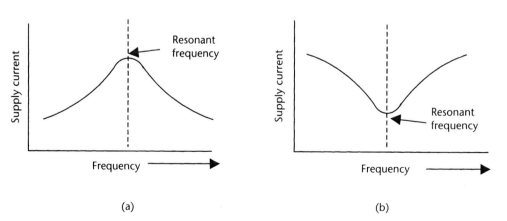

Figure 3.5 Supply current versus frequency in (a) series circuit, and (b) parallel circuit.

3.2 Circuits with R, L, and C

Example 3.4

What is the resonant frequency of the series circuit in Figure 3.3(a)? What is the current at this frequency? Assume the same values as in Example 3.3.

Using Equation (3.9), we get:

$$f = \frac{1}{2\pi\sqrt{LC}} Hz = \frac{1}{2 \times 3.142 \times \sqrt{2.2 \times 10^{-9} \times 33 \times 10^{-3}}} Hz$$

$$f = \frac{1}{2 \times 3.142 \times 8.5 \times 10^{-6}} Hz = \frac{10^6}{53.4} Hz = 1.87 \times 10^4 Hz$$

$$f = 18.7 kHz$$

The current can be obtained using Equation 3.7, at resonance $Z = R$.

$$I = E/Z = 20/18.7 \times 10^3 = 1.07 \text{ mA}$$

When R, L, and C are connected in parallel, as shown in Figure 3.6(a), each component will see the same voltage but not the same current, as shown by the waveforms in Figure 3.6(b).

The source current (I_s) is the vector sum of the currents in each component, and is given by:

$$I_S^2 = I_R^2 + (I_L - I_C)^2 \tag{3.10}$$

The impedance of the circuit (Z) as seen by the input is given by:

$$\frac{1}{Z^2} = \frac{1}{R^2} + \frac{1}{(X_L - X_C)^2} \tag{3.11}$$

At the resonant frequency, I_L and I_C become equal and cancel, and the current (I) is given by:

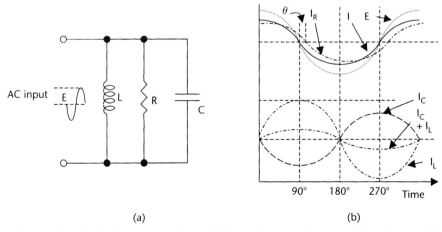

Figure 3.6 (a) Parallel R, C, and L circuits, and (b) waveforms and phase relations in a parallel circuit.

$$I = I_L = I_C = V \times 2\pi fC = V/2\pi fL \qquad (3.12)$$

Thus, at resonance, $E = IR$, as seen from (3.10). Below the resonant frequency, the circuit is inductive, and above the resonant frequency, the circuit is capacitive. Plotting the current against frequency shows the current is a minimum at the resonant frequency, as shown in the frequency plot in Figure 3.5(b). The frequency at resonance is given by (3.9), and the current is given by (3.7).

3.3 RC Filters

Passive networks using resistors and capacitors are extensively used [2], and sometimes small inductors are used in instrumentation circuits for filtering out noise, frequency selection, and frequency rejection, and so forth. Filters can be either passive [3] or active (using amplifiers), and are divided into the following types.

- *High-pass.* Allows high frequencies to pass but blocks low frequencies;
- *Low-pass.* Allows low frequencies to pass but blocks high frequencies;
- *Band-pass.* Allows a specific range of frequencies to pass;
- *Band reject.* Blocks a specific range of frequencies;
- *Twin – T.* Form of band reject filter, but with a sharper response characteristic.

Figure 3.7(a) shows a first order and second order low-pass RC filter [4], which passes low frequencies and rejects high frequencies, as shown in the frequency response curve in Figure 3.7(b). The cutoff frequency (f_c) of a filter is defined as the frequency at which the output power is reduced to one-half of the input power (-3 dB, or the output voltage is $0.707 V_{in}$). In the case of the first order low-pass filter, the cutoff frequency is given by:

$$f_c = 1/2\pi RC \qquad (3.13)$$

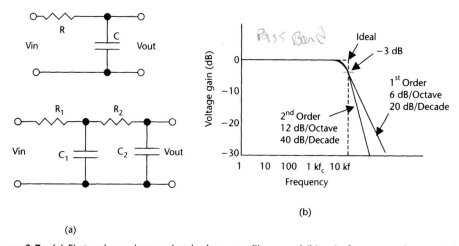

Figure 3.7 (a) First order and second order low-pass filters, and (b) gain-frequency characteristics of first order and second order filters.

In the case of the second order low-pass filter, the cutoff frequency is given by:

$$f_c = \frac{1}{2\pi\sqrt{R_1 R_2 C_1 C_2}} \, Hz \qquad (3.14)$$

The output voltage drops off at 20 dB/decade with a first order filter, 40 dB/decade for a second order filter, 60 dB/decade for a third order filter, and so forth. There is also a phase change associated with filters: −45° for a first order filter, −90° for a second order filter, −135° for a third order filter, and so forth. The phase change can cause instability in active circuits [5].

The input to output voltage ratio for any frequency (f) for a first order low-pass filter is given by:

$$\frac{V_{out}}{V_{in}} = \frac{1}{\left[1 + (f/f_c)^2\right]^{1/2}} \qquad (3.15)$$

The circuit for a high-pass, band-pass, and twin-T passive filters are shown in Figure 3.8. The number of resistive and capacitive elements determines the order of the filter (e.g., first order, second order, and so forth). The circuit configuration determines the characteristics of the filters, such as Butterworth, Bessel, Chebyshev, and Legendre. Figure 3.8(d) shows the frequency characteristics of the filters shown in Figure 3.8(a–c) [6].

These are examples of the use of resistors and capacitors in RC networks. Further descriptions of filters can be found in electronic textbooks [7].

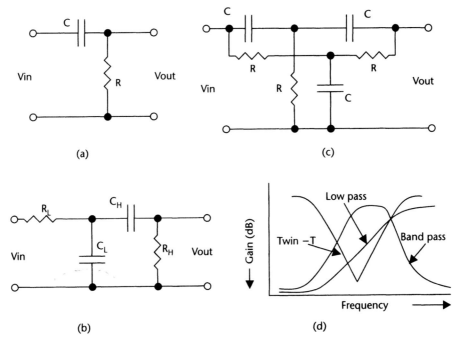

Figure 3.8 (a) High-pass filters, (b) band-pass filters, (c) twin-T band reject filters, and (d) their frequency characteristics.

3.4 Bridge Circuits

Resistor temperature compensation can be achieved in voltage divider circuits or bridge circuits. However, when trying to measure small changes in resistance, as required in strain gauges (see Section 11.3.4), the signal resolution in bridge circuits is much higher than with voltage dividers. Bridge circuits are used to convert small changes in impedance into voltages. These voltages are referenced to zero, so that the signals can be amplified to give high sensitivity to impedance changes.

3.4.1 Voltage Dividers

The two resistive elements R_1 and R_2 in a strain gauge sensor can be connected in series to form a voltage divider, as shown in Figure 3.9. The elements are driven from a supply, V_s. Since the temperature coefficient of resistance is the same in both elements, the voltage at the junction of the elements, V_R, is independent of temperature, and is given by:

$$V_R = \frac{R_2 V_s}{R_1 + R_2} \tag{3.16}$$

Example 3.5

The resistive elements in a strain gauge are each 5kΩ. A digital voltmeter with ranges of 10V, 1V, and 0.1V, and a resolution of 0.1% of FSD, is used to measure the output voltage. If R_2 is the fixed element and R_1 is the element measuring strain, what is the minimum change in R_1 that can be detected? Assume a supply of 10V.

To measure the output voltage ≈ 5V, the 10V range is required, giving a resolution or sensitivity of 10 mV.

$V_R = 5.01V = 5,000 \times 10/(R_1 + 5,000)V$
$5.01(R_1 + 5,000) = 50,000$
$R_1 = (50,000/5.01) - 5,000 = 4,980 Ω$
Resolution = $5,000 - 4,980 = 20 Ω$

3.4.2 dc Bridge Circuits

The simplest and most common bridge network is the DC Wheatstone bridge. The basic bridge is shown in Figure 3.10; this bridge is used in sensor applications, where small changes in resistance are to be measured. The basic bridge is modified for use in many other specific applications. In the basic bridge four resistors are connected

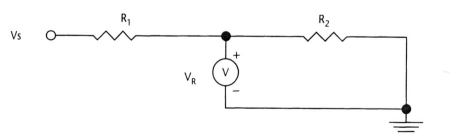

Figure 3.9 Voltage divider.

3.4 Bridge Circuits

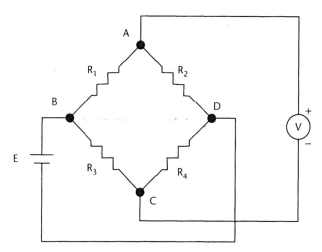

Figure 3.10 Circuit of a basic Wheatstone bridge.

in the form of a diamond with the supply and measuring instruments connected across the bridge as shown. When all the resistors are equal the bridge is balanced; that is, the bridge voltage at A and C are equal (E/2), and the voltmeter reads zero. Making one of the resistors a variable resistor the bridge can be balanced.

The voltage at point C referenced to D = E × $R_4/(R_3 + R_4)$
The voltage at point A referenced to D = E × $R_2/(R_1 + R_2)$
The voltage (V) between A and C = E $R_4/(R_3 + R_4)$ − E $R_2/(R_1 + R_2)$ (3.17)

When the bridge is balanced V = 0,

and $R_3R_2 = R_1R_4$ (3.18)

It can be seen from (3.17) that if R_1 is the resistance of a sensor whose change in value is being measured, the voltage at A will increase with respect to C as the resistance value decreases, so that the voltmeter will have a positive reading. The voltage (V) will change in proportion to small changes in the value of R_1, making the bridge very sensitive to small changes in resistance. Resistive sensors such as strain gauges are temperature sensitive and are often configured with two elements that can be used in a bridge circuit to compensate for changes in resistance due to temperature changes, for instance, if R_1 and R_2 are the same type of sensing element. Then resistance of each element will change by an equal percentage with temperature, so that the bridge will remain balanced when the temperature changes. If R_1 is now used to sense a variable, the voltmeter will only sense the change in R_1 due to the change in the variable, not the change due to temperature [8].

Example 3.6

Resistors R_1 and R_2 in the bridge circuit shown in Figure 3.9 are the strain gauge elements used in Example 3.5. Resistors R_3 and R_4 are fixed resistors, with values of 4.3 kW. The digital voltmeter is also the same meter. What is the minimum change in R_1 that can be detected by the meter? Assume the supply E is 10V.

The voltage at point C will be 5.0V, because $R_3 = R_4$, and the voltage at C equals one-half of the supply voltage. The voltmeter can use the 0.1V range, because the offset is 0V, giving a resolution of 0.1 mV. The voltage that can be sensed at A is given by:

$$E_{AD} = 10 \times R_2/(R_1 + R_2) = 5.00 + 0.0001 \text{V}$$

$$R_1 = (50,000/50,001) - 5,000 \Omega$$

$$R_1 = -49,999$$

$$\text{Resolution} = 1 \Omega$$

This example shows the improved resolution obtained by using a bridge circuit over the voltage divider in Example 3.5.

Lead Compensation

In many applications, the sensing resistor (R_2) may be remote from a centrally located bridge. The resistance temperature device (RTD) described in Sections 10.3.2 and 15.3.4 is an example of such a device. In such cases, by using a two-lead connection as in Figure 3.10, adjusting the bridge resistor (R_4) can zero out the resistance of the leads, but any change in lead resistance due to temperature will appear as a sensor value change. To correct this error, lead compensation can be used. This is achieved by using three interconnecting leads, labeled (a), (b), and (c) in Figure 3.11. A separate power lead (c) is used to supply R_2. Consider point D of the bridge as now being at the junction of the supply and R_2. The resistance of lead (b) is now a part of resistor R_4 and the resistance of lead (a) is now a part of remote resistor R_2. Since both leads have the same length with the same resistance and in the same environment, any changes in the resistance of the leads will cancel, keeping the bridge balanced.

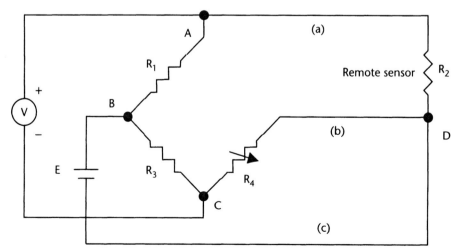

Figure 3.11 Circuit of a Wheatstone bridge with compensation for lead resistance used in remote sensing.

3.4 Bridge Circuits

Current Balanced Bridge

A current feedback loop can be used to automatically null the Wheatstone bridge, as shown in Figure 3.12. A low-value resistor R_s is connected in series with resistor R_2. The output from the bridge is amplified and converted into a current I, which is fed through R_s to develop an offset voltage, keeping the bridge balanced. The current through R_s can then be monitored to measure any changes in R_2. The bridge is balanced electronically to give a fast response time, and there are no null potentiometers to wear out. Initially, when the bridge is at null with zero current from the feedback loop:

$$R_3 (R_2 + R_s) = R_1 \times R_4$$

assume R_2 changes to $R_2 + \delta R_2$.
Then, to rebalance the bridge;

$$R_3 (R_2 + \delta R_2 + R_s) - IR_s = R_1 \times R_4$$

Subtracting the equations

$$\delta R_2 = IR_s/R_3 \tag{3.19}$$

showing a linear relationship between changes in the sensing resistor and the feedback current, with R_s and R_3 having fixed values.

Due to the two features of high sensitivity to small changes in resistance and correction for temperature effects, bridges are extensively used in instrumentation with strain gauges, piezoresistive elements, and magnetoresistive elements. The voltmeter should have a high resistance, so that it does not load the bridge circuit. Bridges also can be used with ac supply voltages and ac meters, not only for the

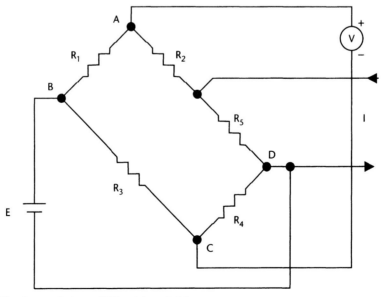

Figure 3.12 Current balanced Wheatstone bridge.

3.4.3 ac Bridge Circuits

The basic concept of dc bridges can also be extended to ac bridges, the resistive elements are replaced with impedances, as shown in Figure 3.13(a). The bridge supply is now an ac voltage. This type of set up can be used to measure small changes in capacitance as required in the capacitive pressure sensor shown in Figure 7.5 in Chapter 7. The differential voltage δV across S is then given by:

$$\delta V = E \frac{Z_2 Z_3 - Z_1 Z_4}{(Z_1 + Z_3)(Z_2 + Z_4)} \qquad (3.20)$$

where E is the ac supply electromotive force (EMF).

When the bridge is balanced, $\delta V = 0$, and (3.20) reduces to:

$$Z_2 Z_3 = Z_1 Z_4 \qquad (3.21)$$

Or

$$R_2(R_3 + j/\omega C_1) = R_1(R_4 + j/\omega C_2)$$

Because the real and imaginary parts must be independently equal

$$R_3 \times R_2 = R_1 \times R_4 \qquad (3.22)$$

and

$$R_2 \times C_2 = R_1 \times C_1 \qquad (3.23)$$

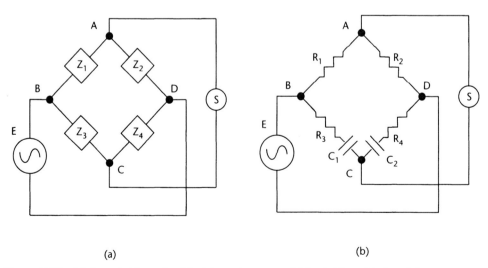

Figure 3.13 (a) An ac bridge using block impedances, and (b) a bridge with R and C components, as described in Example 3.7.

Example 3.7

If the bridge circuit in Figure 3.13(b) is balanced, with $R_1 = 15$ kΩ, $R_2 = 27$ kΩ, $R_3 = 18$ kΩ, and $C_1 = 220$ pF, what are the values of R_4 and C_2?

$$R_2 R_3 = R_1 R_4$$

$$R_4 = 27 \times 18/15 = 32.4 \text{ k}\Omega$$

Additionally,

$$C_2 R_2 = C_1 R_1$$

$$C_2 = 220 \times 15/27 = 122 \text{ pF}$$

Bridge null voltage does not vary linearly with the amount the bridge is out of balance, as can be seen from (3.16). The variation is shown in Figure 3.14(a), so that ΔV should not be used to indicate out-of-balance unless a correction is applied. The optimum method is to use a feedback current to keep the bridge in balance, and measure the feedback current as in (3.19). Figure 3.14(b) shows that for an out-of-balance of less than 1%, the null voltage is approximately linear with ΔR.

3.5 Summary

The effects of applying step voltages to passive components were discussed. When applying a step waveform to a capacitor or inductor, it is easily seen that the currents and voltages are 90° out of phase. Unlike in a resistor, these phase changes give rise to time delays that can be measured in terms of time constants. The phase changes also apply to circuits driven from ac voltages, and give rise to impedances, which can be combined with circuit resistance to calculate circuit characteristics and frequency dependency. Their frequency dependence makes them suitable for frequency selection and filtering.

It is required to measure small changes in resistance in many resistive sensors, such as strain gauges. The percentage change is small, making accurate measurement difficult. To overcome this problem, the Wheatstone Bridge circuit is used. By

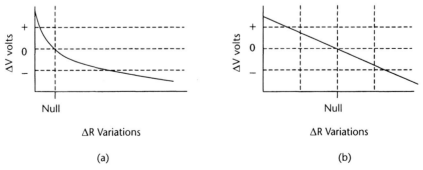

Figure 3.14 Bridge out-of-null voltage plotted against resistor variations, with (a) large variations, and (b) small variations.

comparing the resolution in a voltage divider and a bridge circuit, a large improvement in resolution of the bridge circuit can be seen. Bridge circuits also can be used for temperature compensation, compensation of temperature effects on leads in remote sensing, and with feedback for automatic measurement of resistive changes. The use of bridge networks also can be extended to measure changes in reactive components, as would be required in a capacitive sensor by the use of ac bridge configurations.

References

[1] Sutko, A., and J. D. Faulk, *Industrial Instrumentation,* 1st ed., Delmar Publishers, 1996, pp. 59–61.

[2] Bamble, S., "Demystifying Analog Filter Design," *Sensors Magazine,* Vol. 19, No. 2, February 2002.

[3] Ramson, E., "An Introduction to Analog Filters," *Sensors Magazine,* Vol. 18, No. 7, July 2001.

[4] Johnson, C. D., *Process Control Instrumentation Technology,* 7th ed., Prentice Hall, 2003, pp. 70–81.

[5] Humphries, J. T., and L. P. Sheets, *Industrial Electronics,* 4th ed., Delmar, 1993, pp. 7–11.

[6] Schuler, C. A., *Electronics Principles and Applications,* 5th ed., McGraw-Hill, 1999, pp. 243–245.

[7] Groggins, B., E. Jacobins, and C. Miranda, "Using an FPAA to Design a Multiple-Pole Filter for Low-g Sensing," *Sensors Magazine,* Vol. 15, No. 2, February 1998.

[8] Anderson, K., "Looking Under the (Wheatstone) Bridge," *Sensors Magazine,* Vol. 18, No. 6, June 2001.

[9] Johnson, C. D., *Process Control Instrumentation Technology,* 7th ed., Prentice Hall, 2003, pp. 58–69.

CHAPTER 4
Analog Electronics

4.1 Introduction

Process control electronic systems use both analog and digital circuits. The study of electronic circuits, where the input and output current and voltage amplitudes are continually varying, is known as analog electronics. In digital electronics, the voltage amplitudes are fixed at defined levels, such as 0V or 5V, which represent high and low levels, or "1"s and "0"s. This chapter deals with the analog portion of electronics.

In a process control system, transducers are normally used to convert physical process parameters into electrical signals, so that they can be amplified, conditioned, and transmitted to a remote controller for processing and eventual actuator control or direct actuator control. Measurable quantities are analog in nature; thus, sensor signals are usually analog signals. Consequently, in addition to understanding the operation of measuring and sensing devices, it is necessary to understand analog electronics, as applied to signal amplification, control circuits, and the transmission of electrical signals.

4.2 Analog Circuits

The basic building block for analog signal amplification and conditioning in most industrial control systems is the operational amplifier (op-amp). Its versatility allows it to perform many of the varied functions required in analog process control applications.

4.2.1 Operational Amplifier Introduction

Op-amps, because of their versatility and ease of use, are extensively used in industrial analog control applications. Their use can be divided into the following categories.

1. Instrumentation amplifiers are typically used to amplify low-level dc and low frequency ac signals (in the millivolt range) from transducers. These signals can have several volts of unwanted noise.
2. Comparators are used to compare low-level dc or low frequency ac signals, such as from a bridge circuit, or transducer signal and reference signal, to produce an error signal.

3. Summing amplifiers are used to combine two varying dc or ac signals.
4. Signal conditioning amplifiers are used to linearize transducer signals with the use of logarithmic amplifiers and to reference signals to a specific voltage or current level.
5. Impedance matching amplifiers are used to match the high output impedance of many transducers to the signal amplifier impedance, or amplifier output impedance to that of a transmission line.
6. Integrating and differentiating are used as waveform shaping circuits in low frequency ac circuits to modify signals in control applications.

4.2.2 Basic Op-Amp

The integrated circuit (IC) made it possible to interconnect multiple active devices on a single chip to make an op-amp, such as the LM741/107 general-purpose op-amp. These amplifier circuits are small—one, two, or four can be encapsulated in a single plastic dual inline package (DIP) or similar package. The IC op-amp is a general purpose amplifier that has high gain and low dc drift, so that it can amplify dc as well as low frequency ac signals. When the inputs are at 0V, the output voltage is 0V, or can easily be adjusted to be 0V with the offset null adjustment. Op-amps require a minimal number of external components. Direct feedback is easy to apply, giving stable gain characteristics, and the output of one amplifier can be fed directly into the input of the next amplifier. Op-amps have dual inputs, one of which is a positive input (i.e., the output is in phase with the input), the other a negative input (i.e., the output is inverted from the input). Depending on the input used, these devices can have an inverted or a noninverted output, and can amplify differential sensor signals, or can be used to cancel electrical noise, which is often the requirement with low-level sensor signals. Op-amps are also available with dual outputs (i.e., both positive and negative). The specifications and operating characteristics of bipolar operational amplifiers, such as the LM 741/107 and MOS general purpose and high performance op-amps, can be found in semiconductor manufacturers' catalogs [1].

4.2.3 Op-Amp Characteristics

The typical op-amp has very high gain, high input impedance, low output impedance, low input offset, and low temperature drift. The parameters, although not ideal, give a good basic building block for signal amplification. Typical specifications from a manufacturer's data sheets for a general-purpose integrated op-amp are as follows:

- Voltage gain, 200,000;
- Output impedance, 75Ω;
- Input impedance bipolar, 2 MΩ;
- Input impedance MOS, 10^{12} Ω;
- Input offset voltage, 5 mV;
- Input offset current, 200 nA.

These are just the basic parameters in several pages of specifications [2].

The very high gain of the basic amplifier is called the *open loop gain*, and is difficult to use, because vary small changes at the input to the amplifier (e.g., noise, drift due to temperature changes, and so forth) will cause the amplifier output to saturate. High gain amplifiers also tend to be unstable. The gain of the amplifier can be reduced using feedback, see Section 4.3.1, normally providing a stable fixed gain amplifier. This configuration is called the *closed loop gain*.

Input offset is due to the input stage of op-amps not being ideally balanced. The offset is due to leakage current, biasing current, and transistor mismatch, and is defined as follows.

1. *Input offset voltage.* The voltage that must be applied between the inputs to drive the output voltage to zero.
2. *Input offset current.* The input current required to drive the output voltage to zero.
3. *Input bias current.* Average of the two input currents required to drive the output voltage to zero.

Op-amps require an offset control when amplifying small signals to null any differences in the inputs, so that the dc output of the amplifier is zero when the dc inputs are at the same voltage. In the case of the LM 741/107, provision is made to balance the inputs by adjusting the current through the input stages. This is achieved by connecting a 47k potentiometer between the offset null points, and taking the wiper to the negative supply line, as shown in Figure 4.1. Shown also are the pin numbers and connections (top view) for a DIP.

The ideal signal operating point for an op-amp is at one-half the supply voltage. The LM 741 is normally supplied from +15V and −15V (using a 30V supply), and the signals are referenced to ground [3].

Slew Rate (SR) is a measure of the op-amp's ability to follow transient signals, and affects the amplifier's large signal response. The slew rate is the rate of rise of the output voltage, expressed in millivolts per microsecond, when a step voltage is applied to the input. Typically, a slew rate may range from 500 to 50 mV/μs. The rate is determined by the internal and external capacitances and resistance. The slew rate is given by:

$$SR = \Delta V_o / \Delta t \tag{4.1}$$

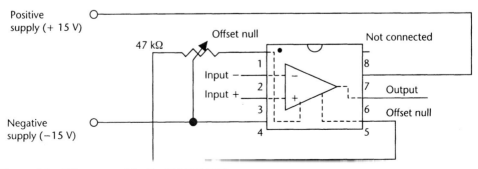

Figure 4.1 Offset control for the LM 741/107 op-amp.

where ΔV_o is the output voltage change, and Δt is the time.

Unity gain frequency of an amplifier is the frequency at which the small signal open loop voltage gain is 1, d or a gain of 0 dB, which in Figure 4.2 is 1 MHz.

The *bandwidth* of the amplifier is defined as the point at which the small signal gain falls 3 dB. In Figure 4.2 and the open loop bandwidth is about 5 Hz, whereas the closed loop bandwidth is approximately 100 kHz, showing that feedback increases the bandwidth.

Gain bandwidth product (GBP) is similar to the slew rate, but is the relation between small signal open loop voltage gain and frequency. It can be seen that the small gain of an op-amp decreases with frequency, due to capacitances and resistances; capacitance is often added to improve stability. The gain versus frequency or Bode plot for a typical op-amp is shown in Figure 4.2. It can be seen from the graph that the GBP is a constant. The voltage gain of an op-amp decreases by 10 (20 dB) for every decade input frequency increase.

The GBP of an op-amp is given by:

$$\text{GBP} = \text{BW} \times A_v \qquad (4.2)$$

where BW is the bandwidth and A_v is the closed loop voltage gain of the amplifier.

Example 4.1

What is the bandwidth of the amplifier whose GBP is 1.5 MHz, if the voltage gain is 180?

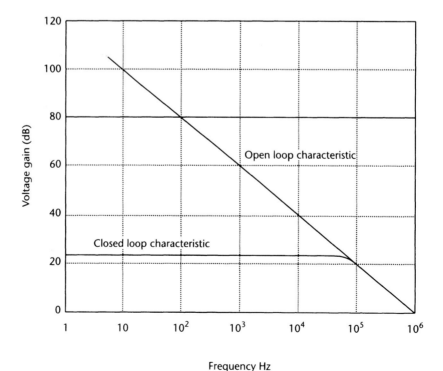

Figure 4.2 Typical frequency response or Bode plot of a compensated op-amp.

4.3 Types of Amplifiers

$$BW = GBP/A_v = 1.5 \text{ MHz}/180$$
$$BW = 8.3 \text{ kHz}$$

4.3 Types of Amplifiers

The op-amp can be configured for voltage or current signal amplification, conversion of voltage signals to current signals and vice versa, impedance matching, and comparison and summing applications in process control.

4.3.1 Voltage Amplifiers

Inverting Amplifier

Consider the ideal op-amp shown in Figure 4.3, which is configured as an inverting voltage amplifier. Resistors R_1 and R_2 provide feedback; that is, some of the output signal is fed back to the input (see Section 15.2.1). The large amplification factor in op-amps tends to make some of them unstable, and causes dc drift of the operating point with temperature. Feedback stabilizes the amplifier, minimizes dc drift, and sets the gain to a known value.

When a voltage input signal is fed into the negative terminal of the op-amp, as in Figure 4.3, the output signal will be inverted. Because of the high input impedance of the op-amp, no current flows into the input, or $I_3 = 0$. Then, since the sum of the currents at the negative input is zero:

$$I_1 + I_2 = 0 \tag{4.3}$$

The voltage at the junction of R_1 and R_2 is zero, the same as the positive input, and is termed a virtual ground from which:

$$V_{out} = \frac{-V_{in} R_2}{R_1} \tag{4.4}$$

The negative sign is because the output signal is negative. In this configuration, the closed loop voltage gain of the stage is:

$Z_{in} = R_1$

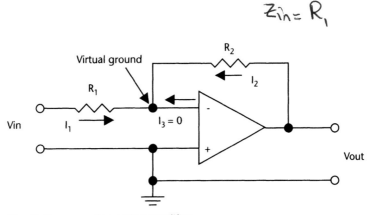

Figure 4.3 Circuit diagrams of inverting amplifier.

$$Gain = \frac{-E_{out}}{E_{in}} = \frac{-R_2}{R_1} \quad (4.5)$$

Unfortunately, ideal op-amps do not exist. Op-amps have finite open loop gain, an output impedance, and finite input impedance. Circuit analysis shows that in a typical amplifier circuit, the effect of these parameters on (4.5) is less than 0.1%, so that in most practical cases, the effects can be ignored.

Example 4.2

In Figure 4.2, if resistor $R_1 = 2{,}700\Omega$ and resistor $R_2 = 470\ k\Omega$, what is the gain, and what is the output voltage amplitude if the ac input voltage is 1.8 mV?

$$Gain = \frac{R_2}{R_1} = \frac{470}{2.7} = 174$$

ac Output Voltage $= -1.8 \times 174$ mV $= -313.5$ mV $= -0.31$V

Summing Amplifier

A summing amplifier is a common use of the inverting amplifier, which is used to sum two or more voltages (see Section 16.6.6). The summing circuit is shown in Figure 4.4. The transfer function is given by:

$$V_{out} = -\left(\frac{V_1 R_2}{R_3} + \frac{V_2 R_2}{R_1}\right) \quad (4.6)$$

Example 4.3

In Figure 4.3, $R_3 = R_1 = 5.6\ k\Omega$, and $R_3 = 220\ k\Omega$. (a) If $V_1 = 27$ mV and $V_2 = 0$V, what is V_{out}? (b) If V_2 is changed to 43 mV, what is the new V_{out}?

$$(a)\ V_{out} = \left(\frac{27 \times 220}{5.6} + 0\right) mV = 1.06V$$

$$(b)\ V_{out} = \left(\frac{27 \times 220}{5.6} + \frac{43 \times 220}{5.6}\right) mV = 1.06V + 1.69V = 2.75V$$

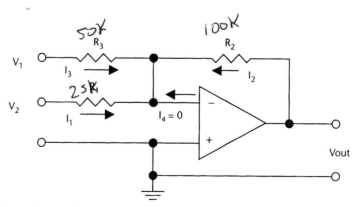

Figure 4.4 Summing amplifier.

4.3 Types of Amplifiers

Noninverting Amplifier

A noninverting amplifier configuration is shown in Figure 4.5. When the input signal is fed into the positive terminal, the circuit is noninverting [4]. Since the negative input is referenced to the positive input, the voltage at the negative terminal is V_{in}. Using Ohm's Law, this gives;

$$\frac{V_{in}}{R_1} = \frac{V_{out} - V_{in}}{R_2}$$

From which the voltage gain is given by:

$$Gain = \frac{E_{out}}{E_{in}} = 1 + \frac{R_2}{R_1} \quad (4.7)$$

In this configuration, the amplifier gain is 1 plus the resistor ratio, so that the gain does not directly vary with the resistor ratio. However, this configuration does give a high input impedance (i.e., that of the op-amp), and a low output impedance.

Example 4.4

In Figures 4.3 and 4.5, $R_1 = 3.9$ kΩ and $R_2 = 270$ kΩ. If a dc voltage of 0.03V is applied to the inputs of each amplifier, what will be the output voltages?

From Figure 4.3:

$$V_{out} = \frac{-270 \times 0.03V}{3.9} = -2.077V$$

$V_{out} = -V_{in}\left(\frac{R_2}{R_1}\right)$

From Figure 4.5:

$$V_{out} = \left(1 + \frac{270}{3.9}\right)0.03V = +2.11V$$

An alternative noninverting amplifier using two inverters is shown in Figure 4.6. The second inverter has unity gain. In this case, the gain is the ratio of R_2 to R_1, as in the case of the inverting amplifier.

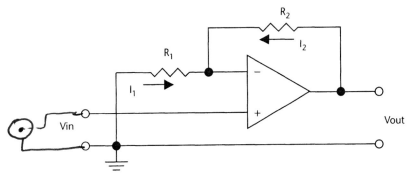

Figure 4.5 Noninverting amplifier circuit using single op-amp.

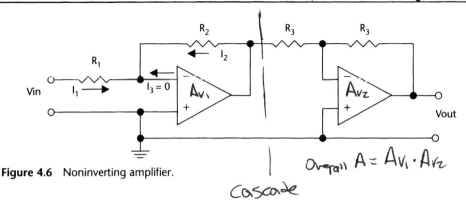

Figure 4.6 Noninverting amplifier.

Voltage Follower

The voltage follower circuit, as shown in Figure 4.7, is an impedance matching device with unity gain (see Section 5.4.1). The configuration has a very high input impedance (>100 MΩ with MOS devices), and low output impedance (<50Ω). The output drivers of the op-amp limit the maximum output current.

Differencing Amplifier

There are many applications for differencing amplifiers in instrumentation, such as amplifying the output from a Wheatstone bridge. Ideally, the output voltage is given by:

$$V_{out} = A(V_1 - V_2) \tag{4.8}$$

where A is the gain of the amplifier.

The basic differencing amplifier is shown in Figure 4.8. The resistors should be matched as shown. Then, for an ideal op-amp, the transfer function is given by:

$$V_{out} = \frac{R_2}{R_1}(V_2 - V_1) \tag{4.9}$$

This equation indicates that the output voltage depends only on the resistor ratio and the input voltage difference $\Delta V(V_2 - V_1)$, and is independent of the input operating point. That is, V_2 can vary from ±10V, and if ΔV does not change, then

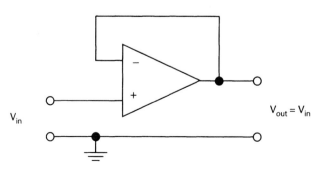

Figure 4.7 Voltage follower.

4.3 Types of Amplifiers

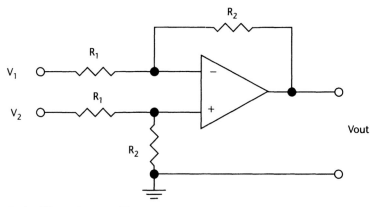

Figure 4.8 Basic differencing amplifier.

the output will not change. In practice, real differential amplifiers only can approach the ideal case.

The Common-Mode (CM) input voltage is defined as the average of the applied input voltages:

$$CM = (V_1 + V_2)/2 \qquad (4.10)$$

The CM gain (A_{CM}) is the ratio of change in output voltage for a change in CM input voltage (ideally zero) [5].

The gain (A) of the op-amp is the ratio of the change in output voltage for a differential input voltage; see (4.8).

The Common-Mode Rejection Ratio (CMRR) is the ratio of the differential gain to the CM gain.

$$CMRR = A/A_{CM} \qquad (4.11)$$

The CMRR is normally expressed as the Common-Mode Rejection (CMR) in decibels.

$$CMR = 20 \log_{10} (CMRR) \qquad (4.12)$$

The higher the CMR value, the better the circuit performance. Typically, values of CMR range from 60 to 100 dB.

Example 4.5

In the dc amplifier shown in Figure 4.8, an input of 130 mV is applied to terminal A and −85 mV is applied to terminal B. What is the output voltage, assuming the amplifier was zeroed with 0V at the inputs? $R_2 = 120k$, $R_1 = 4.7k$

$$E_{out} = \frac{\pm \Delta V_{in} \times 120}{4.7} = [-130 + (85)]mV \times \frac{120}{4.7} = -0.215 \times 25.5V = +5.5V$$

The disadvantage of this circuit is that the input impedance is not very high, and is different for the two inputs. To overcome these drawbacks, voltage followers are

$(V_2 - V_1) = (130mV - -85mV) = 215mV$

normally used to buffer the inputs, resulting in the circuit shown in Figure 4.9, which is used in instrument differencing amplifiers (see Section 15.3.2).

Example 4.6

In the amplifier shown in Figure 4.9, R_1 = 4.7 k?, and R_2 = 560 k?. If V_1 and V_2 are both taken from 0V to 3.7V, and the output changes from 0V to 0.19V, then what is the CMR of the circuit?

$$A_{CM} = 0.19/3.7 = 0.051$$

$$A = 560/4.7 = 119.15$$

$$CMR = 20 \log 119.15/0.051 = 20 \log 2340 = 67.37 \text{ dB}$$

4.3.2 Converters

The circuits shown above were for voltage amplifiers. Op-amps also can be used as current amplifiers, voltage to current converters, current to voltage converters, and special purpose amplifiers.

A *current to voltage converter* is shown in Figure 4.10. When used as a converter, the relation between input and output is called the *Transfer function μ* (or *Ratio*). These devices do not have gain as such, because of the different input and output units. In Figure 4.10, the transfer ratio is given by:

$$\mu = \frac{-E_{out}}{I_{in}} = R_1 \qquad (4.13)$$

Example 4.7

In Figure 4.10, the input current is 83 μA and the output voltage is 3.7V. What is the transfer ratio, and the value of R_1?

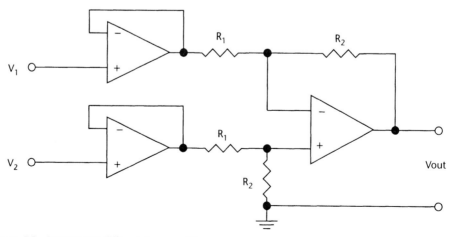

Figure 4.9 Instrument differencing amplifier.

4.3 Types of Amplifiers

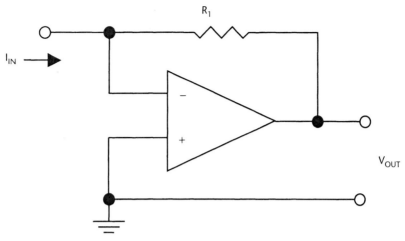

Figure 4.10 Current to voltage converter.

$$\mu = \frac{3.7V}{83\mu A} = 44.5 V/mA = 44.5 kV/A$$

$$R_1 = \frac{3.7V}{83\mu A} = 44.5 k\Omega$$

A *voltage to current converter* is shown in Figure 4.11. The op-amp converts a voltage into a current. In this case, the transfer ratio is given by:

$$\frac{I_{out}}{E_{in}} = \frac{-R_2}{R_1 R_3} \text{ mhos} \tag{4.14}$$

In this case, the units are in mhos (1/ohms), and the resistors are related by the equation:

$$R_1 (R_3 + R_5) = R_2 R_4 \tag{4.15}$$

However, in industrial instrumentation, a voltage to current converter is sometimes referred to as a current amplifier.

Example 4.8

In Figure 4.11, $R_1 = R_4 = 6.5$ kΩ, and $R_2 = 97$ kΩ. What is the value of R_3 and R_5 if the op-amp is needed to convert an input of 2.2V to an output of 15 mA?

$$\frac{I_{out}}{E_{in}} = \frac{R_2}{R_1 R_3} = \frac{15 \times 10^{-3}}{2.2} = 6.8 \times 10^{-3}$$

$$\frac{97 \times 10^3}{6.5 \times 10^3 \times R_3} = 6.8 \times 10^{-3}$$

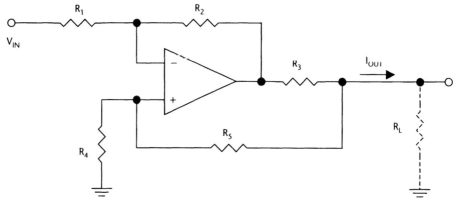

Figure 4.11 Voltage to current converter.

$$R_3 = \frac{14.9}{6.8 \times 10^{-3}} = 2.2 k\Omega$$

$$R_5 = \frac{R_2 R_4}{R_1 - R_3} = R_2 - R_3 = (97 - 2.2)k\Omega = 94.8 k\Omega$$

4.3.3 Current Amplifiers

Devices that amplify currents are referred to as current amplifiers. Figure 4.12 shows a basic current amplifier. The gain is given by:

$$\frac{I_{out}}{I_{in}} = \frac{R_3 R_1}{R_2 R_4} \quad I_{out} = \frac{R_3 R_1}{R_2 R_4} \times I_{in} \quad (4.16)$$

where the resistors are related by the equation:

$$R_2(R_4 + R_6) = R_3 R_5 \quad (4.17)$$

$$\frac{R_3}{R_2} = \frac{(R_4 + R_6)}{R_5}$$

Example 4.9

In Figure 4.12, $I_1 = 1.23 \mu A$, $R_1 = 12 k\Omega$, $R_4 = 3.3 k\Omega$, $R_5 = 4.6 k\Omega$, and $R_6 = 120 k\Omega$. What is the gain and output current?

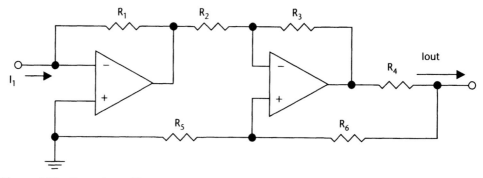

Figure 4.12 Current amplifier.

From (4.17) $R_3/R_2 = (R_4 + R_6)/R_5 = (3.3 + 120)/4.6 = 26.8$

From (4.16) $I_{out} = I_{in} R_3 \times R_1/R_2 \times R_4 = 1.23 \times 26.8 \times 12/3.3\ \mu A$
  $I_{out} = 119.9\ \mu A = 0.12\ mA$

Gain = 119.9/1.23 = 97.45

4.3.4 Integrating and Differentiating Amplifiers

Two important functions that are used in process control (see Section 16.6.5) are integration and differentiation [6].

An *integrating amplifier* is shown in Figure 4.13(a). In this configuration, the feedback resistor is replaced with an integrating capacitor. Using the ideal case, the currents at the input to the amplifier can be summed as follows:

$$V_{in}/R_1 + C\, dV_{out}/dt = 0 \tag{4.18}$$

Solving the equation gives:

$$V_{out} = -\frac{1}{RC}\int V_{in}\, dt \tag{4.19}$$

This shows that the output voltage is the integral of the input voltage, with a scale factor of $-1/RC$. If V_{in} is a step function, then the circuit can be used as a linear ramp generator. Assume V_{in} steps E volts and stays at that value, the equation reduces to:

$$V_{out} = -(E/RC)t \tag{4.20}$$

which is a linear ramp with a negative slope of E/RC. This circuit in practice is used with outer circuits for limiting and resetting.

A *differentiating circuit* is shown in Figure 4.13(b). In this configuration, the input resistor is replaced with a capacitor. The currents can be summed in the ideal case as follows:

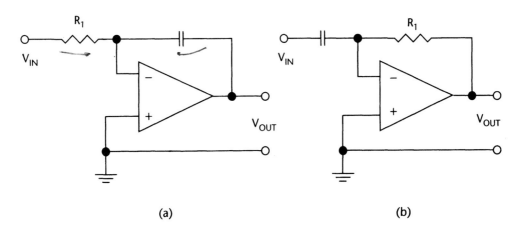

Figure 4.13 Circuits showing (a) integrating amplifier and (b) differentiating amplifier.

$$C = \frac{dV_{in}}{dt} + \frac{V_{out}}{R_1} = 0 \qquad (4.21)$$

Solving for the output voltage, the response is;

$$V_{out} = -RC\frac{dV_{in}}{dt} \qquad (4.22)$$

Showing that V_{out} is the derivative of V_{in}, the circuit tends to be unstable, but in practice is easily stabilized when used with other circuits.

4.3.5 Nonlinear Amplifiers

Many sensors have a logarithmic or nonlinear transfer characteristic. Such devices require signal linearization. This can be implemented using amplifiers with nonlinear characteristics. These are achieved by the use of nonlinear elements, such as diodes or transistors in the feedback loop [7].

Two examples of *logarithmic amplifiers* are shown in Figure 4.14. Figure 4.14(a) shows a logarithmic amplifier using a diode in the feedback loop, and Figure 4.14(b) shows a logarithmic amplifier using a bipolar junction transistor in the feedback loop. The summation of the currents in an ideal case gives:

$$V_{in}/R_1 = I_D \qquad (4.23)$$

where I_D is the current flowing through the diode, or the collector/emitter current of the transistor.

The relation between the diode voltage (V_D) and the diode current is given by:

$$V_D = A \log_c I_D/I_R \qquad (4.24)$$

where A is the proportionality constant depending on c the base of the logarithm, and I_R is the diode leakage current.

Because the voltage across the diode is also the output voltage (V_{out}), (4.23) and (4.24) can be combined, giving:

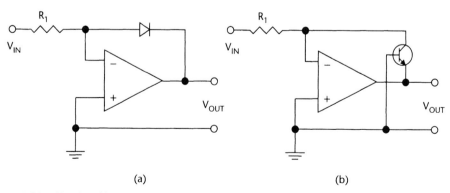

Figure 4.14 Circuits of logarithmic amplifiers using (a) a diode feedback, and (b) an NPN transistor.

4.3 Types of Amplifiers

$$V_D = V_{out} = A \log_C V_{in}/R_1 I_R = A \log_C V_{in} - A \log_C R_1 I_R \quad (4.25)$$

which shows that the logarithmic relationship between V_{out} and V_{in}, and $A \log_C R_1 I_R$ is an offset constant.

The relation between the transistor's base emitter voltage V_{BE} and the collector/emitter current (I_C) is given by:

$$V_{BE} = A \log_C I_C/I_R \quad (4.26)$$

Because the V_{BE} of the transistor is also V_{out}, (4.23) and (4.26) can be combined to give:

$$V_{BE} = V_{out} = A \log_C V_{in}/R_1 I_R = A \log_C V_{in} - A \log_C R_1 I_R \quad (4.27)$$

which shows that the logarithmic relationship between V_{out} and V_{in}, and $A \log_C R_1 I_R$ is an offset constant.

Combinations of resistors and nonlinear elements can be used in multiple feedback loops to match the characteristics of many sensors for linearization of the output from the sensor (see Section 15.2.2). Logarithmic amplifiers are commercially available from some device manufacturers.

Antilogarithmic amplifiers perform the inverse function of the logarithmic amplifier. Two versions of the antilogarithmic amplifier are shown in Figure 4.15. The circuit equations can be obtained in a similar manner to the equations for the logarithmic amplifiers.

4.3.6 Instrument Amplifiers

Because of the very high accuracy requirements in instrumentation, op-amp circuits are not ideally suited for low-level instrument signal amplification, but require impedance matching. Op-amps can have different input impedances at the two inputs. The input impedances can be relatively low, tend to load the sensor output, and can have different gains at the inverting and noninverting inputs. Common mode noise can be a problem. An op-amp configured for use as an instrument amplifier is shown in Figure 4.16. This amplifier has balanced inputs with very high input impedance, low output impedance, and high common mode noise reduction.

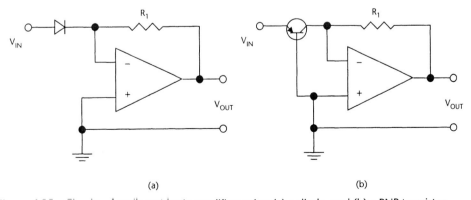

Figure 4.15 Circuits of antilogarithmic amplifiers using (a) a diode, and (b) a PNP transistor.

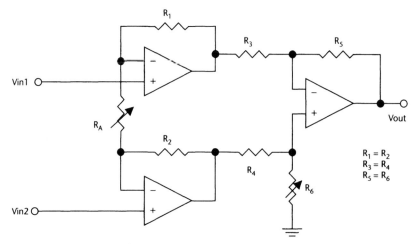

Figure 4.16 Circuit schematic of an instrumentation amplifier.

Gain is set by R_A. Devices similar to the one shown in Figure 4.16 are commercially available for instrumentation. R_A is normally an external component, so that the end user can set the gain [8].

The output voltage is given by:

$$V_{out} = \frac{R_5}{R_3}\left(\frac{2R_1}{R_A} + 1\right)(V_{IN2} - V_{IN1}) \quad (4.28)$$

Figure 4.17 shows a practical circuit using an instrumentation amplifier, which is used to amplify the output signal from a resistive bridge. R_6 is used to adjust for any zero signal offset. This circuit also can be used as a differential transmission line receiver, as shown in Section 14.4.2.

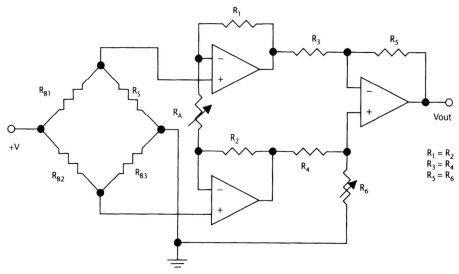

Figure 4.17 Instrumentation amplifier used for offset adjustment and amplification of a signal from a bridge.

4.3.7 Input Protection

Amplifiers, like all ICs, are susceptible to damage from excessive input voltages, such as from input voltages that are larger than the supply voltages, electrostatic discharge (ESD), or EMI pickup [9].

The inputs of ICs are internally protected. However, the protection can give rise to leakage currents, so that the protection is limited. Typically, the overvoltage protection is limited to approximately ±8V greater than the supply voltages. That is, with a ±15V supply, the protection is ± 23V. The protection can be improved by the use of external resistors and clamps, if this is practicable.

Electrostatic discharge is the biggest problem, particularly for CMOS devices, because of their high input impedance. There are two main sources of ESD. The human body can generate a large amount of static energy, which can destroy a device just by handling it. Hence, there is the need for ground straps when handling ICs. The other source of ESD, which can be as high as 16 kV, is from equipment. The human body and equipment ESD model is given in MIL-STD 883B (military standard), and similar models are given in the IEC 1000.4.2 standard.

EMI normally can be reduced to an acceptable level by capacitive filtering at the input to the IC.

4.4 Amplifier Applications

The stage gain of an op-amp with feedback is limited to about 500, primarily by feedback resistor values, as well as from amplifier considerations. For instance, an amplifier with a 5 kΩ input resistor would have a feedback resistor of 2.5 MΩ, which is comparable to the input impedance of the bipolar op-amp.

In process control, amplifiers are used in many applications other than signal amplification, filtering, and linearization [10]. Some of these applications are as follows:

- Capacitance Multiplier;
- Gyrator;
- Sine Wave Oscillator;
- Power Supply Regulator;
- Level Detection;
- Sample and Hold;
- Voltage Reference;
- Current Mirror;
- Voltage to Frequency Converter;
- Voltage to Digital Converter;
- Pulse Amplitude Modulation.

More information on the design and use of these circuits can be found in analog electronic textbooks.

4.5 Summary

This chapter introduced and discussed integrated op-amps, and how their low drift characteristics make them a suitable building block for both low frequency ac and dc small signal amplification. However, op-amps are not ideal amplifiers because of the mismatch at the inputs, input impedance, and different gain at the inputs. The high open loop gain characteristics of the op-amp make it necessary to use feedback for stabilization, and the use of a set zero is required for adjustment of the input mismatch. The op-amp is a very versatile device, and can be used in many configurations for amplifying low-level voltages or currents, summing voltages, and converting between voltage and current. They also may be used as nonlinear amplifiers, as comparators, and for waveform shaping. The use of op-amps is not limited to signal amplification in process control; they have many other applications. Op-amps are susceptible to excessive supply voltage and ESD, so that care and protection is needed in handling.

References

[1] Mancini, R., *Op-Amps for Everyone*, 1st ed., Elsevier Publishing, 2003.
[2] Schuler, C. A., *Electronics Principles and Applications*, 5th ed., McGraw-Hill, 1999, pp. 221–238.
[3] Sutko, A., and J. D. Faulk, *Industrial Instrumentation*, 1st ed., Delmar Publishers, 1996, pp. 80–89.
[4] Wurcer, S. A., and L. W. Counts, "A Programmable Instrument Amplifier for 12 Bit Resolution Systems," *Proc. IEEE Journal of Solid State Circuits*, Vol. SC17, No. 6, December 1982, pp. 1102–1111.
[5] Nash, E., "A Practical Review of Common Mode and Instrumentation Amplifiers," *Sensors Magazine*, Vol. 15, No. 7, July 1998.
[6] Johnson, C. D., *Process Control Instrumentation Technology*, 7th ed., Prentice Hall, 2003, pp. 97–99.
[7] Humphries, J. T., and L. P. Sheets, *Industrial Electronics*, 4th ed., Delmar, 1993, pp. 5–8.
[8] Harrold, S., "Designing Sensor Signal Conditioning with Programmable Analog ICs," *Sensors Magazine*, Vol. 20, No. 4, April 2003.
[9] Bryant, J., et al., "Protecting Instrumentation Amplifiers," *Sensors Magazine*, Vol. 17, No. 4, April 2000.
[10] Humphries, J. T., and L. P. Sheets, *Industrial Electronics*, 4th ed., Delmar, 1993, pp. 46–70.

Digital Electronics

5.1 Introduction

Digital electronics has given us the power to accurately control extremely complex processes that were beyond our wildest dreams a few years ago [1]. It would take many volumes to cover the subject of digital technology, so in this text we can only scratch the surface. There is a place for both analog and digital circuits in instrumentation. Sensors and instrumentation functions are analog in nature. However, digital circuits have many advantages over analog circuits. Analog signals are easily converted to digital signals using commercially available analog to digital converters (ADC). In new designs, digital circuits will be used wherever possible.

Some of the advantages of digital circuits are:

- Lower power requirements;
- Increased cost effectiveness;
- Ability to control multivariable systems simultaneously;
- Ability to transmit signals over long distances without loss of accuracy and elimination of noise;
- Higher speed signal transmission;
- Memory capability for data storage;
- Compatibility with controllers and alphanumeric displays.

5.2 Digital Building Blocks

The basic building blocks used in digital circuits are called gates. The types of gates are Buffer, Inverter, AND, NAND, OR, NOR, XOR, and XNOR [2]. These basic blocks are interconnected to build functional blocks, such as encoders, decoders, adders, counters, registers, multiplexers, demultiplexers, memory, and so forth. The functional blocks are then interconnected to make systems, such as calculators, computers, microprocessors, clocks, function generators, transmitters, receivers, digital instruments, telephone systems, ADCs, and Digital to Analog Converters (DAC), to name a few.

Figure 5.1 shows the traditional logic symbols used together with the Boolean equation describing the gate function. The gates were originally developed using bipolar technology, but are now made using CMOS technology, which has the

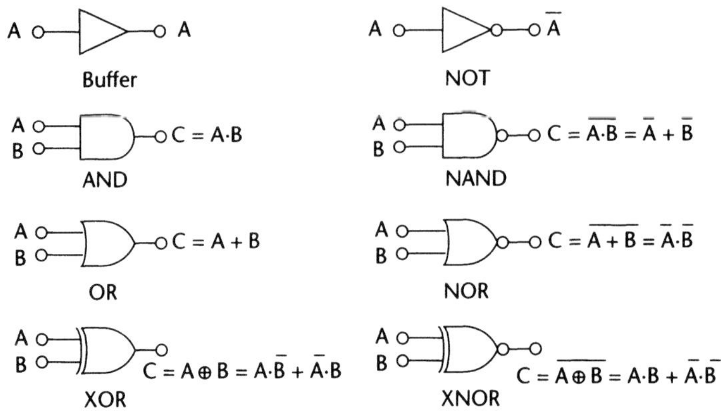

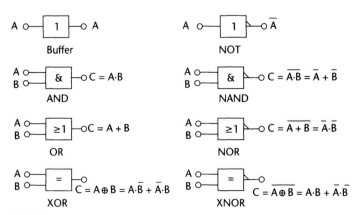

Figure 5.1 Traditional digital logic gate symbols.

advantages of low power requirements, small size, high speed, and high fanout, or the ability to drive a large number of gates.

The American National Standards Institute (ANSI) and the IEEE have developed a set of standard symbols for gates, which they are pushing very hard for acceptance. These symbols are given in Figure 5.2. Either set of logic symbols may be encountered in practice, so it is necessary to be familiar with both. A lesser-known third set of logic symbols was developed by the National Electrical Manufacturers Association (NEMA) [3]

To understand how a problem is analyzed to obtain a Boolean expression, consider the system shown in Figure 5.3. A storage tank is being filled with a liquid, and it is necessary to sound an alarm when certain parameters exceed specifications. We label the variables as follows: the flow rate is A, the humidity B, and the temperature C. Set points have been established for these variables, and depending on whether the variables are above or below the set points, a 1 or 0 is assigned to each variable, in order to develop the Boolean expression for the system. When the Boolean variable Y goes to the 1 state, an alarm will be activated. The conditions for turning on the alarm are:

- High flow with low temperature;
- High flow with high humidity;

Figure 5.2 ANSI/IEEE standard logic symbols.

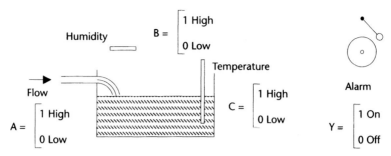

Figure 5.3 Setup to illustrate the development of a Boolean equation.

- Low flow with high humidity and low temperature.

Using these conditions, we can now define a Boolean expression that will give Y = 1 for each equation, as follows:

- $Y = 1 = A \cdot \overline{C}$ condition 1
- $Y = 1 = A \cdot B$ condition 2
- $Y = 1 = \overline{A} \cdot B \cdot \overline{C}$ condition 3

These three equations can now be combined with the OR operation, so that if any of these conditions exist, the alarm will be activated. This gives:

$$Y = A \cdot \overline{C} + A \cdot B + \overline{A} \cdot B \cdot \overline{C} \tag{5.1}$$

This equation can now be used to define the digital logic required to activate the alarm system.

Example 5.1

Develop a digital circuit to implement the alarm conditions discussed for Figure 5.3, using the gate symbols given in Figure 5.1.
   The starting point is the Boolean expression developed for Figure 5.3, which is (5.1).

$$Y = A \cdot \overline{C} + A \cdot B + \overline{A} \cdot B \cdot \overline{C}$$

There are many gate combinations that can be used to implement this logic equation. One possible solution using AND and OR gates is given in Figure 5.4. Any equations or Boolean expressions should be simplified before implementation.

5.3 Converters

We live in an analog world and sensor measurements are analog in nature. All of our computational functions, signal transmission, data storage, signal conditioning, and so forth, derive many benefits from the digital world. It is therefore necessary to convert our analog signals into a digital format for processing, and then back to

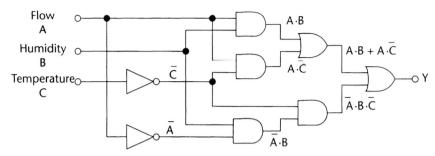

Figure 5.4 Solution for Example 5.1.

analog for final control. The interface between the analog world and digital world uses converters. An ADC changes the analog signal into a digital format, and a DAC changes the digital signal back to analog. The characteristics of these converters must be precise and accurately known, to establish the relationship between the analog and digital signal.

5.3.1 Comparators

The simplest form of information transfer between an analog signal and a digital signal is a comparator. This device is simply a high gain amplifier that is used to compare two analog voltages, and depending on which voltage is larger, will give a digital "0" or "1" signal. This device is shown in Figure 5.5(a), with the input and output waveforms in Figure 5.5(b). The comparator is an integral part of ADCs and DACs, and of many monitoring devices.

One of the input voltages in this case, V_a, is known as a fixed trigger level, a set level, or a reference voltage. The other voltage is the variable, which when compared to V_a in a comparator, will give a digital "1" or "0" signal, depending on whether it is greater than or less than the reference voltage.

Example 5.2

A "1" signal is required to trigger an alarm, if the fluid level in a tank is more than 3m deep. The level sensor gives an output of 9.3 mV for every centimeter increase in depth. What is the required alarm voltage?

In this case, the sensor is connected to terminal V_b, and a reference level of 2.79V (9.3 × 100 × 3 mV) is connected to terminal V_a in Figure 5.2(a).

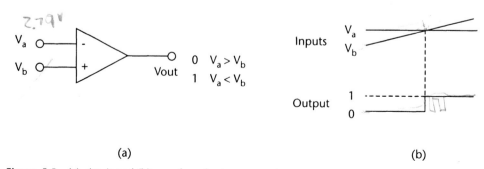

Figure 5.5 (a) circuit and (b) waveform basic comparators.

5.3 Converters

Hysteresis is obtained with positive feedback, as shown in Figure 5.6(a), and is often used in comparators to minimize or overcome noise problems. Some noise can be filtered out, but it is difficult to completely eliminate all of it. Noise can cause the comparator to switch back and forth, giving uncertainty in the trigger point. This is shown in Figure 5.6(b), where input V_b is varying as it increases due to noise. This input, when used in a comparator without feedback, gives several "1" level outputs as shown, which may cause problems when trying to interpret the signal. If positive feedback (or hysteresis) is used, as shown in Figure 5.6(a), then a clean output is obtained, as shown in the lower waveform in Figure 5.6(b). Positive feedback produces a dead band. Once the comparator has been triggered, the trigger point is lowered, so that the varying input must drop to below the new reference point before the comparator output will go low.

In Figure 5.6(a), the condition for the output to go high is given by:

$$V_b \geq V_a \tag{5.2}$$

After the output has been driven high, the condition for the output to return to low is given by:

$$V_b \leq V_a - (R_1/R_2)V_h \tag{5.3}$$

where V_h is the voltage of the output when high.

The dead band or hysteresis is given by $(R_1/R_2)V_h$, and therefore can be selected by the choice of resistors.

Example 5.3

In Example 5.2, if there are waves with amplitude of 35 cm due to pumping, what is the value of R_2 to give a dead band with a 5-cm safety margin to prevent the comparator output from going low? Assume $R_1 = 5$ kΩ, and output high = 5V.

$$\text{The dead band is } (35 + 5) \times 9.3 \text{ mV} = 37.2 \text{ mV}$$
$$(R_1/R_2)V_h = 37.2 \text{ mV}$$
$$R_2 = 5 \times 5/0.0372 \text{ k}\Omega = 672 \text{ k}\Omega$$

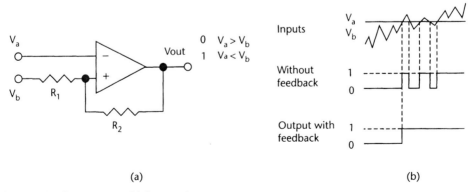

Figure 5.6 Comparator with hysteresis.

5.3.2 Digital to Analog Converters

There are two basic methods of converting digital signals to analog signals: DACs, which are normally used to convert a digital word into a low power voltage reference level or waveform generation; and pulse width modulation (PWM), which is used to convert a digital word into a high power voltage level for actuator and motor control [4].

DACs change digital information into analog voltages using a resistor network or a current mirror method. Using either of these methods, the analog signals are low power and are normally used as a low power voltage level, but can be amplified and used for control. Using a resistor network, a DAC converts a digital word into an analog voltage by using the resistors to scale a reference voltage, resulting in a voltage value proportional to the value of the binary word. For instance, when the binary value is zero, the output voltage is zero, and when the binary number is at a maximum, the output is a fraction less than the reference voltage, which may be scaled up to give discrete output voltage levels. The output voltage (V_{out}) from a DAC is given by:

$$V_{out} = V_{ref}\left(2^{n-1} + ----+2^1 + 2^0\right)/2^n \quad (5.4)$$

For an 8-bit device, the maximum output voltage is $V_{ref} \times 255/256 = 0.996\ V_{ref}$, and for a 10-bit DAC, the maximum output voltage is $V_{ref} \times 1023/1024 = 0.999\ V_{ref}$, showing that the maximum output voltage is slightly less than the reference voltage. In the 8-bit DAC, the reference voltage may be scaled up by 256/255 to give discrete output voltage steps.

Example 5.4

An 8-bit DAC has a reference voltage of 5V. What would be the voltage corresponding to a binary word of 10010011?

$$V_{out} = 5(128 + 16 + 2 + 1)/256\text{V}$$
$$V_{out} = 2.871\text{V}$$

A DAC is typically an IC in a black box, but it can be constructed from discrete components. It is usually more cost effective to use an IC, but it can be useful in some cases to understand the structure, which may have uses in other applications. A DAC uses either a resistive ladder network, which can be resistor ratios, such as R, $2R$, $4R$, or $8R$, but in this method the resistors can get very large. For instance, if the lowest value resistor is 5 kΩ, then the spread for 8 bits is up to 1.28 MΩ. This wide spread in resistors leads to inaccuracies, due to the different coefficients of resistance with temperature at high and low resistor values. A more practical resistor network is the R-$2R$ ladder, where only two values of resistor are required. A 4-stage ladder network is shown in Figure 5.7, where CMOS switches are used to switch the reference voltage (V_R), and ground to the resistor network.

In an R-$2R$ network, a Thévenin voltage source can be used to obtain the relation between the output voltage (V_{out}) and the voltage applied to the resistor (V_R or 0). The output voltage is given by:

5.3 Converters

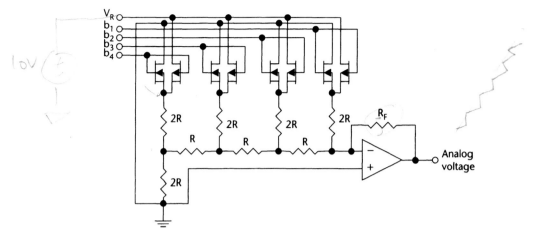

Figure 5.7 Typical DAC using an R/2R resistor network.

$$V_{out} = (-R_F)\left(\frac{V^n}{2R} - - \frac{V_2}{2^{n-1}R} + \frac{V1}{2^n R}\right) \quad (5.5)$$

where R_F is the amplifier feedback resistor, $V_n, --V_2, V_1$ (V_R or 0) (V_n being the MSB and V_1 the LSB) are the voltages applied to the 2R resistors in the R-2R network.

Example 5.5

What is the output voltage from a 4-bit R-2R DAC, if the feedback resistor is 10 k? and the network R = 5 kΩ? Assume a reference voltage of 4.8V, and a binary input of 1011.

$$V_{out} = (-10k\Omega)\left(\frac{4.8V}{2 \times 5k\Omega} + \frac{0V}{4 \times 5k\Omega} + \frac{4.8V}{8 \times 5k\Omega} + \frac{4.8V}{16 \times 5k\Omega}\right)$$

$$V_{out} = (-10)(0.48 + 0.12 + 0.06)V = -6.6V$$

or using (5.4),

V_{out} = 4.8 × 11/16V × −2 (amplifier gain) = −6.6V

The amplifier gain is 2, as the network impedance is 5 kΩ, and the amplifier uses 10 kΩ in the feedback.

The linear transfer function of a 3-bit DAC is shown in Figure 5.8. In this case, there are eight steps with the maximum voltage equal to seven-eighths of the reference voltage [5].

An alternative method to resistor ladders used in integrated circuit DACs is a current mirror technique to provide the transfer function. A 4-bit DAC using a current mirror is shown in Figure 5.9. The ratios of the sizes of the P-MOS devices to the reference P-MOS device are binary, to give binary current ratios when the N-MOS devices in series with the P devices are turned On. Thus, the current through R_2 is proportional to the value of the binary input. Large ratios of device

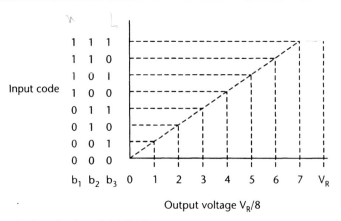

Figure 5.8 Transfer function for a 3-bit DAC.

sizes are not required as are required with binary resistors, because different reference currents can be generated with mirroring techniques. The reference current for the mirror is set by R_1, which is normally an integral part of a voltage reference, such as a bandgap reference. The advantages of the mirroring technique include the following: the devices are smaller than resistors; they can be mirrored with an accuracy equal to or greater than with resistor ratios; and the impedance of the N-MOS switches is not critical, because of the high output impedance of the P-MOS current mirrors.

Commercial DACs, such as the DAC 0808, are shown in Figure 5.10. The DAC 0808 is an 8-bit converter using an R-2R ladder network, which will give an output resolution or accuracy of 1 in $(2^8 - 1)$. The -1 is necessary, because the first number is 0, leaving 255 steps. This shows that an 8-bit DAC can reproduce an analog voltage to an accuracy of $\pm 0.39\%$. For higher accuracy analog signals, a 12-bit commercial DAC would be used, which would give an accuracy of $\pm 0.025\%$.

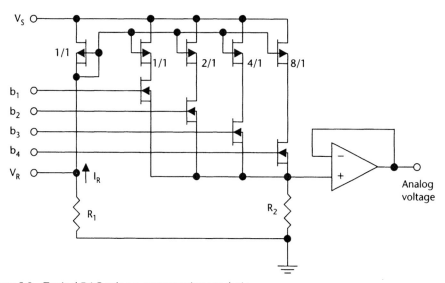

Figure 5.9 Typical DAC using a current mirror technique.

5.3 Converters

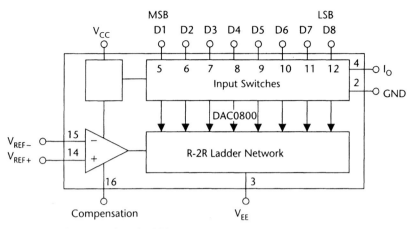

Figure 5.10 Block diagram of DAC 0800.

Example 5.6

A 4-bit DAC with a conversion frequency of 20 kHz and a reference voltage of 1.6V is used to generate a 1 kHz sine wave. Show the DAC voltage steps and output waveform.

Figure 5.11 shows the generation of the 1 kHz sine wave. In this example, the p-p voltage of the sine wave is generated by the 16 steps of the digital signal, thus giving peak voltages of 0V and 1.5V (1.6 × 15/16). With a 20 kHz conversion rate, an output voltage is obtained every 50 μs or 18°, so that using sin θ, the voltage of the sine wave can be calculated every 18°. This is given in column 2. The closest DAC voltage is then selected, and is given in column 3. These step voltages are then plotted as shown with the resulting sine wave. In practice, the conversion rate could be higher, giving a better approximation to a complete sine wave, and/or the resolution of the DAC could be increased. Shown also is the binary code from the DAC (4 bits only). A simple RC filter can smooth the step waveform to get the sine wave. The example is only to give the basic conversion idea.

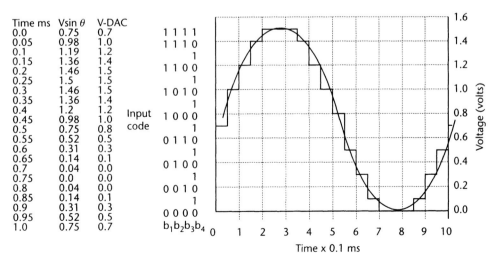

Figure 5.11 1-kHz sine waveform reproduced from a 4-bit DAC.

PWM switches the supply voltage varying the duration that the voltage is applied to reproduce an analog signal, and is shown in Figure 5.12. The width of the output pulses shown are modulated, going from narrow to wide and back to narrow. If the voltage pulses shown are averaged with time, then the width modulation shown will give a sawtooth waveform. The other half of the sawtooth is generated using the same modulation, but with a negative supply, or with the use of a bridge circuit to reverse the current flow. The load limits the current. This type of width modulation is normally used for power drivers for ac motor control or actuator control from a dc supply. The output devices are input controlled power devices, such as a BJT or IGBT (see Section 13.4.1). They are used as switches, since they are either On or Off, and can control more than 100 kW of power. This method of conversion produces low internal dissipation with high efficiency, which can be as high as 95% of the power going into the load; whereas, analog power drivers are only 50% efficient at best, and have high internal power dissipation [6].

5.3.3 Analog to Digital Converters

Sensors are devices that measure analog quantities, and normally give an analog output, although techniques are available to convert some sensor outputs directly into a digital format. The output from most sensors is converted into a digital signal using an ADC. A digital number can represent the amplitude of an analog signal, as previously stated. For instance, an 8-bit word can represent numbers up to 256, so that it can represent an analog voltage or current with an accuracy of 1 in 255 (one number being zero). This assumes the conversion is accurate to 1 bit, which is normally the case, or 0.4 % accuracy. Similarly, 10-bit and 12-bit words can represent analog signals to accuracies of 0.1% and 0.025%, respectively.

Commercial integrated ADCs are available for instrumentation applications. Several techniques are used for the conversion of analog to digital signals, including: flash, successive approximation, resistor ladders, ramp, and dual slope techniques.

Flash converters are the fastest technique for converting analog voltages into digital signals. The device basically consists of a series of comparators (typically 255), biased to decreasing reference voltages as they go lower down the chain. This concept is shown in Figure 5.13 for an 8-bit converter, where only seven comparators are required, since a converter is not needed for 0V. The comparators give a "0" output when the analog voltage is less than its reference voltage, and a "1" output

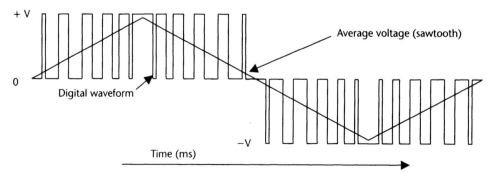

Figure 5.12 PWM signal to give a 1 kHz sawtooth waveform, using positive and negative supplies.

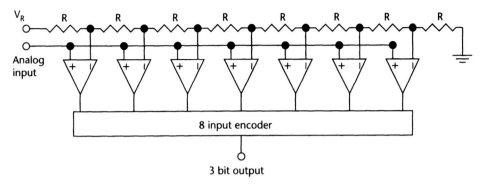

Figure 5.13 3-bit flash converter.

when the analog voltage is higher than its reference input voltage. These outputs are then encoded to give a 3-bit digital word. The devices are very fast, but expensive and with limited applications. A number of flash converters are commercially available, including the 8-bit flash converter manufactured by Maxim (MAX 104), which gives a resolution of ±0.39% with an output sample rate of 1 GSPS (10^9 samples per second).

A *Ramp-up* or *Stepped-up ADC* is a low-speed device that compares the analog voltage to a ramp voltage generated by an integrator or a resistor network, as in a DAC. The voltage is ramped up, or stepped up, until the voltage from the DAC is within the resolution of the converter. When it equals the input voltage, the steps are counted, and the digital word count represents the analog input voltage. This method is slow, due to the time required for high counts, and is only used in low-speed applications. The device is a medium cost converter. A 12-bit device has a conversion time of approximately 5 ms.

Successive approximation is a parallel feedback ADC that feeds back a voltage from a DAC, as shown in Figure 5.14. A comparator compares the analog input voltage to the voltage from the DAC. The logic in the successive approximation method sets and compares each bit, starting with the MSB (b_1). Starting with V_x at 0, the first step is to set b_1 to 1, making $V_x = V_R/2$, and then to compare it to the input voltage. If the input is larger, b_1 remains at 1, and b_2 is set to 1, making $V_x = (3V_R/4)$.

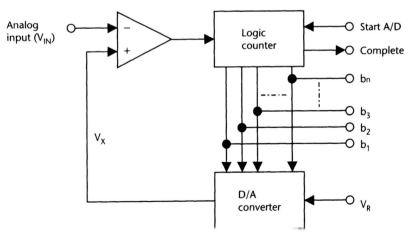

Figure 5.14 Successive approximation block diagram.

This new value is again compared to the input voltage, and so on. If V_x is greater than the input, b_1 is reset to 0, and b_2 is set to 1, and then compared to the input. This process is repeated to the LSB of the DAC.

This is a high-speed technique, since each bit is compared only once to the input. For instance, an 8-bit device needs to make only eight comparisons. On the other hand, if the DAC were ramped up, it would require 256 comparisons. This technique makes a high-speed, medium-cost DAC with good accuracy. This type of device can convert an analog voltage to 12-bit accuracy in 20 μs, and a less expensive device can convert an analog signal to 8-bit accuracy in 30 μs [7].

Example 5.7

Use the sequence given above for a 4-bit successive approximation ADC to measure 3.7V, if the converter has a 4.8V reference (V_R).

From the sequence above:

- Step 1
  - Set b_1 to 1 V_x = 4.8 × 8/16V = 2.4V
  - V_{in} > 2.4V b_1 is not reset and stays at 1 (MSB).
- Step 2
  - Set b_2 to 1 V_x = 2.4 + 4.8 × 4/16V = 3.6V
  - V_{in} > 3.6V b_2 is not reset and stays at 1.
- Step 3
  - Set b_3 to 1 V_x = 3.6 + 4.8 × 2/16V = 4.2V
  - V_{in} < 4.2V b_3 is reset and returns to 0.
- Step 4
  - Set b_4 to 1 V_x = 3.6 + 4.8 × 1/16V = 3.9V
  - V_{in} < 3.9V b_4 is reset and returns to 0 (LSB).

This shows that a binary word 1100 represents 3.7V.

Dual slope converters are low-cost devices with good accuracy, and are very tolerant of high noise levels in the analog signal, but are slow compared to other types of converters. It is the most common type of ramp converter, and is normally the choice for multimeters and applications where high speed is not required. A 12-bit conversion takes approximately 20 ms. A block diagram of the device is shown in Figure 5.15. The input voltage is fed to a comparator for a fixed period of time, charging up an integrating capacitor, which also averages out the noise in the input voltage. After the fixed time period, the input to the integrator is switched to a negative voltage reference, which is used to discharge the capacitor. During the charging time (T_s), the output voltage from the integrator (V_{out}) is given by:

$$V_{out} = \frac{1}{RC} \int V_{in} dt \qquad (5.6)$$

because V_{in} is constant

5.3 Converters

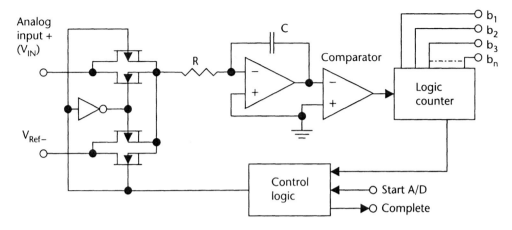

Figure 5.15 Dual slope ADC.

$$V_{out} = \frac{1}{CR} \times T_s V_{in} \qquad (5.7)$$

where T_s is the time taken for the output voltage from the integrator to reach V_{out}.

After the charging time, the integrator capacitor is discharged by electronically switching the input voltage to the integrator to a negative reference using the CMOS switches. The voltage to the comparator now decreases until it reaches the initial voltage, which is sensed by the comparator. This discharge is given by:

$$V_{out} = -\frac{1}{CR} T_R V_{Ref} \qquad (5.8)$$

where T_R is the time taken to discharge the capacitor.

Combining (5.6) and (5.7), we get:

$$\frac{1}{CR} \times T_S V_{in} = \frac{1}{CR} T_R V_{Ref}$$

$$V_{in} = \frac{T_R}{T_S} V_{ref} \qquad (5.9)$$

This shows that if T_S is a set time, then there is a linear relation between V_{in} and T_R. It should be noted that the conversion is independent of the values of R and C.

Example 5.8

What is the conversion time for the dual slope converter shown in Figure 5.15? If R = 1 MΩ, C = 5 nF, V_{ref} = 5V, T_s = 50 ms, and the input voltage is 8.2V, what is the capacitor discharge time?

The first step is to find the time the capacitor discharges after the 50-ms set time.

$$T_R = T_S \times V_{in}/V_{ref} = 50 \times 10^{-3} \times 8.2V/5V = 82 \text{ ms}$$

The total conversion time is 50 ms + 82 ms = 132 ms, and 82 ms is the capacitor discharge time. Because of the linear relation between V_{in} and T_R, V_{ref} and V_s can be chosen to give a direct reading of V_{in}.

Figure 5.16 shows the block diagram of the ADC0804, which is a commercial 8-bit ADC designed using CMOS technology with TTL-level compatible outputs. The device uses successive approximation to convert the analog input voltage to digital signal. The device has a very flexible design, and is microprocessor compatible. The 8-bit conversion time is 100 μs, and the manufacturer's data sheets should be consulted for the device parameters [8].

5.3.4 Sample and Hold

Analog signals are constantly changing, so for a converter to accurately measure the voltage at a specific time instant, a sample and hold technique is used to capture the voltage level and hold it long enough for the measurement to be made. Such a circuit is shown in Figure 5.17(a), with the waveforms shown in Figure 5.17(b). The CMOS switch in the sample and hold circuit has a low impedance when turned On, and a very high impedance when Off. The voltage across capacitor C follows the input analog voltage when the FET is On, and holds the dc level of the analog voltage when the FET is turned Off. During the Off period, the ADC measures the dc level of the analog voltage and converts it into a digital signal. Since the sampling frequency of the ADC is much higher than the frequency of the analog signal, the varying amplitude of the analog signal can be represented in a digital format during each sample period, and then stored in memory. The analog signal can be regenerated from the digital signal using a DAC. There is a time delay when using a sample and hold circuit, because of the time it takes to convert the analog voltage into its digital equivalent. This can be seen from the waveforms [9].

5.3.5 Voltage to Frequency Converters

An alternative to the ADC is the voltage to frequency converter. After the analog voltage is converted to a frequency, it is then counted for a fixed interval of time, giv-

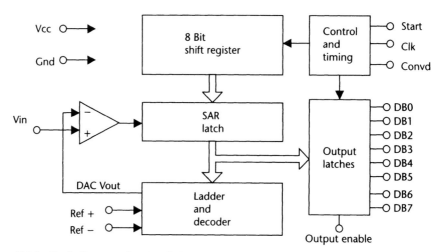

Figure 5.16 Block diagram of an LM 0804 ADC.

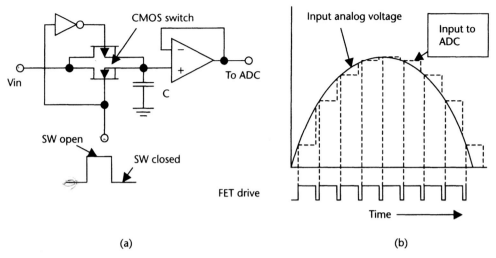

Figure 5.17 (a) Sample and hold circuit, and (b) waveforms for the circuit.

ing a count that is proportional to the frequency and the analog signal. Commercial units, such as the LM331 shown in Figure 5.18, are available for this conversion. These devices have a linear relation between voltage and frequency. The operating characteristics of the devices are given in the manufacturer's data sheets.

The comparator compares the input voltage to the voltage across capacitor C_2. If the input voltage is larger, then the comparator triggers the one-shot timer. The output of the timer will turn On the current source, charging C_2 for a period of $1.1C_1R_2$, making the voltage across C_2 higher than the input voltage. At the end of the timing period, the current source is turned Off, the timer is reset, and C_1R_2 is discharged. The capacitor C_2 will now discharge through R_3 until it is equal to the input voltage, and the comparator again triggers the one-shot timer, starting a new cycle. The switched current source can be adjusted by R_1.

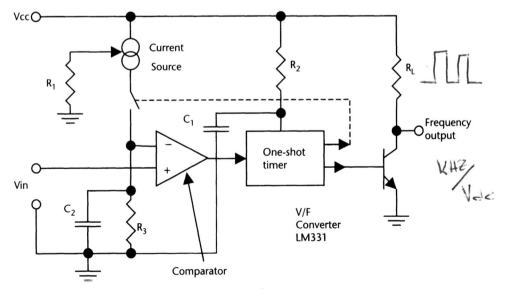

Figure 5.18 Block diagram of an LM331 voltage to frequency converter.

5.4 Data Acquisition Devices

The central processor is required to interface with a large number of sensors and to drive a number of actuators. The processor has a limited number of input and output ports, so that the data has to be channeled into the input ports via external units, such as multiplexers, and channeled out via demultiplexers to be distributed to the external actuators. The multiplexers work by time division multiplex. For instance, an 8-bit multiplexer will accept eight inputs, and output the signals one at a time under process control [10].

5.4.1 Analog Multiplexers

A 4-bit analog multiplexer is shown in Figure 5.19. The analog input signals can be alternately switched by CMOS analog switches to the output buffer, similar to a rotary switch. A decoder controls the switches. The decoder has input enables and address bits. When the enables are 0, all the outputs from the decoder are 0, holding all of the analog switches Off. When the enables are 1, the address bits are decoded, so that only one of the control lines to the analog switches is a 1, turning the associated switch On and sending the signal to the output buffer. The output from the multiplexer is fed to the controller via an ADC. Analog multiplexers are commercially available with 4, 8, or 16 input channels to 1 output channel, such a device is the CD 4529. Analog demultiplexers are also commercially available.

5.4.2 Digital Multiplexers

Figure 5.20 shows the block diagram of a 4-bit digital multiplexer. The operation is similar to an analog multiplexer, except that the inputs are digital. When the enable is 0, the outputs from the decoder are all 0, inhibiting data from going through the input NAND gates, and the data output is 0. When the enable is 1, the addresses are decoded, so that one output from the decoder is 1, which opens an input gate to allow the digital data on that channel to be outputted. Digital multiplexers are commercially available with 4, 8, or 16 input channels. An example of a 16-channel digital input device is the SN74150, and its companion device, the SN 74154, a 16-channel output demultiplexer.

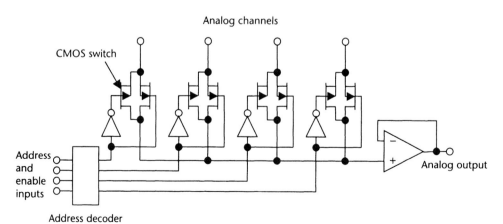

Figure 5.19 Analog multiplexer.

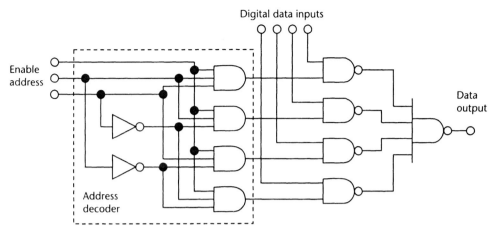

Figure 5.20 Digital multiplexer.

5.4.3 Programmable Logic Arrays

Many systems have large blocks of gates to perform custom logic and sequential logic functions. These functions were constructed using the SN 74 family of logic gates. The logic gates can now be replaced with a programmable logic array (PLA). One of these devices replaces many gate devices, requires less power, and can be configured (programmed) by the end user to perform all of the required system functions. The devices also have the flexibility to be reprogrammed if an error in the logic is found or there is a need to upgrade the logic. Because they can be programmed by the end user, there is no wasted time, unlike with the Read Only Memory (ROM), which had to be programmed by the manufacturer. However, the Electrically Erasable Programmable ROM (EEPROM) is available as a reprogrammable device, and technology-wise is similar to the PLA.

5.4.4 Other Interface Devices

A number of controller peripheral devices, such as timing circuits, are commercially available. These circuits are synchronized by clock signals from the controller that are referenced to very accurate crystal oscillators, which are accurate to within less than 0.001%. Using counters and dividers, the clock signal can be used to generate very accurate delays and timing signals. Compared to RC-generated delays and timing signals, which can have tolerances of more than 10%, the delays and timing signals generated by digital circuits in new equipment are preferred.

Other peripheral devices are digital comparators, encoders, decoders, display drivers, counters, shift registers, and serial to parallel converters.

5.5 Basic Processor

Figure 5.21 shows a simplified diagram of a digital processor. The system typically consists of the central processing unit (CPU) with arithmetic unit (ALU), random access memory (RAM), read only memory (ROM), and data input/output ports. Communication between the units uses three buses: (1) a two-way data bus that

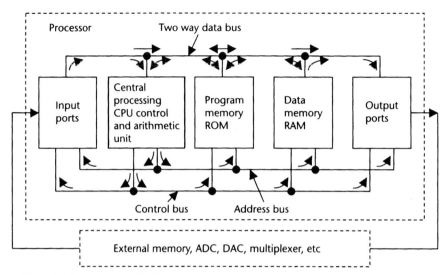

Figure 5.21 Digital processor block diagram.

passes data between all of the individual units; (2) a one-way address bus from the CPU that gives the address to which the data is to be sent or retrieved; and (3) a one-way control bus that selects the unit to which the data and address is to be sent, or the address from which data is to be retrieved. The input/output ports are used for communication with other computers or peripheral units. The processor forms the heart of the controller, as shown in Chapter 14, Figure 14.3. The input port can receive analog sensor data via ADCs and multiplexers. Once received, the data is first conditioned by stored equations or lookup tables to correct for linearity, offset, span, or temperature. The new data is then compared to the preset value, and the appropriate signal sent via the output port and a DAC to control an actuator. The input port can receive serial or parallel formatted data. In the case of serial data, it is converted to parallel data by a serial to parallel converter for internal use. The controller has the ability to serially scan a large number of sensors via the input port, and send data to a number of actuators via the output port. The sensors are scanned every few milliseconds, and the output data to the actuators updated, giving continuous monitoring and control. The controller can be used for sequential control, as well as continuous monitoring and control.

5.6 Summary

Digital electronics were introduced in this chapter not only as a refresher, but also to extend digital concepts to their applications in process control. Physical variables are analog and the central processor is digital; therefore, various methods of converting analog to digital and digital to analog were discussed. Analog data can be converted to digital using successive approximation for low-speed applications, or by using flash conversion for high speed, when converting digital to analog weighted resistor techniques for an analog voltage output, or pulse width modulation for power control. Data acquisition devices are used to feed data to the central processor. The use of analog and digital multiplexers in data acquisition systems is shown with demultiplexers, and in the basic processor block diagram.

The use of comparators in analog to digital and digital to analog converters was discussed. The various methods of conversion and their relative merits were given, along with a discussion on analog to frequency converters.

References

[1] Jones, C. T., *Programmable Logic Controllers*, 1st ed., Patrick-Turner Publishing Co., 1996, pp. 17–22.

[2] Tokheim, R. L., *Digital Electronics Principles and Applications*, 6th ed., Glencoe/McGraw-Hill, 2003, pp. 77–113.

[3] Humphries, J. T., and L. P. Sheets, *Industrial Electronics*, 4th ed., Delmar, 1993, pp. 566–568.

[4] Miller, M. A., *Digital Devices and Systems with PLD Applications*, 1st ed., Delmar, 1997, pp. 452–486.

[5] McGonigal, J., "Integrated Solutions for Calibrated Sensor Signal Conditioning," *Sensor Magazine*, Vol. 20, No. 10, September 2003.

[6] Madni, A., J. B. Vuong, and P. T. Vuong, "A Digital Programmable Pulse-Width-Modulation Converter," *Sensors Magazine*, Vol. 20, No. 5, May 2003.

[7] Banerjee, B., "The 1 Msps Successive Approximation Register A/D Converter," *Sensors Magazine*, Vol. 18, No. 12, December 2001.

[8] Contadini, F., "Demystifying Sigma-Delta ADCs," *Sensors Magazine*, Vol. 19, No. 8, August 2002.

[9] Khalid, M., "Working at High Speed: Multi-megahertz 16-Bit A/D Conversion," *Sensors Magazine*, Vol. 15, No. 5, May 1998.

[10] Johnson, C. D., *Process Control Instrumentation Technology*, 7th ed., Prentice Hall, 2003, pp. 147–158.

CHAPTER 6
Microelectromechanical Devices and Smart Sensors

6.1 Introduction

The development of new devices in the microelectronics industry has over the past 50 years been responsible for producing major changes in all industries. The technology developed has given cost-effective solutions and major improvements in all areas. The microprocessor is now a household word, and is embedded in every appliance, entertainment equipment, most toys, and every computer in the home. In the process control industry, processes and process control have been refined to a level only dreamed of a few years ago. The major new component is the microprocessor, but other innovations in the semiconductor industry have produced accurate sensors for measuring temperature, time, and light intensity; along with microelectromechanical devices for measuring pressure, acceleration, and vibration. This has certainly brought about a major revolution in the process control industry. Silicon has been the semiconductor of choice. Silicon devices have a good operating range ($-50°$ to $+150°C$), a low leakage, and can be mass-produced with tight tolerances. The processing of silicon has been refined, so that many millions of devices can be integrated and produced on a single chip, which enables a complete electronic system to be made in a package. Several unique properties of silicon make it a good material for use in sensing physical parameters. Some of these properties are as follows:

- The piezoresistive effect can be used in silicon for making strain gauges;
- The Hall Effect or transistor structures can be used to measure magnetic field strength;
- Linear parametric variation with temperature makes it suitable for temperature measurement;
- Silicon has light-sensitive parameters, making it suitable for light intensity measurements;
- Silicon does not exhibit fatigue, and has high strength and low density, making it suitable for micromechanical devices.

Micromechanical sensing devices can be produced as an extension of the standard silicon device process, enabling the following types of sensors:

- Pressure;
- Force and strain;
- Acceleration;
- Vibration;
- Flow;
- Angular rate sensing;
- Frequency filters.

Micromechanical devices are very small, have low mass, and normally can be subjected to high overloads without damage.

6.2 Basic Sensors

Certain properties of the semiconductor silicon crystal can be used to sense physical properties, including temperature, light intensity, force, and magnetic field strength.

6.2.1 Temperature Sensing

A number of semiconductor parameters vary linearly with temperature, and can be used for temperature sensing. These parameters are diode or transistor junction voltages, zener diode voltages, or polysilicon resistors. In an integrated circuit, the fact that the differential base emitter voltage between two transistors operating at different current densities (bandgap) is directly proportional to temperature is normally used to measure temperature. The output is normally adjusted to give a sensitivity of 10 mV per degree (C, F, or K). These devices can operate over a very wide supply voltage range, and are accurate to $\pm 1°C$ or K and $\pm 2°F$, over the temperature range $-55°$ to $+150°C$ [1]. The device families manufactured by National Semiconductor are as follows:

- LM35, which gives 10 mV/°C;
- LM34, which gives 10 mV/°F;
- LM135, which gives 10 mV/K (the output voltage = 2.73V at 0°C).

A simplified block diagram of the LM75 is shown in Figure 6.1. This device is calibrated in Celsius, has a 10-bit DAC, and a register for digital readout. The analog voltage signal from the temperature sensor is digitized in a 9-bit Delta-Sigma ADC with a 10-bit decimation filter. Three address pins (A0, A1, A2) are available, so that any one of eight devices can be selected when connected to a common bus. An overtemperature alarm pin is available that can be programmed from the two-way data bus [2].

6.2.2 Light Intensity

Semiconductor devices are in common use as photointensity sensors. Photodiodes, phototransistors, and integrated photosensors are commercially available [3]. Integrated devices have on-chip temperature compensation and high sensitivity, and can

6.2 Basic Sensors

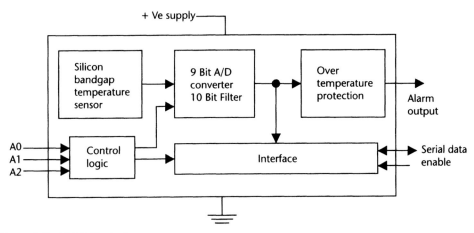

Figure 6.1 LM75 Block diagram.

be configured to have a voltage, digital, or frequency output that is proportional to intensity. The devices also can be made sensitive to visual or infrared frequency spectra. Photons cause junction leakage in photodiodes that is proportional to light intensity; thus, the reverse leakage is a measure of light intensity. In photodiodes, the effect is to increase the base current of the device giving an amplified collector-emitter current. Devices are available with sensitivities of 80 mV per $\mu W/cm^2$ (880 nm) and a linearity of 0.2%. Programmable light to frequency converters can sense light intensities of 0.001 to 100 k$\mu W/cm^2$ with low temperature drift, and an absolute frequency tolerance of ±5%. Some typical devices families manufactured by Texas Instrument are as follows:

- TSL230 Light to frequency converter;
- TSL235 Light to frequency converter;
- TSL250 Light to voltage converter;
- TSL260 IR Light to voltage converter.

Figure 6.2 shows the circuit of the front end of an integrated temperature-compensated phototransistor. The circuit is similar to that of the front end of an op-amp. In the op-amp, the bases of the differential input pair of transistors are driven from the differential input signals. In this case, the bases are joined together and internally biased. One of the input transistors is typically screened with metallization, while the other is not screened, so that incident light from a window in the package can fall on the transistor, which will then become a phototransistor. The light intensity is converted into an electrical signal and amplified. Because the input stage is a differential stage, it is temperature-compensated. The current supply is typically from a bandgap regulator [4].

6.2.3 Strain Gauges

In a resistive type of strain gauge, the gauge factor (GF) is the fractional change in resistance divided by the fractional change in length, and is given by:

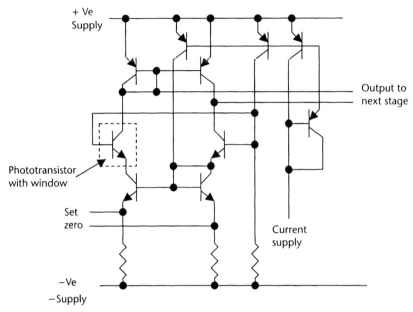

Figure 6.2 Temperature-compensated integrated phototransistor circuit.

$$GF = \Delta R/R/\Delta l/l = \Delta R/R/\text{strain} \qquad (6.1)$$

where $\Delta R/R$ is the fractional change in resistance, and $\Delta l/l$ is the fractional change in length or strain.

The semiconductor strain gauge is more sensitive than the deposited resistive strain gauge, since it uses the piezoresistive effect, which can be very large. The physical dimension change then can be ignored. The gauge factor can be either positive or negative, depending upon doping, and can be 100 times higher than for foil gauges. The gauge factor in silicon depends on the crystal orientation, and the type of doping of the gauge, which can be n or p. Strain gauges are normally p-doped resistors, since these have the highest gauge factor. Lightly doped or higher valued resistors have higher gauge factors, but are more temperature-sensitive. Thus, the resistance is a compromise between gauge factor and temperature sensitivity. The semiconductor gauge is very small (0.5 × 0.25 mm), does not suffer from fatigue, and is commercially available with a compensating device perpendicular to the measuring device, or with four elements to form the arms of a bridge. The strain gauge also can be integrated with compensating electronics and amplifiers to give conditioning and high sensitivity [5].

6.2.4 Magnetic Field Sensors

Magnetic fields can be sensed using the Hall Effect, magnetoresistive elements (MRE), or magnetotransistors [6]. Some applications for magnetic field sensors are given in Section 11.2.1.

The *Hall Effect* occurs when a current flowing in a carrier (semiconductor) experiences a magnetic field perpendicular to the direction of the current flow. The interaction between the two causes the current to be deflected perpendicular to the direction of both the magnetic field and the current. Figure 6.3 shows the effect of a magnetic field on the current flow in a Hall Effect device. Figure 6.3(a) shows the current flow without a magnetic field, and Figure 6.3(b) shows the deflection of the current flow with a magnetic field, which produces the Hall voltage. Table 6.1 gives the characteristics of some common materials used as Hall Effect devices.

The MRE is the property of a current-carrying ferromagnetic material to change its resistance in the presence of a magnetic field. As an example, a ferromagnetic (permalloy) element (20% iron and 80% nickel) will change its resistivity approximately 3% when the magnetic field is rotated 90°. The resistivity rises to a maximum when the current and magnetic field are coincident to each other, and are at a minimum when the magnetic field and current are perpendicular to each other. The effect of rotating an MRE is shown in Figure 6.4(a). The attribute is known as the anisotropic magnetoresistive effect. The resistance R, or an element, is related to the angle q between the current and the magnetic field directions by the expression:

$$R = R_{11} \cos^2 q + R_{\perp} \sin^2 q \qquad (6.2)$$

where R_{11} is the resistance when the current and magnetic fields are parallel, and $R_{\perp}$ is the resistance when the current and magnetic fields are perpendicular.

MRE devices give an output when stationary, which makes them suitable for zero speed sensing, or position sensing. For good sensitivity and to minimize temperature effects, four devices are normally arranged in a Wheatstone bridge configuration. In an MRE device, aluminum strips are put 45° across the permalloy element to linearize the device, as shown in Figure 6.4(b). The low resistance aluminum strips cause the current to flow 45° to the element, which biases the element into a linear operating region. Integrated MRE devices can typically operate from −40 to +150°C at frequencies up to 1 MHz.

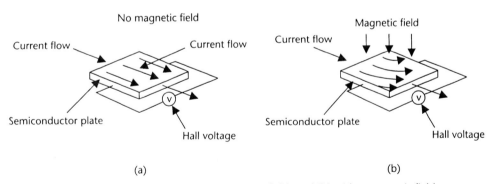

Figure 6.3 Hall Effect device (a) without a magnetic field, and (b) with a magnetic field.

Table 6.1 Hall Effect Sensitivities

Material	Temperature Range	Supply Voltage	Sensitivity @ 1 kA/m	Frequency Range
Indium	−40° to +100°C	1V	7 mV	0 to 1 MHz
GaAs	−40° to +150°C	5V	1.2 mV	0 to 1 MHz
Silicon	−40° to +150°C	12V	94 mV	0 to 100 kHz

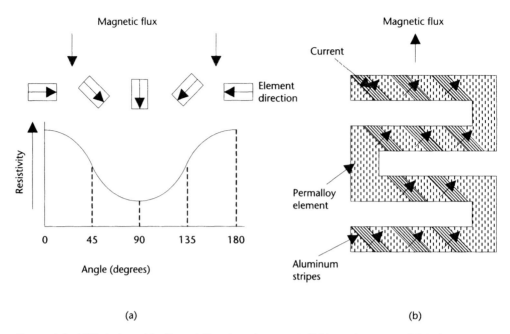

Figure 6.4 MRE devices: (a) effect of direction of magnetic field on element, and (b) element layout.

Magnetotransistors can be made using bipolar or CMOS technology. Figure 6.5(a) shows the topology of a PNP bipolar magnetotransistor. The electron flow without a magnetic field is shown in Figure 6.5(a), and the electron flow in the presence of a magnetic field is shown in Figure 6.5(b), and the junction cross section in shown Figure 6.5(c). The device has two collectors as shown. When the base is forward biased, the current from the emitter is equally divided between the two collectors when no magnetic field is present. When a magnetic field is present, the current flow is deflected towards one of the collectors, similar to the Hall Effect device. This gives an imbalance between the current in the two collectors, which is proportional to the magnetic field strength, can be amplified as a differential signal, and can be used to measure the strength of the magnetic field [7]. A comparison of the sensitivities of magnetic field sensors is given in Table 6.2.

6.3 Piezoelectric Devices

The piezoelectric effect is the coupling between the electrical and mechanical properties of certain materials. If a potential is applied across piezoelectric material, then a mechanical change occurs. This is due to the nonuniform charge distribution

Table 6.2 Comparison of the Sensitivities of Magnetic Field Sensors

Sensor Type	Temperature Range	Sensitivity @ 1 kA/m	Frequency Range	Mechanical Stress
Hall Effect (Si)	−40° to +150°C	90 mV	0 to 100 kHz	High
MRE	−40° to +150°C	140 mV	0 to 1 MHz	Low
Magnetotransistor	−40° to +150°C	250 mV	0 to 500 kHz	Low

6.3 Piezoelectric Devices

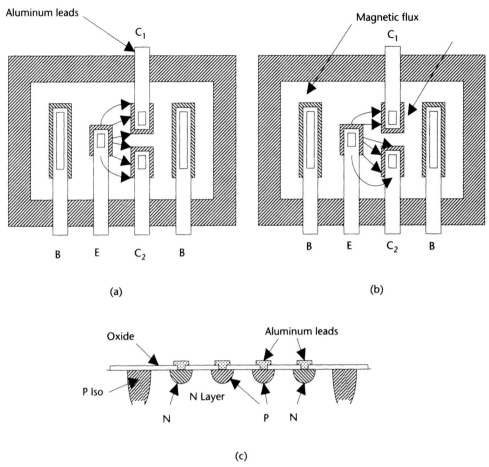

Figure 6.5 (a) Electron flow in a PNP magnetotransistor without a magnetic field, (b) electron flow in a PNP magnetotransistor with a magnetic field, and (c) cross section of a PNP magnetotransistor.

within the crystal structure of the material. When the material is exposed to an electric field, the charges try to align themselves with the electric field, causing a change in shape of the crystal. The same polarization mechanism causes a voltage to develop across the crystal in response to mechanical stress, which makes piezoelectric devices suitable for use in the measurement of force. Some naturally occurring crystals exhibit the piezoelectric effect, such as quartz, Rochelle salt, lithium sulphate, and tourmaline. Another important group of piezoelectric materials are the piezoelectric ceramics, such as lead-zirconate-titanate (PZT), lead-titanate, lead-zirconate, and barium-titanate.

The differences between the characteristics of crystalline quartz and PZT make them suitable for use in widely differing application areas. Quartz, because of its stability, minimal temperature effects, and high Q, is an ideal timing device. PZT, due to its low Q, but much higher dielectric constant, higher coupling factor, and piezoelectric charge constant, gives a much higher performance as a transducer (see Table 6.3). These factors also make PZT useful for micromotion actuators, or micropositioning devices.

Table 6.3 Piezoelectric Material Characteristics

	Symbol	Units	Quartz	PZT
Dielectric constant	K^T		4.5	1,800
Coupling factor	k33		0.09	0.66
Charge constant	d33	$C/N \times 10^{-12}$	2.0	460
Voltage constant	g33	$V\,m/N \times 10^{-3}$	50	28
Quality factor	Q		10^5	80

6.3.1 Time Measurements

The measurement of time is so common that it is not considered as sensing. However, many process control operations have critical timing requirements. Highly accurate timing can be obtained using atomic standards, but quartz crystal–controlled timing devices have an accuracy of better than 1 in 10^6, which is far more accurate than any other type of parameter measurement. Such system can be integrated onto a single silicon chip with an external crystal.

Because of its stability and high Q, quartz makes an excellent reference oscillator. Thin sections of the crystal can be used as the frequency selective element in an oscillator. The crystal slice is lapped and etched down until the desired resonant frequency is reached. Thin metal electrodes are then deposited on both sides of the crystal. A voltage applied to the electrodes induces mechanical movement and vibration at the natural resonant frequency of the crystal, producing a voltage, which in an active circuit can be used to produce a sustained frequency. Above 15 MHz, the slice becomes very thin and brittle. However, a harmonic of the fundamental frequency can be used to extend the frequency range, or a phase locked loop system can be used to lock a high frequency to the crystal frequency. This type of circuit would be used in a high frequency transmitter, such as that used to transmit sensor data from a remote location. The orientation of the crystal slice with respect to the crystal axis is of prime importance for temperature stability. The pyroelectric effect can be reduced to 1 p/m over 20°C with the correct orientation, such as the AT cut referenced to the z-axis. The quartz crystal used in watch circuits has a low frequency (32.768 kHz), and is etched in the shape of a tuning fork. The tolerance is ±20 p/m with a stability of −0.042 p/m.

Timing falls into two types of measurement: first, generating a window of known duration, such as to time a process operation or to measure an unknown frequency, and second, timing an event, as would be required in distance measurement.

Window generation is shown in Figure 6.6, which shows the block diagram of a system for measuring an unknown frequency. The output from a 1 MHz crystal oscillator is shaped and divided by 2×10^6, which generates a 1 second window to open the AND gate, letting the unknown frequency be counted for 1 second and displayed. The divider can be a variable divider, and set to give a variable gate width for timing a process operation for any required duration.

Figure 6.7 shows a block diagram of a circuit for measuring an unknown *time duration*, such as the time for a radar pulse to reach an object and return to the receiver. The unknown signal is used to gate the output from the 1 MHz oscillator to

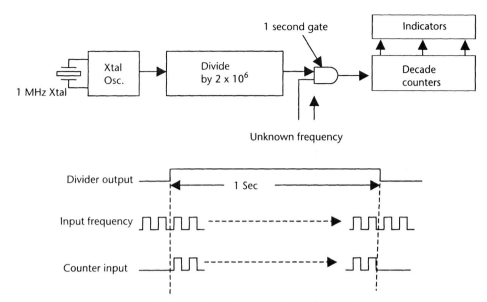

Figure 6.6 Block schematic of 1 second gate for measuring unknown frequency.

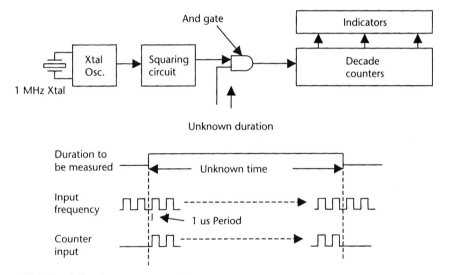

Figure 6.7 Circuit for time measurement.

the counter, and the indicators will give the time in microseconds. Oscillator frequencies can be chosen and divided down, so that the display gives a distance measurement directly related to the time being measured.

6.3.2 Piezoelectric Sensors

PZT devices are commonly used for sensors. Because of the unique relation between force and voltage, a force is readily converted into a voltage. Size for size, PZT devices are approximately 100 times more sensitive than quartz. They can be used for dynamic measurements from approximately 1 Hz to well over 10 kHz, but they are not good for static measurements. PZT devices are small and cost-effective, and are used for vibration, shock, and acceleration sensing.

6.3.3 PZT Actuators

When a voltage is applied across a PZT element in the longitudinal direction (axis of polarization), it will expand in the transverse direction (perpendicular to the direction of polarization). When the fields are reversed, the motion is reversed. The motion can be of the order of tens of microns, with forces of up to 100N. A two-layer structure with the layers polarized in the same direction is shown in Figure 6.8(a), and the two layers act as a single layer. If two layers are polarized in opposite directions as shown in Figure 6.8(b), then the structure acts like a cantilever. Such a structure can have a movement of up to 1 millimeter and produce forces of several hundred Newtons. Multilayer devices can be made to obtain different kinds of motion. Piezoelectric actuators are normally specified by free deflection and blocked force. Free deflection is the displacement at maximum operating voltage when the actuator is free to move and does not exert any force. Blocked force is the maximum force exerted when the actuator is not free to move. The actual deflection depends on the opposing force.

Piezoelectric actuators are used for ultraprecise positioning, for generation of acoustical and ultrasonic waves, in alarm buzzers, and in micropumps for intravenous feeding of medication.

6.4 Microelectromechanical Devices

The techniques of chemical etching have been extended to make semiconductor micromachined devices a reality, and make miniature mechanical devices possible. Micromachining silicon can be divided into bulk or surface micromachining. With bulk micromachining, the silicon itself is etched and shaped, but with surface micromachining, layers of material are deposited or grown on the surface of the silicon and shaped. Sacrificial layers then are etched to fabricate micromechanical structures [8]. The advantage of surface micromachining technology is that it produces smaller structures. This approach shares many common steps with current IC

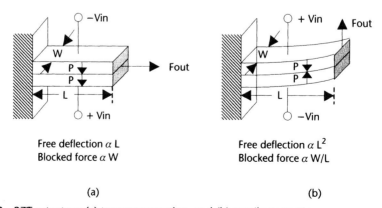

Figure 6.8 PZT actuators: (a) transverse motion, and (b) cantilever motion.

technology, but can introduce strain into the structures, which can cause warping after the structures are made, and may require careful annealing. Bulk devices do not suffer from this problem.

6.4.1 Bulk Micromachining

A unique property of crystalline material is that it can be etched along the crystal planes using a wet anisotropic etch, such as potassium hydroxide. This property is used in the etching of bulk silicon to make pressure sensors, accelerometers, micropumps, and other types of devices.

Pressure sensors are made by etching the backside of the wafer, which can contain over 100 dies or sensors. A photolithographic process is used to define the sensor patterns. The backside of the wafer is covered with an oxide and a layer of light-sensitive resist. The resist is selectively exposed to light through a masked patterned and then developed. The oxide can now be wet etched using buffered hydrofluoric acid. The resist will define the pattern in the oxide, which in turn is used as the masking layer for the silicon etch. This sequence of events is shown in Figure 6.9.

A single die is shown in Figure 6.10(a). The silicon is etched along the crystal plane at an angle of 54.7°, and after etching, a thin layer of silicon is left, as shown in the cross section. The silicon wafer is then reversed, and the topside is masked and etched using a process similar to that used on the backside of the wafer.

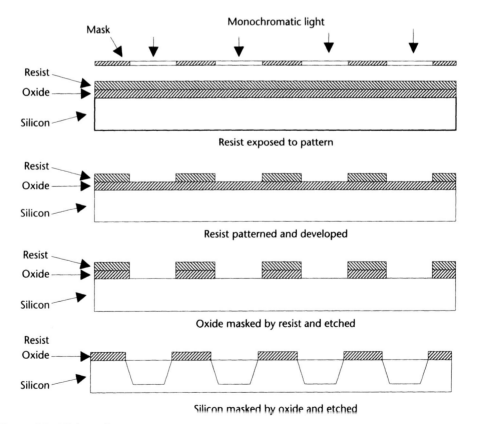

Figure 6.9 Wafer etch process.

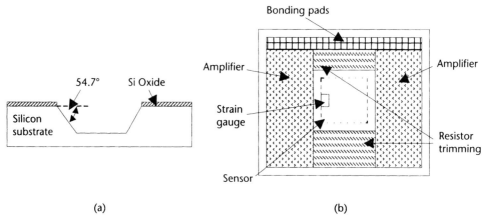

Figure 6.10 (a) Backside etch of silicon pressure sensor die, and (b) position of conditioning circuits on the top side of the die.

Diffusions are made for strain gauge and transistors, interconnections, and metal resistors are then deposited on the top surface. Figure 6.10(b) shows the topological layout of a single die on the topside of the wafer, showing the following: the location of the conditioning circuits; the strain gauge, which is four piezoresistors forming a bridge; bonding pads; and nickel-chrome resistors that are laser trimmed to correct for offset and sensitivity.

After processing is complete, each die is tested and trimmed for offset and span. The wafer is then bonded to a second constraint wafer, which gives strain relief when the wafer is assembled in a plastic package. If the pressure sensor is an absolute pressure sensor, then the cavity is sealed with a partial vacuum If the sensor is to be used to gauge differential pressures, then the constraint wafer will have holes etched through to the cavity, as shown in Figure 6.11. The diaphragm is typically $3,050 \times 3,050\,\mu m$ in a medium range pressure sensor, and the signal compensated die size will be approximately $3,700 \times 3,300\,\mu m$. The diaphragm thickness and size both will vary with the pressure range being sensed. After testing, the wafer is cut into individual dies and the good dies are assembled in a plastic package, as shown in Figure 6.11. Plastic packages come in many different shapes and sizes, depending upon the application of the pressure sensor.

The strain gauge can be four individual piezoresistive devices forming a conventional dc Wheatstone bridge for improved sensitivity, or can be an X-ducer (Motorola patent), which is effectively four piezoresistive devices in a bridge. The integrated amplifier used to amplify the sensor signal, and the trimming resistors, are shown in Figure 6.12. The amplifiers are connected to form an instrument amplifier circuit. The circuit shown is typical of piezoelectric strain gauge elements in pressure sensors. A number of resistors are trimmed to adjust for temperature, offset, and span. After trimming, the operating temperature range is from $-50°$ to $+100°C$, giving an accuracy of better than 1% of reading.

Gas flow sensors have been developed using bulk micromachining techniques. The mass air flow sensors utilize temperature-resistive films laminated within a thin film of dielectric (2 to 3 μm thick), suspended over a micromachined cavity, as shown in Figure 6.13(a). The heated resistor also can be suspended over the cavity. Heat is transferred from one resistor to another by the mass of the gas flowing. The

6.4 Microelectromechanical Devices

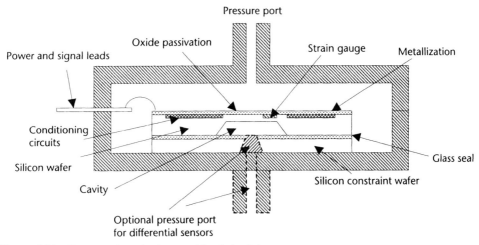

Figure 6.11 Cross section of micromachined absolute pressure sensor.

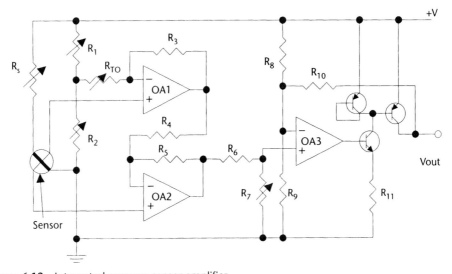

Figure 6.12 Integrated pressure sensor amplifier.

imbalance in the resistance caused by heat transfer is directly proportional to heat flow. The advantages of this type of anemometer are its small size and low thermal mass. It does not impede gas flow, and its low thermal mass reduces the response time to approximately 3 ms. However, the sensor is somewhat fragile and can be damaged by particulates. Figure 6.13(b) gives an example of the control circuit used with the mass flow sensor.

6.4.2 Surface Micromachining

In surface micromachining technology, layers of material are deposited (e.g., polysilicon) or grown (e.g., silicon dioxide) on the surface of the silicon, and then shaped using a photolithographic process. The sacrificial oxide layers are then etched [9]. Leaving freestanding structures.

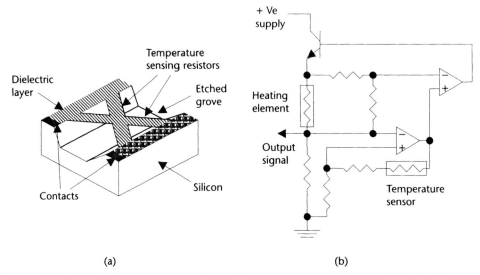

(a) (b)

Figure 6.13 (a) Hot wire anemometer microminiature temperature sensor, and (b) circuit for mass air flow sensor.

Accelerometers using surface micromachining techniques are in volume production. Figure 6.14 shows the cross section of the surface micromachined accelerometer. The topological view of the comb structure is shown in Figure 6.15. The photolithographic process steps are similar to those used in the backside etching of the pressure sensor, but are simplified here. The simplified process sequence shown in Figure 6.14(a) is as follows:

- The contact areas are diffused into the silicon wafer, after resist, pattern exposure, and oxide etch.
- A sacrificial layer of silicon dioxide (approximately $2\,\mu$m thick) is grown over the surface of the wafer, and resist is applied and patterned.

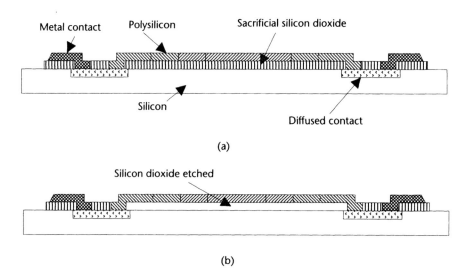

Figure 6.14 Cross section of a surface micromachined accelerometer.

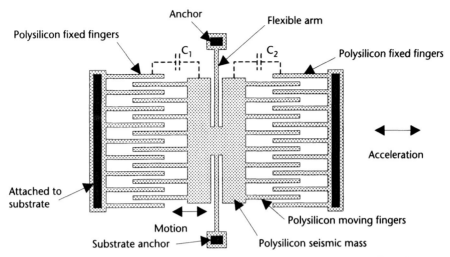

Figure 6.15 Topological view of the comb structure of a surface micromachined accelerometer.

- Holes are then etched in the oxide, so that the next layers can make contact with the diffused areas.
- Lightly doped polysilicon (approximately 2 μm thick) is then deposited on the oxide. Resist is applied and patterned, and the polysilicon is then etched to the pattern shown in Figure 6.15.
- A metal layer (aluminum) is then deposited; resist is applied and patterned; and the metal is etched to form the leads to the control electronics.
- The structure is annealed to remove stresses in the polysilicon, so that when released, the fingers in the polysilicon will lie flat and not bow or buckle.
- The final step is to remove the sacrificial oxide under the polysilicon. This is done by masking the wafer with resist, so that only the polysilicon structure is exposed, and an oxide etchant is used to etch away the exposed oxide, but not the polysilicon. This also will remove the oxide under the polysilicon, leaving a freestanding structure, as shown in Figure 6.14(b).

This is a very simplified process. Since the control electronics around the sensing structure are also included in the processing steps, these steps are omitted. Figure 6.15 shows the direction of movement of the structure during acceleration. Movement of the seismic mass causes its fingers to move with respect to the fixed fingers, changing the capacitance between them. The change in differential capacitance between C_1 and C_2 then can be amplified and used to measure the acceleration of the device. The values of C_1 and C_2 are of the order of 0.2 pF for full-scale deflection. The movement of the seismic mass gives about a 10% change in the value of the capacitance if open loop techniques are used for sensing. The integrated control electronics have temperature compensation, and are trimmed for offset and sensitivity. The control electronics can use open loop or closed loop techniques for sensing acceleration. Using open loop switched capacitor techniques, capacitance changes of approximately 0.1 fF can be sensed. Using closed loop techniques, electrostatic forces can be used to balance the forces produced on the seismic mass by acceleration, thus holding the seismic mass in its central position. At the distances involved, electrostatic forces are very large. Approximately 2V is required for balance at full acceleration. The dc balance voltage is directly proportional to acceleration. This dc balance voltage can be achieved using techniques such as width modulation or delta-sigma modulation, where the width of the driving waveforms applied to the plates is varied, giving an electrostatic force to balance the force of acceleration. The delta-sigma modulator output can be either a serial digital output or an analog output.

An alternative accelerometer layout is shown in Figure 6.16. The seismic mass is made of polysilicon with a polysilicon lower plate, with the spacing between the plates of approximately 2 μm. The processing steps are the same as in the previous structure. The center of mass of the top plate is displaced from the mounting pillar or anchor, the mass has torsion suspension, and the displacement under acceleration is sensed using differential capacitive sensing. A test plate is used to simulate acceleration using electrostatic forces. A difference of approximately 2V between the test plate and the seismic mass will give the same displacement as approximately 20g, for an accelerometer designed to measure up to 50g acceleration.

Filters also have been developed using surface micromachining techniques, and are similar in layout to the accelerometer. Figure 6.17 shows the topology of a comb microresonator, and the circuit used for a bandpass filter. The Q of the filter is controlled through negative feedback, and is the ratio of the feedback MOS devices. These devices typically have a rejection of 35 dB. The microfilters or resonators are small in size and operate in the range from 20 to 75 kHz. Other practical devices using surface micromachining techniques are vibration sensors and gyroscopes.

6.5 Smart Sensors Introduction

The advances in computer technology, devices, and methods have produced vast changes in the methodology of process control systems. These systems are moving away from a central control system, and towards distributed control devices.

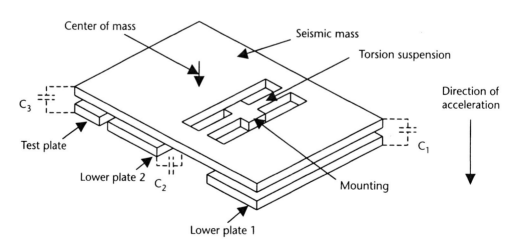

Figure 6.16 Surface micromachined accelerometer.

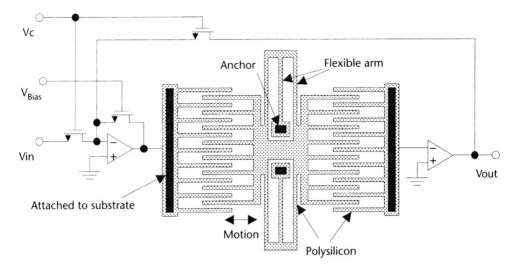

Figure 6.17 Topology of a microresonator.

6.5.1 Distributed System

The distributed system has a microprocessor integrated with the sensor. This allows direct conversion to a digital signal, conditioning of the signal, generation of a signal for actuator control, and diagnostics. The implementation of smart sensors has many advantages over a central control system [10]. These are as follows:

- The smart sensor takes over the conditioning and control of the sensor signal, reducing the load on the central control system, allowing faster system operation.
- Smart sensors use a common serial bus, eliminating the need for discrete wires to all sensors, greatly reducing wiring cost, large cable ducts, and confusion over lead destination during maintenance or upgrades (especially if lead markers are missing or incorrectly placed).
- Smart sensors have powerful built-in diagnostics, which reduces commissioning and startup costs and maintenance.
- Direct digital control provides high accuracy, not achievable with analog control systems and central processing.
- Uniformity in programming means that the program only has to be learned once, and new devices can be added to the bus on a plug-and-play basis.
- Individual controllers can monitor and control more than one process variable.
- The set points and calibration of a smart sensor are easily changed from the central control computer.
- The cost of smart sensor systems is presently higher than that of conventional systems, but when the cost of maintenance, ease of programming, ease of adding new sensors is taken into account, the long-term cost of smart sensor systems is less.

The implementation of smart sensors does have some drawbacks. These are:

- If upgrading to smart sensors, care has to be taken when mixing old devices with new sensors, since they may not be compatible.
- If a bus wire fails, the total system is down, which is not the case with discrete wiring. However, with discrete wiring, if one sensor connection fails, it may be necessary to shut the system down. The problem of bus wire failure can be alleviated by the use of a redundant backup bus.

6.5.2 Smart Sensors

Smart sensor is a name given to the integration of the sensor with an ADC, a proportional integral and derivative (PID) processor, a DAC for actuator control, and so forth. Such a setup is shown in Figure 6.18 for the mixture of two liquids in a fixed ratio, where the flow rates of both liquids are monitored using differential pressure sensors. The temperatures of the liquids are also monitored to correct the flow rates for density changes and any variations in the sensitivity of the DP cells. All of the sensors in this example can be MEM devices. The electronics in the smart sensor contains all the circuits necessary to interface to the sensor, amplify and condition the signal, and apply proportional, integral, and derivative action (PID) (see Chapter 16). When usage is varying, the signals from the sensors are selected in sequence by the multiplexer (Mux), and are then converted by the ADC into a digital format for the internal processor. After signal evaluation by the processor, the control signals are generated, and the DACs are used to convert the signal back into an analog format for actuator control. Communication between the central control computer and the distributed devices is via a common serial bus. The serial bus, or field bus, is a single twisted pair of leads used to send the set points to the peripheral units and to monitor the status of the peripheral units. This enables the processor in the smart sensor to receive updated information on factors such as set points, gain, operating mode, and so forth; and to send status and diagnostic information back to the central computer [11].

Smart sensors are available for all of the control functions required in process control, such as flow, temperature, level, pressure, and humidity control. The distributed control has many advantages, as already noted.

6.6 Summary

Integrated sensors and micromechanical devices were introduced in this chapter. These devices are silicon-based, made using chemical-etching techniques. Properties of integrated silicon devices can be used to accurately measure temperature, light, force, and magnetic field strength. The piezoelectric effect in other materials is used for accurate time generation and in microposition actuators. Integrated micromechanical devices are made using either bulk or surface micromachining techniques. Not only are these devices very small, but conditioning and sensitivity adjustment can be made as an integral part of the sensor, since they are silicon-based. This has the advantage of noise reduction, high sensitivity, improved reliability, and the ability to add features that normally would require extensive

6.6 Summary

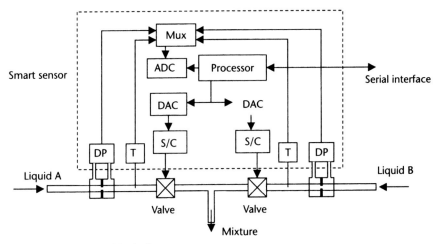

Figure 6.18 Smart sensor block diagram.

external circuits. As electronic devices become more cost effective, conventional systems will be replaced with distributed systems using smart sensors, which have a number of advantages in process control facilities, such as reduced loading on the controller, minimized wiring to peripheral units, and simplified expansion with the plug-and-play concept.

References

[1] Humphries, J. T., and L. P. Sheets, *Industrial Electronics*, 4th ed., Delmar, 1993, pp. 333–336.

[2] Lacanette, K., "Using IC Temperature Sensors to Protect Electronic Systems," *Sensors Magazine*, Vol. 14, No. 1, January 1997.

[3] Johnson, C. D., *Process Control Instrumentation Technology*, 7th ed., Prentice Hall, 2003, pp. 289–297.

[4] Baker, B. C., "Keeping the Signal Clean in Photo-sensing Instrumentation," *Sensors Magazine*, Vol. 14, No. 6, June 1997.

[5] Nagy, M. L., C. Apanius, and J. W. Siekkinen, "A User Friendly High-Sensitivity Strain Gauge," *Sensor Magazine*, Vol. 18, No. 6, June 2001.

[6] Caruso, M., et al., "A New Perspective on Magnetic Field Sensors," *Sensors Magazine*, Vol. 15, No. 12, December 1998.

[7] Lenz, J. E., "A Review of Magnetic Sensors," *Proceedings IEEE*, Vol. 78, No. 6, pp. 973–989.

[8] Markus, K. W., V. Dhuler, and R. Cohen, "Smart MEMs: Flip Chip Integration of MEMs and Electronics," *Proceedings Sensors Expo*, September 1994.

[9] Ristic, L., *Sensor Technology and Devices*, 1st ed., Norwood, MA: Artech House, Inc., 1994, pp. 95–144.

[10] Battikha, N. E., *The Condensed Handbook of Measurement and Control*, 2nd ed., ISA, 2004, pp. 171–173.

[11] Pullen, D., "Overview of Smart Sensor Interfaces," *Proceedings Sensors Expo*, September 1994.

Pressure

7.1 Introduction

Pressure is the force per unit area that a liquid or gas exerts on its surroundings, such as the force or pressure of the atmosphere on the surface of the Earth, and the force that liquids exert on the bottom and walls of a container. Pressure is not only an important parameter for process control, but also as an indirect measurement for other parameters. Not only is it important to select the right device for the required range and accuracy, but the device must be immune to contamination and interaction with the fluid being measured. As technology evolves, new and improved methods of accurately measuring pressures are constantly being developed [1].

7.2 Pressure Measurement

Pressure units are a measure of force acting over unit area. It is most commonly expressed in pounds per square inch (psi) or sometimes pounds per square foot (psf) in English units; or Pascals (Pa) in metric units, which is the force in Newtons per square meter (N/m²).

$$Pressure = \frac{force}{area} \quad \frac{lbs}{in^2} \tag{7.1}$$

Example 7.1

The liquid in a container has a total weight of 152 kN, and the container has a 8.9 m² base. What is the pressure on the base?

$$Pressure = \frac{152}{8.9} kPa = 17.1 kPa$$

$$\frac{152 kN}{8.9 m^2} = 17.07 kPa$$

7.2.1 Hydrostatic Pressure

The pressure at a specific depth in a liquid is termed hydrostatic pressure. The pressure increases as the depth in a liquid increases. This increase is due to the weight of the fluid above the measurement point. The pressure p is given by:

$$p = \gamma h \tag{7.2}$$

$$P_a = \frac{N}{m^2}$$

where γ is the specific weight (lb/ft³ in English units, or N/m³ in SI units), and h is the distance from the surface in compatible units (e.g., ft, in, cm, or m).

Example 7.2

What is the depth in a lake, if the pressure is 0.1 MPa?

$$\text{Depth} = 0.1 \text{ MPa} \div 9.8 \text{ kN/m}^3 = 10.2 \text{m}$$

The pressure at a given depth in a liquid is independent of the shape of the container or the volume of liquid contained. This is known as the Hydrostatic Paradox. The value of the pressure is a result of the depth and density. The total pressure or forces on the sides of the container depend on its shape, but at a specified depth, the pressure is given by (7.2).

Head is sometimes used as a measure of pressure. It is the pressure in terms of a column of a particular fluid (e.g., a head of 1 ft or 1m of water). For example, the pressure exerted by a 1-ft head of water is 62.4 psf, and the pressure exerted by 1-ft head of glycerin is 78.6 psf. Here again, (7.2) applies.

Example 7.3

What is the pressure at the base of a water tower that has 35m of head?

$$p = 9.8 \text{ kN/m}^3 \times 35\text{m} = 343 \text{ kPa}$$

7.2.2 Specific Gravity

The specific gravity (SG) of a liquid or solid is defined as the density of a material divided by the density of water. SG also can be defined as the specific weight of the material divided by the specific weight of water at a specified temperature. The specific weights and specific gravities of some common materials are given in Table 7.1. The specific gravity of a gas is its density (or specific weight) divided by the density (or specific weight) of air at 60°F and 1 atmospheric pressure (14.7 psia). In the SI system, the density in grams per cubic centimeter or megagrams per cubic meter and the SG have the same value. Both specific weight and density are temperature-dependent parameters, so that the temperature should be specified when they are being measured. SG is a dimensionless value, since it is a ratio.

Table 7.1 Specific Weights and Specific Gravities of Some Common Materials

	Temperature	Specific Weight lb/ft³	kN/m³	Specific Gravity
Acetone	60°F	49.4	7.74	0.79
Alcohol (ethyl)	68°F	49.4	7.74	0.79
Glycerin	32°F	78.6	12.4	1.26
Mercury	60°F	846.3	133	13.56
Steel		490	76.93	7.85
Water	39.2°F	62.43	9.8	1.0

Conversion factors: 1ft³ = 0.028m³; 1 lb = 4.448N; 1 lb/ft³ = 0.157 kN/m³.

Example 7.4

What is the specific gravity of glycerin, if the specific weight of glycerin is 12.4 kN/m³?

$$SG = 12.4/9.8 = 1.26$$

7.2.3 Units of Measurement

Many industrial processes operate at pressures that are referenced to atmospheric pressure, and are known as gauge pressures. Other processes operate at pressures referenced to a vacuum, or can be referred to as negative gauge pressure. Atmospheric pressure is not a fixed value, but depends on factors such as humidity, height above sea level, temperature, and so forth. The following terms exist when considering atmospheric constants.

1. Atmospheric pressure is measured in pounds per square inch (psi), in the English system.
2. Atmospheric pressure is measured in Pascals (Pa or N/m²), in the SI system.
3. Atmospheric pressure can be stated in inches or centimeters of water.
4. Atmospheric pressure can be stated in inches or millimeters of mercury.
5. Atmosphere (atm) is the equivalent pressure in atmospheres.
6. 1 torr = 1 mm mercury, in the metric system.
7. 1 bar (1.013 atm) = 100 kPa, in metric system.

Table 7.2 gives the conversions between various pressure measurement units.

Example 7.5

What pressure in psi corresponds to 98.5 kPa?

$$p = 98.5 \text{ kPa } (6.895 \text{ kPa/psi}) = 98.5/6.895 \text{ psi} = 14.3 \text{ psi}$$

Following are the six terms in common use applied to pressure measurements.

1. *Total vacuum* is zero pressure or lack of pressure, as would be experienced in outer space, and is very difficult to achieve in practice. Vacuum pumps can only approach a true vacuum.

Table 7.2 Pressure Conversions

	Water		Mercury**		kPa	psi
	in#	cm*	mm	in		
1 psi	27.7	70.3	51.7	2.04	6.895	1
1 psf	0.19	0.488	0.359	0.014	0.048	0.007
1 kPa	4.015	10.2	7.5	0.295	1	0.145
1 atm	407.2	1034	761	29.96	101.3	14.7
1 torr	0.535	1.36	1	0.04	0.133	0.019
1 millibar	0.401	1.02	0.75	0.029	0.1	0.014

#at 39°F *at 4°C **Mercury at 0°C

2. Atmospheric pressure is the pressure on the Earth's surface, due to the weight of the gases in the Earth's atmosphere (14.7 psi or 101.36 kPa absolute). The pressure decreases above sea level. For example, at an elevation of 5,000 ft, it has dropped to approximately 12.2 psi (84.122 kPa).
3. *Absolute pressure* is the pressure measured with respect to a vacuum, and is expressed in psia or kPa(a). Note the use of a and g when referencing the pressure to absolute and gauge.
4. Gauge pressure is the pressure measured with respect to atmospheric pressure, and is normally expressed in psig or kPa(g). Figure 7.1 shows graphically the relation between atmospheric, gauge, and absolute pressures.
5. *Vacuum* is a pressure between total vacuum and normal atmospheric pressure. Pressures less than atmospheric pressure are often referred to as "negative gauge," and indicated by an amount below atmospheric pressure. As an example, −5 psig corresponds to 9.7 psia.
6. *Differential pressure* is the pressure measured with respect to another pressure, and is expressed as the difference between the two values. This represents two points in a pressure or flow system, and is referred to as the "delta p," or Δp.

Example 7.6

The atmospheric pressure is 14.5 psi. If the absolute pressure is 2,865.6 psfa, what is the gauge pressure?

$$\text{Gauge pressure} = \frac{2865.6\ psfa}{144} - 14.5\ psi = 19.9\ psia - 14.5\ psi = 5.4\ psig$$

Example 7.7

What is the gauge pressure in (a) kPa, and (b) N/cm², at a distance 5.5 ft below the surface of a column of water?

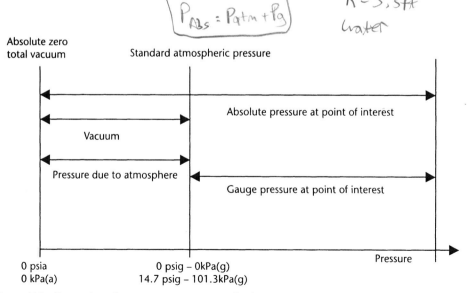

Figure 7.1 Illustration of gauge pressure versus absolute pressure.

7.2 Pressure Measurement

(a) $p = 9.8(5.5/3.28)$ kPa $= 9.8 \times 1.68$ kPa $= 16.4$ kPa(g)
(b) $p = 16.4$ N/m² $= 16.4/10{,}000$ N/cm² $= 1.64 \times 10^{-3}$ N/cm²(g)

The pressure in this case is the gauge pressure [i.e., kPa(g)]. To get the total pressure, the pressure of the atmosphere must be taken into account. The total pressure (absolute) in this case is $9.8 + 101.3 = 111.1$ kPa(a). The g and a should be used where possible to avoid confusion. In the case of psi and psf, this becomes psig and psfg, or psia and psfa. In the case of kPa, use kPa(a) or kPa(g). It also should be noted that if glycerin were used instead of water, then the pressure would be 1.26 times higher, since its specific gravity is 1.26.

7.2.4 Buoyancy

Buoyancy is the upward force exerted on an object immersed or floating in a liquid. The weight is less than it is in air, due to the weight of the displaced fluid. The upward force on the object causes the weight loss, called the buoyant force, and is given by:

$$B = \gamma V \qquad (7.3)$$

where B is the buoyant force in pounds, γ is the specific weight in pounds per cubic foot, and V is the volume of the displaced liquid in cubic feet. If working in SI units, then B is in newtons, γ is in newtons per cubic meter, and V is in cubic meters.

In Figure 7.2, items a, b, c, and d, are the same size, and the buoyancy forces on a and c are the same, although their depths are different. There is no buoyant force on d, since the liquid cannot get under it to produce the buoyant force. The buoyant force on b is one-half that on a and c, since only one-half of the object is submerged.

Example 7.8

What is the buoyant force on a plastic cube with 2.5m sides, floating in water, if three-quarters of the block is submerged?

$$B = 9.8 \text{ kN/m}^3 \times 2.5\text{m} \times 2.5\text{m} \times 2.5\text{m} \times 3/4 = 114.8 \text{ kN}$$

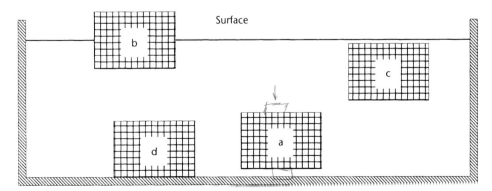

Figure 7.2 Immersed object to demonstrate buoyancy.

Example 7.9

What is the apparent weight of a 3.7m³ block of wood totally immersed in acetone? Assume the specific weight of wood is 8.5 kN/m³.

Weight of wood in air = 3.7 × 8.5 kN = 31.45 kN

Buoyant force on wood = 3.7 × 7.74 kN = 28.64 kN

Apparent weight = 31.45 − 28.64 = 2.81 kN (287 kg)

Pascal's Law states that the pressure applied to an enclosed liquid (or gas) is transmitted to all parts of the fluid and to the walls of the container. This is demonstrated in the hydraulic press in Figure 7.3. A force of F_s exerted on the small piston (ignoring friction) will exert a pressure in the fluid given by:

$$p = \frac{F_S}{A_S} \qquad (7.4)$$

where A_s is the cross-sectional area of the smaller piston.

Since the pressure is transmitted through the liquid to the second cylinder, according to Pascal's Law, the force on the larger piston (F_L) is given by:

$$F_L = pA_L \qquad (7.5)$$

where A_L is the cross-sectional area of the large piston (assuming the pistons are at the same level), from which:

$$F_L = \frac{A_L F_S}{A_S} \qquad (7.6)$$

It can be seen that the force F_L is magnified by the ratio of the piston areas. This principle is used extensively in hoists, hydraulic equipment, and so forth.

Example 7.10

In Figure 7.3, if the area of the small piston A_s is 8.2 in², and the area of the large piston A_L is 2.3 ft², what is the force F_L on the large piston, if the force F_s on the small piston is 25N?

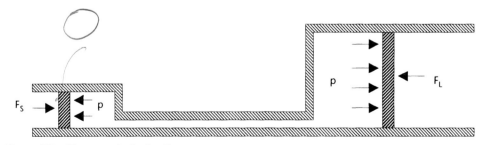

Figure 7.3 Diagram of a hydraulic press.

Force F_L on piston = $\dfrac{25N \times 2.3 \times 144}{8.2} = 1009.7N = 1.0097 kN$

7.3 Measuring Instruments

Several instruments are available for pressure measurement, these instruments can be divided into pressure measuring devices and vacuum measuring devices. U-tube manometers are being replaced with smaller and more rugged devices, such as the silicon diaphragm. Vacuum measuring devices require special techniques for the measurement of very low pressures.

7.3.1 Manometers

Manometers are good examples of pressure measuring instruments, although they are not as common as they previously were, because of the development of new, smaller, more rugged, and easier to use pressure sensors [2].

U-tube manometers consist of "U" shaped glass tubes partially filled with a liquid. When there are equal pressures on both sides, the liquid levels will correspond to the zero point on a scale, as shown in Figure 7.4(a). The scale is graduated in pressure units. When a higher pressure is applied to one side of the U-tube, as shown in Figure 7.4(b), the liquid rises higher in the lower pressure side, so that the difference in height of the two columns of liquid compensates for the difference in pressure. The pressure difference is given by:

$$P_R - P_L = \gamma \times \text{difference in height of the liquid in the columns} \qquad (7.7)$$

where γ is the specific weight of the liquid in the manometer.

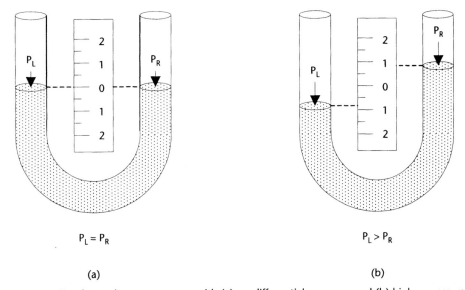

Figure 7.4 Simple U-tube manometers, with (a) no differential pressure, and (b) higher pressure on the left side.

Example 7.11

The liquid in a manometer has a specific weight of 8.5 kN/m³. If the liquid rises 83 cm higher in the lower pressure leg, what is difference in the pressure between the higher and the lower legs?

$$\Delta p = \gamma \Delta h = 8.5 \times 83/100 \text{ kPa} = 7.05 \text{ kPa}$$

Example 7.12

What is the liquid density in a manometer, if the difference in the liquid levels in the manometer tubes is 1.35m, and the differential pressure between the tubes is 7.85 kPa?

$$\gamma = \frac{p}{h} = \frac{7.85 kPa}{1.35m} \times \frac{N/m^2}{9.8 N/m^3} kg/m^3 = 0.59 Mg/m^3$$

7.3.2 Diaphragms, Capsules, and Bellows

Gauges are a major group of sensors that measure pressure with respect to atmospheric pressure. Gauge sensors are usually devices that change their shape when pressure is applied. These devices include diaphragms, capsules, bellows, and Bourdon tubes.

Diaphragms consist of a thin layer or film of a material supported on a rigid frame, as shown in Figure 7.5(a). Pressure can be applied to one side of the film for gauge sensing, with the other inlet port being left open to the atmosphere. Pressures can be applied to both sides of the film for differential sensing, and absolute pressure sensing can be achieved by having a partial vacuum on one side of the diaphragm. A wide range of materials can be used for the sensing film: from rubber to plastic for low pressures, silicon for medium pressures, and stainless steel for high pressures. When a pressure is applied to the diaphragm, the film distorts or becomes slightly spherical, and can be sensed using a strain gauge, piezoelectric, or changes in capacitance techniques. Older techniques included magnetic and carbon pile devices. In the device shown, the position of the diaphragm is sensed using capacitive techniques, and the measurement can be made by using an ac bridge or by using pulse-switching techniques. These techniques are very accurate, and excellent linear correlation between pressure and output signal amplitude can be obtained.

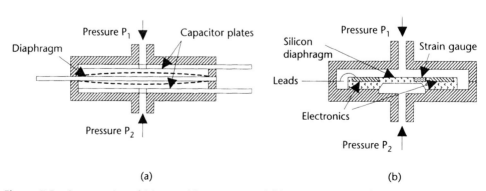

Figure 7.5 Cross section of (a) capacitive sensor, and (b) microminiature silicon pressure sensor.

Silicon diaphragms are now in common use. Since silicon is a semiconductor, a piezoresistive strain gauge and amplifier electronics can be integrated into the top surface of the silicon structure, as shown in Figure 7.5(b) (see also Figure 6.9). These devices have built-in temperature compensation for the strain gauges and amplifiers, and have high sensitivity, giving a high output voltage (5V FSD). They are very small, accurate (<2% FSD), reliable, have a good temperature operating range ($-50°$ to $120°C$), are low cost, can withstand high overloads, have good longevity, and are unaffected by many chemicals. Commercially made devices are available for gauge, differential, and absolute pressure sensing up to 200 psi (1.5 MPa). This range can be extended by the use of stainless steel diaphragms to 10,000 psi (70 MPa) [3].

The cross section shown in Figure 7.5(b) is a differential silicon chip (sensor die) microminiature pressure sensor. The dimensions of the sensing elements are very small, and the die is packaged into a plastic case (0.2 in thick × 0.6 in diameter, approximately). The sensor is used in a wide variety of industrial applications, widely used in automotive pressure sensing applications (e.g., manifold air pressure, barometric air pressure, oil, transmission fluid, break fluid, power steering, tire pressure, and many other applications such as blood pressure monitors) [4].

Capsules are two diaphragms joined back to back. Pressure can be applied to the space between the diaphragms, forcing them apart to measure gauge pressure. The expansion of the diaphragm can be mechanically coupled to an indicating device. The deflection in a capsule depends on its diameter, material thickness, and elasticity. Materials used are phosphor bronze, stainless steel, and iron-nickel alloys. The pressure range of instruments using these materials is up to 50 psi (350 kPa). Capsules can be joined together to increase sensitivity and mechanical movement, or, as shown in Figure 7.6, they can be connected as two opposing devices, so that they can be used to measure differential pressures. The measuring system uses a closed loop technique to convert the movement of the arm (pressure) into an electrical signal, and to maintain it in its neutral position in a force balance system. When there is a pressure change, movement in the arm is sensed by the linear variable differential transformer (LVDT), or other type of position sensor. The signal is amplified and drives an electromagnet to pull the arm back to its neutral position. The current needed to drive the electromagnet is then proportional to the applied pressure, and the output signal amplitude is proportional to the electromagnet's current. The advantage of the closed loop control system is that any nonlinearity in the

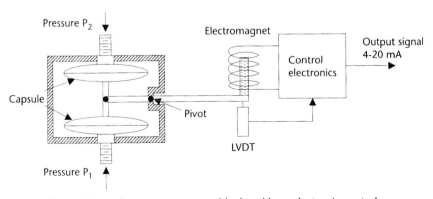

Figure 7.6 Differential capsule pressure sensor with closed loop electronic control.

mechanical system is virtually eliminated. This setup provides a high-level output, high resolution, good accuracy, and stability. The current to the electromagnet can be a varying amplitude dc, or a pulse width modulated current that easily can be converted to a digital signal.

Bellows are similar to capsules, except that instead of being joined directly together, the diaphragms are separated by a corrugated tube or a tube with convolutions, as shown in Figure 7.7. When pressure is applied to the bellows, it elongates by stretching the convolutions, rather than the end diaphragms. The materials used for the bellows type of pressure sensor are similar to those used for the capsule, giving a pressure range for the bellows of up to 800 psi (5 MPa). Bellows devices can be used for absolute, gauge, and differential pressure measurements.

Differential measurements can be made by mechanically connecting two bellows to be opposing each other when pressure is applied to them, as shown in Figure 7.7. When pressures P_1 and P_2 are applied to the bellows, a differential scale reading is obtained. P_2 could be atmospheric pressure for gauge measurements. The bellows is the most sensitive of the mechanical devices for low-pressure measurements (i.e., 0.5 to 210 kPa).

7.3.3 Bourdon Tubes

Bourdon tubes are hollow, flattened, or oval cross-sectional beryllium, copper, or steel tubes, as shown in Figure 7.8(a). The flattened tube is then shaped into a three-quarter circle, as shown in Figure 7.8(b). The operating principle is that the outer edge of the cross section has a larger surface than the inner portion. When pressure is applied, the outer edge has a proportionally larger total force applied because of its larger surface area, and the diameter of the circle increases. The walls of the tube are between 0.01 and 0.05 in thick. The tubes are anchored at one end. When pressure is applied to the tube, it tries to straighten, and in doing so, the free

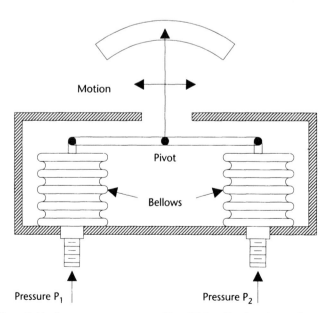

Figure 7.7 Differential bellows pressure gauges ($P_1 - P_2$) for direct scale reading.

end of the tube moves. This movement can be mechanically coupled to a pointer, which will indicate pressure as a line-of-sight indicator, or it can be coupled to a potentiometer, which will give a resistance value proportional to pressure as an electrical signal. The Bourdon tube dates from the 1840s. It is reliable, inexpensive, and one of the most common general-purpose pressure gauges.

Bourdon tubes can withstand overloads of up to 40% of their maximum rated load without damage, but, if overloaded, may require recalibration. The Bourdon tube is normally used for measuring positive gauge pressures, but also can be used to measure negative gauge pressures. If the pressure to the Bourdon tube is lowered, then the diameter of the tube reduces. This movement can be coupled to a pointer to make a vacuum gauge. Bourdon tubes can have a pressure range of up to 10,000 psi (70 MPa) [5].

Bourdon tubes also can be shaped into helical or spiral shapes to increase their range. Figure 7.9(a) shows the Bourdon tube configured as a spiral, and Figure 7.9(b) shows a tube configured as helical pressure gauge. These configurations are more sensitive than the circular Bourdon tube, and extend the lower end of the Bourdon tube pressure range from 3.5 kPa down to 80 Pa.

7.3.4 Other Pressure Sensors

Barometers are used for measuring atmospheric pressure. A simple barometer was the mercury in glass barometer, which is now little used due to its fragility and the toxicity of the mercury. The aneroid (no fluid) barometer is favored for direct reading [e.g., the bellows in Figure 7.7, or the helical Bourdon tube in Figure 7.9(b)], and the solid state absolute pressure sensor is favored for electrical outputs.

A *Piezoelectric pressure gauge* is shown in Figure 7.10. Piezoelectric crystals produce a voltage between their opposite faces when a force or pressure is applied to the crystal. This voltage can be amplified, and the device used as a pressure sensor. To compensate for the force produced by the weight of the diaphragm when the

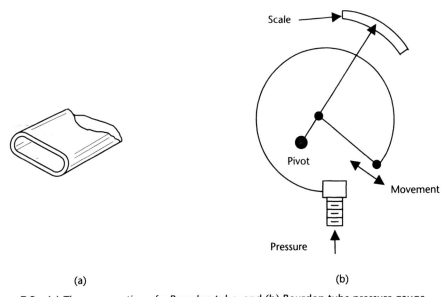

Figure 7.8 (a) The cross section of a Bourdon tube, and (b) Bourdon tube pressure gauge.

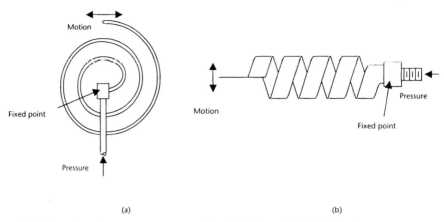

Figure 7.9 Bourdon tube pressure gauges: (a) spiral tube, and (b) helical tube.

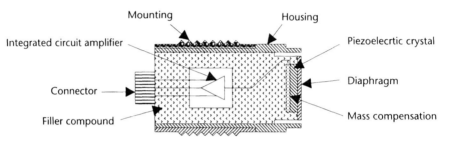

Figure 7.10 Cross section of a piezoelectric sensing element.

sensor is moving or its velocity is changing (e.g., if vibration is present), a seismic weight or mass is sometimes used on the other side of the piezoelectric crystal to the diaphragm [6]. Piezoelectric devices have good sensitivity, a wide operating temperature (up to 300°C), and a good frequency response (up to 100 kHz), but are not well suited for low frequency (less than 5 Hz) or for dc operation, due to offset and drift caused by temperature changes (pyroelectric effect). Piezoelectric devices are better suited for dynamic rather than static measurements.

7.3.5 Vacuum Instruments

Vacuum instruments are used to measure pressures less than atmospheric pressure. Bourdon tubes, diaphragms, and bellows can be used as vacuum gauges, but they measure negative pressures with respect to atmospheric pressure. The silicon absolute pressure gauge has a built-in low-pressure reference, so it is calibrated to measure absolute pressures. Conventional devices can be used down to 20 torr (5 kPa), and this range can be extended down to approximately 1 torr with special sensing devices.

Ionization gauges can be used to measure pressures from 10^{-3} atm down to approximately 10^{-12} atm. The gas is ionized with a beam of electrons and the current is measured between two electrodes in the gas. The current is proportional to the number of ions per unit volume, which is also proportional to the gas pressure. An ionization gauge is shown in Figure 7.11(a) [7].

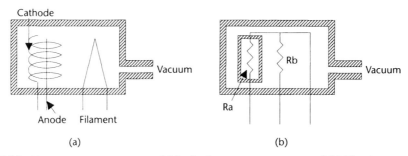

Figure 7.11 Vacuum pressure gauges: (a) Ionization pressure gauge, and (b) Pirani gauge.

The *Pirani gauge* is shown in Figure 7.11(b). With special setups and thermocouples, the Pirani gauge can measure vacuums down to approximately 1 torr (10^{-3} atm). These methods are based on the relation of heat conduction and radiation from a heating element, to the number of gas molecules per unit volume in the low-pressure region, which determines the pressure.

The *McLeod gauge* is a device set up to measure low pressures (1 torr). The device compresses the low-pressure gas to a level at which it can be measured. The change in volume and pressure then can be used to calculate the original gas pressure, providing the gas does not condense.

7.4 Application Considerations

When installing pressure sensors, care should be taken to select the correct pressure sensor for the application. This section gives a comparison of the characteristics of the various types of pressure sensors, installation considerations, and calibration.

7.4.1 Selection

Pressure sensing devices are chosen for pressure range, overload requirements, accuracy, temperature operating range, line-of-sight reading, electrical signaling, and response time. In some applications, there are other special requirements. Parameters such as hystersis and stability should be obtained from the manufacturer's specifications. For most industrial applications that involve positive pressures, the Bourdon tube is a good choice for direct visual readings, and the silicon pressure sensor for the generation of electrical signals [8]. Both types of devices have commercially available sensors to measure from 5 psi FSD, up to 10,000 psi (70 MPa) FSD. Table 7.3 gives a comparison of the two types of devices.

Table 7.3 Comparison of Bourdon Tube Sensor and Silicon Sensor

Device	Maximum Pressure Range, psi (MPa)	Accuracy, FSD	Response Time, Seconds	Overload
Bourdon tube (stainless steel)	10,000 (70)	1%	1	40%
Silicon sensor with stainless steel diaphragm	10,000 (70)	1%	1×10^{-3}	400%

Table 7.4 lists the operating range for several types of pressure sensors. The range shown is the full range, and may involve several devices made of different types of materials. The accuracy may be misleading, since it depends on the range of the device. The values given are typical, and may be exceeded with new materials and advances in technology [9].

7.4.2 Installation

The following should be taken into consideration when installing pressure-sensing devices.

1. The distance between the sensor and the source should be kept to a minimum.
2. Sensors should be connected via valves for ease of replacement.
3. Overrange protection devices should be included at the sensor.
4. To eliminate errors due to trapped gas in sensing liquid pressures, the sensor should be located below the source.
5. To eliminate errors due to trapped liquid in sensing gas pressures, the sensor should be located above the source.
6. When measuring pressures in corrosive fluids and gases, an inert medium is necessary between the sensor and source, or the sensor must be corrosion-resistant.
7. The weight of liquid in the connecting line of a liquid pressure sensing device located above or below the source will cause errors at zero, and a correction must be made by the zero adjustment, or otherwise compensated for in measurement systems.
8. Resistance and capacitance can be added to electronic circuits to reduce pressure fluctuations and unstable readings.

7.4.3 Calibration

Pressure sensing devices are calibrated at the factory. In cases where a sensor is suspect and needs to be recalibrated, the sensor can be returned to the factory for recalibration, or it can be compared to a known reference. Low-pressure devices can be calibrated against a liquid manometer. High-pressure devices can be calibrated with a deadweight tester, using weights on a piston to accurately reproduce high

Table 7.4 Approximate Pressure Ranges for Pressure Sensing Devices

Device	Pressure Range, psi (MPa)	Temperature Range	Accuracy ±
U-tube manometer	0.1–120 (0.7 kPa–1 MPa)	ambient	0.02 in
Diaphragm	0.5–400 (3.5 kPa–0.28 MPa)	90°C maximum	0.1% FSD
Solid state diaphragm	0.2–200 (1.4 kPa–0.14 MPa)	−50° to +120°C	1.0% FSD
Stainless steel diaphragm	20–10,000 (140 kPa–70 MPa)	−50° to +120°C	1.0% FSD
Capsule	0.5–50 (3.5 kPa–0.031 MPa)	90°C maximum	0.1% FSD
Bellows	0.1–800 (0.7 kPa–0.5 MPa)	90°C maximum	0.1% FSD
Bourdon tube	0.5–10,000 (3.5 kPa–70 MPa)	90°C maximum	0.1% FSD
Spiral Bourdon	0.01–4,000 (70 Pa–25 MPa)	90°C maximum	0.1% FSD
Helical Bourdon	0.02–8,000 (140 Pa–50 MPa)	90°C maximum	0.1% FSD
Piezoelectric stainless steel	20–10,000 (140 kPa–70 MPa)	−270° to +200°C	1.0% FSD

pressures. Accurately calibrated standards can be obtained from the National Institute of Standards and Technology, which also can perform recalibration of standards.

7.5 Summary

Pressure measurements can be made in either English or SI units. Pressures can be referenced to atmospheric pressures, where they are termed gauge pressures; or referenced to vacuum, where they are referred to as absolute pressures. Hydrostatic pressures are the pressure at a depth in a liquid due to the weight of liquid above, which is determined by its density or specific weight. Similarly, when an object is placed in a liquid, a force is exerted (buoyancy) on the object equal to the weight of liquid displaced.

A number of pressure measuring instruments are available, such as the Bourdon tube and bellows, although some of the older types, such as the U-tube manometer, are being replaced by smaller, easier to use, and less fragile devices, such as the silicon pressure sensor. All of the pressure sensors can be configured to measure absolute, differential, or gauge pressures. However, to measure very low pressures close to true vacuum, special devices, such as the ionization and Pirani gauges, are required.

Pressures are often used as an indirect measure of other physical parameters. Care has to be taken not to introduce errors when mounting pressure gauges. The choice of the pressure gauge depends on the needs of the application. A list of characteristics was given, along with the precautions that should be used when installing pressure gauges.

Definitions

Absolute pressure   is the pressure measured with respect to a vacuum, and is expressed in psia or kPa(a).

Atmospheric pressure   is the pressure on the Earth's surface due to the weight of the gases in the Earth's atmosphere, and is normally expressed at sea level as 14.7 psi or 101.36 kPa.

Density (ρ)   is the mass per unit volume of a material (e.g., lb (slug)/ft^3, or kg/m^3).

Differential pressure   is the pressure measured with respect to another pressure.

Dynamic pressure   is the pressure exerted by a fluid or gas when it impacts on a surface or an object due to its motion or flow.

Gauge pressure   is the pressure measured with respect to atmospheric pressure, and is normally expressed as psig or kPa(g).

Impact pressure (total pressure)   is the sum of the static and dynamic pressures on a surface or object.

Specific gravity (SG) of a liquid or solid is a ratio of the density (or specific weight) of a material, divided by the density (or specific weight) of water at a specified temperature. The specific gravity of a gas is its density (or specific weight), divided by the density (or specific weight) of air, at 60°F and 1 atmosphere pressure (14.7 psia). In the SI system, the density in grams per cubic centimeter or megagrams per cubic meter and SG have the same value.

Specific weight (γ) is the weight per unit volume of a material (e.g., lb/ft^3, or N/m^3).

Static pressure is the pressure of a fluid or gas that is stationary or not in motion.

Total vacuum which is zero pressure or lack of pressure, as would be experienced in outer space.

Vacuum is a pressure measurement made between total vacuum and normal atmospheric pressure (14.7 psi).

References

[1] Wilson, J. S., "Pressure Measurement Principles and Practice," *Sensors Magazine*, Vol. 20, No. 1, January 2003.

[2] Jurgen, R. K., *Automotive Electronics Handbook*, 2nd ed., McGraw-Hill, 1999, pp. 2.3–2.25.

[3] Bryzek, J., "Signal Conditioning for Smart Sensors and Transducers," *Proceedings Sensor Expo*, 1993, pp. 151–160.

[4] Burgess, J., "Tire Pressure Monitoring: An Industry Under Pressure," *Sensors Magazine*, Vol. No. 7, July 2003.

[5] Humphries, J. T, and L. P. Sheets, *Industrial Electronics*, 4th ed., Delmar, 1993, pp. 356–359.

[6] Bicking, R. E., "Fundamentals of Pressure Sensor Technology," *Sensor Magazine*, Vol. 15, No. 11, November 1998.

[7] Johnson, C. D., *Process Control Instrumentation Technology*, 7th ed., Prentice Hall, 2003, p. 256.

[8] Bitko, G., A. McNeil, and R. Frank, "Improving the MEMS Pressure Sensor," *Sensors Magazine*, Vol. 17, No. 7, July 2000.

[9] Battikha, N. E., *The Condensed Handbook of Measurement and Control*, 2nd ed., ISA, 2004, p. 122.

CHAPTER 8
Level

8.1 Introduction

This chapter discusses the measurement of the level of liquids and free-flowing solids. There are many widely varying methods for the measurement of liquid level. Level measurement is an important part of process control. Level sensing can be single point, continuous, direct, or indirect. Continuous level monitoring measures the level of the liquid on an uninterrupted basis. In this case, the level of the material will be constantly monitored, and hence the volume can be continuously monitored, if the cross-sectional area of the container is known.

Examples of direct and indirect level measurements are using a float technique, or measuring pressure and calculating the liquid level. Accurate level measurement techniques have been developed. New measurement techniques are constantly being introduced and old ones improved [1]. Level measuring devices should have easy access for inspection, maintenance, and replacement. Free-flowing solids include dry powders, crystals, rice, grain, and so forth.

8.2 Level Measurement

Level sensing devices can be divided into four categories: (1) direct sensing, in which the actual level is monitored; (2) indirect sensing, in which a property of the liquid, such as pressure, is sensed to determine the liquid level; (3) single point measurement, in which it is only necessary to detect the presence or absence of a liquid at a specific level; and (4) free-flowing solid level sensing [2].

8.2.1 Direct Level Sensing

A number of techniques are used for direct level sensing, such as direct visual indication using a sight glass or a float. Ultrasonic distance measuring devices also may be considered.

The *Sight glass* or gauge is the simplest method for direct visual reading. As shown in Figure 8.1, the sight glass is normally mounted vertically adjacent to the container. The liquid level then can be observed directly in the sight glass. The ends of the glass are connected to the top and bottom of the tank via shutoff valves, as would be used with a pressurized container (boiler) or a container with volatile, flammable, hazardous, or pure liquids. In cases where the tank contains inert liquids, such as water, and pressurization is not required, the tank and sight glass

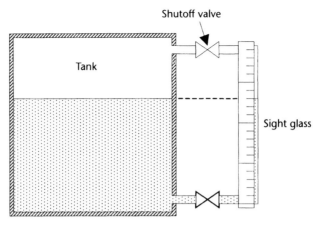

Figure 8.1 Sight glass for visual observation of liquid levels.

both can be open to the atmosphere. Glass gauges are cheap but easily broken, and should not be used with hazardous liquid. For safety reasons, they should not be longer than 4 ft. Cold liquids also can cause condensation on the gauge. Gauges should have shutoff valves in case of breakage (sometimes automatic safety shutoff valves are used) and to facilitate replacement. In the case of high pressure or hazardous liquids, a nonmagnetic material can be used for the sight glass with a magnetic float that can rotate a graduated disk, or can be monitored with a magnetic sensor, such as a Hall Effect device [3].

Float sensors are shown in Figure 8.2. There are two types of floats shown: the angular arm and the pulley. The float material is less dense than the density of the liquid, and floats up and down on top of the material being measured. In Figure 8.2(a), a float with a pulley is used. This method can be used with either liquids or free-flowing solids. With free-flowing solids, agitation is sometimes used to help level the solids. The advantages of the float sensor are that they are almost independent of the density of the liquid or solid being monitored, are accurate and robust,

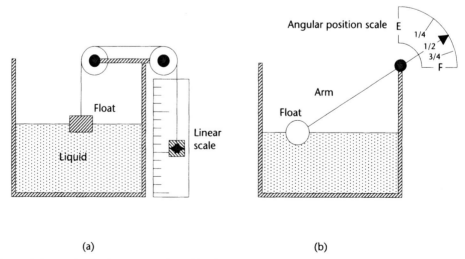

Figure 8.2 Methods of measuring liquid levels, using (a) a simple float with level indicator on the outside of the tank, and (b) an angular arm float.

and have a linear output with level height. However, accuracy can be affected by material accumulation on the float, corrosion, chemical reactions, and friction in the pulleys. If the surface of the material being monitored is turbulent, causing the float reading to vary excessively, some means of damping might be required, such as a stilling well. In Figure 8.2(b), a ball float is attached to an arm, and the angle of the arm is measured to indicate the level of the material. A spherical float shape is used to provide maximum buoyancy, and it should be one-half submerged for maximum sensitivity, and to have the same float profile independent of angle. An example of this type of sensor is the fuel level monitor in the tank of an automobile. Due to lack of headroom in this application, the angle of the float arm can go only from approximately 0° to 90°. The fuel gauge shows the output voltage from a potentiometer driven by the float. Although very simple and cheap to manufacture, the angular float sensor has the disadvantage of nonlinearity, as shown by the line-of-sight scale in Figure 8.2(b).

Figure 8.3(a) shows a pulley-type float sensor with a linear radial scale that can be replaced with a potentiometer to obtain an electrical signal. Figure 8.3(b) shows an angular arm float. The travel of the arm on this float is ±30°, giving a scale that is more linear than in the automotive application, and which can be linearized for industrial use. The scale can be replaced by a potentiometer to obtain an electrical signal.

Ultrasonic or sonic devices can be used for single point or continuous level measurement of a liquid or a solid. A setup for continuous measurement is shown in Figure 8.4. A pulse of sonic waves (approximately 10 kHz) or ultrasonic waves (more than 20 kHz) from the transmitter is reflected from the surface of the liquid to the receiver, and the time for the echo to reach the receiver is measured. The time delay gives the distance from the transmitter and receiver to the surface of the liquid, from which the liquid level can be calculated, knowing the velocity of ultrasonic waves (approximately 340 m/s). Since there is no contact with the liquid, this method can be used for solids, and corrosive and volatile liquids. Sonic and ultrasonic devices

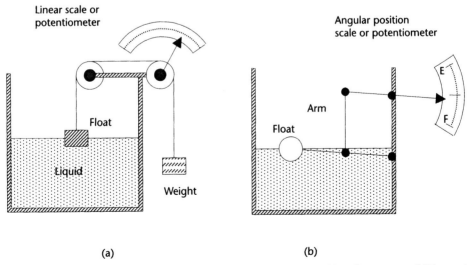

Figure 8.3 Float level sensors with radial scales or potentiometers: (a) pulley type, and (b) angular arm type with ±30° angle.

are reliable, accurate, and cost-effective. They can be used in high humidity, have no moving parts, and are unaffected by material density or conductivity. Vibration or high noise levels can affect the devices. Dust can give false signals or attenuate the signals by deposit buildup on the transmitting and receiving devices. Care should be taken not to exceed the operating temperature of the devices, and correction may be required for the change in velocity of the sonic waves with humidity, temperature, and pressure [4].

In a liquid, the transmitter and receiver can be placed on the bottom of the container, and the time measured for an echo to be received from the surface of the liquid back to the receiver can be used to calculate the depth of the liquid.

8.2.2 Indirect Level Sensing

A commonly used method of indirectly measuring a liquid level is to measure the hydrostatic pressure at the bottom of the container. The level can be extrapolated from the pressure and the specific weight of the liquid. The level of liquid can be measured using displacers, capacitive probes, bubblers, resistive tapes, or by weight measurements.

Pressure is often used as an indirect method of measuring liquid levels. Pressure increases as the depth increases in a fluid. The pressure is given by:

$$p = \gamma h \tag{8.1}$$

where p is the pressure, γ is the specific weight, and h is the depth. Note that the units must be consistent, the specific weight is temperature-dependent, and temperature correction is required.

Example 8.1

A pressure gauge located at the base of an open tank containing a liquid with a specific weight of 13.6 kN/m³ registers 1.27 MPa. What is the depth of the fluid in the tank?

From (8.1)

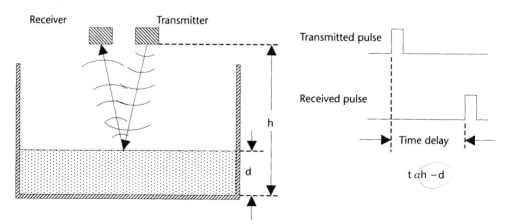

Figure 8.4 Use of ultrasonic devices for continuous liquid level measurements made by timing reflections from the surface of the liquid.

8.2 Level Measurement

$$h = \frac{p}{\gamma} = \frac{1.27 MPa}{13.6 kN/m^3} = 93.4 m$$

The pressure can be measured by any of the methods given in the section on pressure. In Figure 8.1, a differential pressure sensor can replace the sight glass. These devices are affected by liquid density and are susceptible to dirt. It is sometimes necessary to mount the pressure sensor above or below the zero liquid level, as shown in Figure 8.5, in which case an adjustment to the zero point is required. When the sensor is below the zero level, as shown in Figure 8.5(a), "zero suppression" is required to allow for H; the distance of the measuring instrument above or below the bottom of the container when it is above the zero level, as shown in Figure 8.5(b), "zero elevation" is required to allow for H. Shutoff valves should be used for maintenance and replacement, and cleanout plugs to remove solids. The dial on the pressure gauge can be calibrated directly in liquid depth.

A *displacer* with force sensing is shown in Figure 8.6. This device uses the change in buoyant force on an object to measure the changes in liquid level. The displacers must have a higher specific weight than that of the liquid whose level is being measured, and has to be calibrated for the specific weight of the liquid. A force or strain gauge measures the excess weight of the displacer. There is only a small movement in this type of sensor, compared to the movement in a float sensor. Displacers are simple, reliable, and accurate, but are affected by the (temperature-dependent) specific weight of the liquid, and buildup on the dispenser of coatings and depositions from the liquid. A still well may be required where turbulence is present in the liquid [5].

The buoyant force on the cylindrical displacer shown in Figure 8.6 is given by:

$$\text{Buoyant Force}(F) = \frac{\gamma \pi d^2 L}{4} \quad (8.2)$$

where γ = specific weight of the liquid, d is float diameter, and L is the length of the displacer submerged in the liquid.

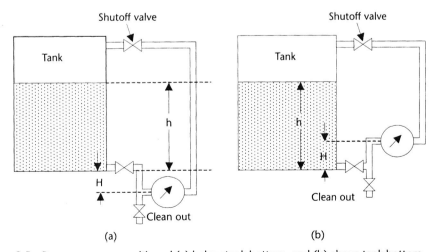

Figure 8.5 Pressure sensors positioned (a) below tank bottom, and (b) above tank bottom.

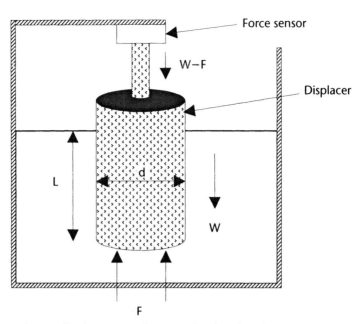

Figure 8.6 Displacer with a force sensor for measuring liquid level.

The weight, as seen by the force sensor, is given by:

$$\text{Weight on force sensor} = \text{Weight of displacer } (W) - F \quad (8.3)$$

It should be noted that the units must be in the same measurement system, the liquid must not rise above the top of the displacer, and the displacer must not touch the bottom of the container.

Example 8.2

A 13-in diameter displacer is used to measure changes in water level. If the water level changes by 1.2m, what is the change in force sensed by the force sensor?

From (8.3):

$$\text{Change in force} = (W - F_1) - (W - F_2) = F_2 - F_1$$

From (8.2):

$$F_2 - F_1 = \frac{9.8 kN/m^3 \times \pi (0.13m)^2 \times 1.2}{4} = 156 N$$

Example 8.3

A 7.3-in diameter displacer is used to measure acetone levels. What is the change in force sensed if the liquid level changes by 2.3 ft?

$$F_2 - F_1 = \frac{49.4 lb/ft^3 \times \pi \times 7.3^2 \, m^2 \times 2.3 ft}{4 \times 144 in/ft} = 33 lb$$

Capacitive probes can be used in liquids and free-flowing solids for continuous level measurement [6]. Materials placed between the plates of a capacitor increase the capacitance by a factor (μ), known as the dielectric constant of the material. For instance, air has a dielectric constant of 1, and water has a dielectric constant of 80. When two capacitor plates are partially immersed in a nonconductive liquid, the capacitance (C_d) is given by:

$$C_d = C_a \mu \frac{d}{r} + C_a \qquad (8.4)$$

Where C_a is the capacitance with no liquid, μ is the dielectric constant of the liquid between the plates, r is the height of the plates, and d is the depth or level of the liquid between the plates.

The dielectric constants of some common liquids are given in Table 8.1. There are large variations in dielectric constant with temperature, so that temperature correction may be needed. From (8.4), the liquid level is given by:

$$d = \frac{(C_d - C_a)}{\mu C_a} r \qquad (8.5)$$

Example 8.4

A 1.3m-long capacitive probe has a capacitance of 31 pF in air. When partially immersed in water with a dielectric constant of 80, the capacitance is 0.97 nF. What is the length of the probe immersed in water?

From (8.5):

$$d = \frac{(0.97 \times 10^3 \, pf - 31 pf) 1.3m}{80 \times 31 pf} = 0.49m = 49cm$$

The capacitive probe shown in Figure 8.7(a) is used to measure the level in a nonconducting liquid, and consists of an inner rod with an outer shell. The capacitance is measured between the two using a capacitance bridge. In the portion of the probe that is out of the liquid, air serves as the dielectric between the rod and outer shell. In the section of the probe immersed in the liquid, the dielectric is that of the liquid, which causes a large capacitive change. Where the tank is made of metal, it can serve as the outer shell. The capacitance change is directly proportional to the level of the liquid. The dielectric constant of the liquid must be known for this type of measurement. The dielectric constant varies with temperature, as can be seen from Table 8.1, so that temperature correction is required [7]. If the liquid is conductive, then one of the plates is enclosed in an insulator, as shown in Figure 8.7(b).

Table 8.1 Dielectric Constant of Some Common Liquids

Water	80 @ 20°C	Acetone	20.7 @ 25°C
Water	88 @ 0°C	Alcohol (ethyl)	24.7 @ 25°C
Glycerol	42.5 @ 25°C	Gasoline	2.0 @ 20°C
Glycerol	47.2 @ 0°C	Kerosene	1.8 @ 20°C

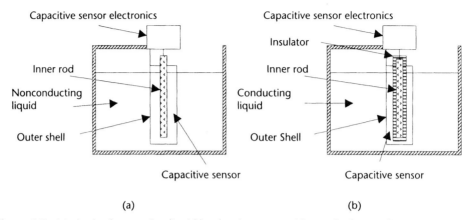

Figure 8.7 Methods of measuring liquid levels using a capacitive probe for continuous monitoring in (a) nonconducting liquid, and (b) conducting liquid.

The dielectric constant is now that of the insulator, and the liquid level sets the area of the capacitor plate.

Bubbler devices require a supply of clean air or inert gas to prevent interaction with the liquid, as shown in Figure 8.8. Gas from a pressure regulator is forced through a tube via a flow regulator, and the open end of the tube is close to the bottom of the tank. The specific weight of the gas is negligible compared to the specific weight of the liquid, and can be ignored. The pressure required to force the liquid out of the tube is equal to the pressure at the end of the tube due to the liquid, which is the depth of the liquid multiplied by the specific weight of the liquid (requiring temperature correction). This method can be used with corrosive liquids, since the material of the tube can be chosen to be corrosion-resistant. Electrical power is not required, and variations in specific weight will affect the readout [8].

Example 8.5

How far below the surface of the water is the end of a bubbler tube, if bubbles start to emerge from the end of the tube when the air pressure in the bubbler is 263 kPa?
From (8.1):

$$h = \frac{p}{\gamma} = \frac{263 kPa \times 10^{-4}}{1 gm/cm^3} = 263 cm$$

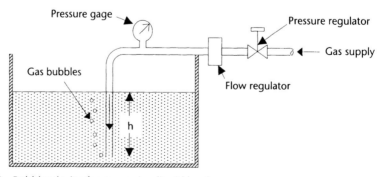

Figure 8.8 Bubbler device for measuring liquid level.

8.2 Level Measurement

Resistive tapes can be used to measure liquid levels, as shown in Figure 8.9. A resistive element is placed in close proximity to a conductive strip in an easily compressible nonconductive sheath. The pressure of the liquid pushes the conductive strip against the resistive element, shorting out a length of the resistive element that is proportional to the depth of the liquid. The sensor can be used in corrosive liquids or slurries. It is cheap, but is not rugged or accurate. It is prone to humidity problems, measurement accuracy is dependent on material density, and is not recommended for use with explosive or flammable liquids.

Load cells can be used to measure the weight of a tank and its contents. The weight of the container is subtracted from the total reading, leaving the weight of the contents of the container. Knowing the cross-sectional area of the tank and the specific weight of the material, the volume and/or depth of the contents can be calculated. This method is well-suited for continuous measurement, and the material being weighed does not come into contact with the sensor. The level (depth) depends on the density of the material [9].

The weight of a container can be used to calculate the level of the material in the container. From Figure 8.10, the volume (V) of the material in the container is given by:

$$V = \text{area} \times \text{depth} = \pi r^2 \times d \qquad (8.6)$$

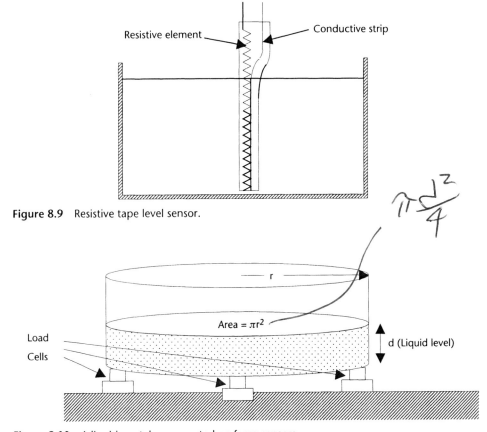

Figure 8.9 Resistive tape level sensor.

Figure 8.10 A liquid container mounted on force sensors.

where r is the radius of the container, and d is the depth of the material.

The weight of material (W) in a container is given by:

$$W = \gamma V \tag{8.7}$$

Example 8.6

What is the depth of the liquid in a container if the specific weight of the liquid is 56 lb/ft³, the container weighs 33 lb, and has a diameter of 63 in? A load cell measures the total weight to be 746 lb.

Using (8.6) and (8.7), we get the following:

Weight of liquid = 746 − 33 = 713 lb

$$\text{Volume of liquid} = \frac{3.14 \times 63 \times 63 \times d}{12 \times 12 \times 4} ft = \frac{713 lb}{56 lb/ft^3}$$

$$\text{Depth}(d) = \frac{12.73 \times 576}{12470} ft = 0.588 ft = 7 in$$

8.2.3 Single Point Sensing

In On/Off applications, single point sensing can be used with conductive probes, thermal probes, and beam breaking probes.

Conductive probes are used for single point measurements in liquids that are conductive and nonvolatile, since a spark can occur. Conductive probes are shown in Figure 8.11. Two or more probes can be used to indicate set levels. If the liquid is in a metal container, then the container can be used as the common probe. When the liquid is in contact with two probes, the voltage between the probes causes a current to flow, indicating that a set level has been reached. Thus, probes can be used to indicate when the liquid level is low, and when to operate a pump to fill the container. A third probe can be used to indicate when the tank is full, and to turn off the filling pump. The use of ac voltages is normally preferred to the use of dc voltages, to prevent electrolysis of the probes.

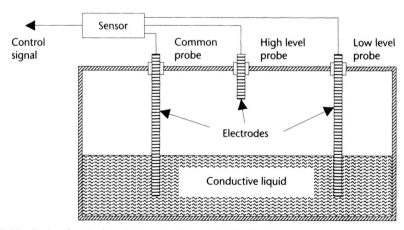

Figure 8.11 Probes for single point sensing in conductive liquids.

Thermal probes consist of a heating element adjacent to a temperature sensor. When the liquid rises above the probe, the heat is dissipated and the temperature at the sensor drops. The probe is a simple, low-cost, and reliable device for single point sensing.

Beam breaking methods are sometimes used for pressurized containers. For single point measurement, as shown in Figure 8.12(a), only one transmitter and one detector are required. The beams can be light, sonic or ultrasonic waves, or radiation. The devices are low-cost and of simple construction, but can be affected by deposits. If several single point levels are required, a detector will be required for each level measurement, as shown in Figure 8.12(b). The disadvantages of this radiation system are the cost, the need for special engineering, and the need to handle radioactive material. However, this system can be used with corrosive or very hot liquids. High-pressure containers are used where conditions would be detrimental to the installation of other types of level sensors.

8.2.4 Level Sensing of Free-Flowing Solids

Paddle wheels driven by electric motors can be used for sensing the level of solids that are in the form of power, grains, or granules. When the material reaches and covers the paddle wheel, the torque needed to turn the paddles greatly increases. The torque can be an indication of the depth of the material. Such a setup is shown in Figure 8.13(a). Some agitation may be required to level the solid particles. This is an inexpensive device and is good for most free-flowing materials, but is susceptible to vibration and shock. If the density of the material is greater than 0.9 lb/ft^3 (12.8 kg/m^3), then a vibration device can be used, as shown in Figure 8.13(b). The probe vibrates at the natural frequency of a tuning fork, and the frequency changes when in contact with a material. The change in resonant frequency is detected. The probe, which has no moving parts, is rugged, reliable, and only requires low maintenance, but its operation can be affected by other vibration sources. These devices may need protection from falling materials, and the proper location of the probe is essential for correct measurement [10].

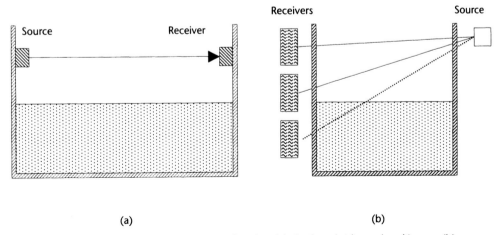

Figure 8.12 Liquid level measurements made using (a) single point beam breaking, or (b) multipoint beam breaking.

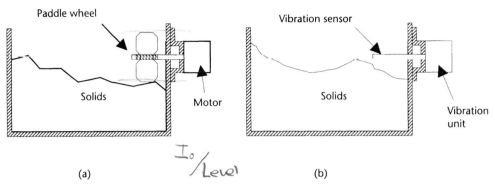

Figure 8.13 Sensors for measuring free-flowing solids: (a) paddle wheel, and (b) vibration device.

8.3 Application Considerations

A number of factors affect the choice of sensor for level measurement, such as pressure on the liquid, temperature of the liquid, turbulence, volatility, corrosiveness, level of accuracy required, single point or continuous measurement, direct or indirect, particulates in a liquid, free-flowing solids, and so forth [11]. Table 8.2 gives a comparison of the characteristics of level sensors.

When considering the choice of level sensor, temperature effects are a major consideration. Density and dielectric constants are affected by temperature, making indirect level measurements temperature-sensitive (see Chapter 10). This makes it necessary to monitor temperature as well as level, so that the level reading can be compensated. The types of sensors needing temperature compensation are: capacitive, bubblers, pressure, displacers, ultrasonic distance-measuring devices, and load cells.

Floats are often used to sense fluid levels, since they are unaffected by particulates, can be used for slurries, can be used with a wide range of liquid specific

Table 8.2 Level Measuring Devices

Type	Continuous/ Point	Liquid/ Solid	Temperature Range	Pressure Range, psi	Accuracy	Measuring Range, ft
Sight glass	P,C	L	260°C	6,000	±0.25 in	4
Differential pressure	P,C	L	650°C	6,000	±0.5%	Depends on cell
Pulley float	P,C	L	150°C	300	±0.12 in	60
Displacer	P,C	L	260°C	300	±0.25%	10
Bubbler	P,C	L	Dew point	1 atm	±1.0%	Unlimited
Capacitive	P,C	L,S	−30° to +980°C	5,000	±1.0%	20
Resistive	P	L	−30° to +80°C	3,000	±0.12 in	100
Sonic	P,C	L,S	−40° to +150°C	100	±1.0%	3–150
Ultrasonic	P,C	L,S	−25° to +60°C	300	±1.0%	3–12
Radiation	P,C	L,S	60°C	Unlimited	±1.0%	15
Paddle	P	S	175°C	30	±1.0 in	

weights. Due to their shape, flat floats are less susceptible to turbulence on the surface of the liquid. When the float is used to measure more than 50 cm of liquid depth, any change in float depth will have minimal effect on the measured liquid depth.

Displacers must never be completely submerged when measuring liquid depth, and must have a specific weight greater than that of the liquid. Care must also be taken to ensure that the liquid does not corrode the displacer, and the specific weight of the liquid is constant over time. The temperature of the liquid also may have to be monitored to make corrections for density changes. Displacers can be used to measure depths up to approximately 3m with an accuracy of ± 0.5 cm.

Capacitive device accuracy can be affected by the placement of the device. The dielectric constant of the liquid also should be regularly monitored. Capacitive devices can be used in containers that are pressurized up to 30 MPa, can be used in temperatures up to 1,000°C, and can measure depths up to 6m with an accuracy of $\pm 1\%$

Pressure gauge choice for measuring liquid levels can depend on the following considerations:

1. The presence of particulates, which can block the line to the gauge;
2. Damage caused by excessive temperatures in the liquid;
3. Damage due to peak pressure surges;
4. Corrosion of the gauge by the liquid;
5. If the liquid is under pressure a differential pressure gauge is needed;
6. Distance between the tank and the gauge;
7. Use of manual valves for gauge repair.

Differential pressure gauges can be used in containers with pressures up to 30 MPa and temperatures up to 600°C, with an accuracy of $\pm 1\%$. The liquid depth depends on its density and the pressure gauge used, and temperature correction is required.

Bubbler devices require certain precautions to ensure a continuous supply of clean dry air or inert gas (i.e., the gas used must not react with the liquid). It may be necessary to install a one way valve to prevent the liquid being sucked back into the gas supply lines if the gas pressure is lost. The bubbler tube must be chosen so that the liquid does not corrode it. Bubbler devices are typically used at atmospheric pressure. An accuracy of approximately 2% can be obtained, and the depth of the liquid depends on gas pressure available.

Ultrasonic devices can be used in containers with pressures up to 2 MPa, temperatures up to 100°C, and depths up to 30m, with an accuracy of approximately 2%.

Radiation devices are used for point measurement of hazardous materials. Due to the hazardous nature of the materials, personnel should be trained in its use, transportation, storage, identification, and disposal, observing rules set by OSHA.

Other considerations are that liquid level measurements can be affected by turbulence, readings may have to be averaged, and/or baffles may have to be used to reduce the turbulence. Frothing in the liquid also can be a source of error, particularly with resistive or capacitive probes.

8.4 Summary

This chapter introduced the concepts of level measurement. Level measurements can be direct or indirect continuous monitoring, or single point detection. Direct reading of liquid levels using ultrasonic devices is noncontact, and can be used for corrosive and volatile liquids and slurries.

Indirect measurements involve the use of pressure sensors, bubblers, capacitance, or load cells, which are all temperature-sensitive and will require temperature data for level correction. Of these sensors, load cells do not come into contact with the liquid, and are therefore well suited for the measurement of corrosive, volatile, and pressurized liquids and slurries.

Single point monitoring can use conductive probes, thermal probes, or ultrasonic or radioactive devices. Of these devices, the ultrasonic and radioactive devices are noncontact, and can be used with corrosive and volatile liquids, and in pressurized containers. Care has to be taken in handling radioactive materials.

The measurement of the level of free-flowing solids can be made with capacitive probes, a paddle wheel, or with a vibration-type of device.

References

[1] Harrelson, D., and J. Rowe, "Multivariable Transmitters, A New Approach for Liquid Level," *ISA Expo.* 2004.

[2] Vass, G., "The Principles of Level Measurement," *Sensors Magazine*, Vol. 17, No. 10, October 2000.

[3] Battikha, N. E., *The Condensed Handbook of Measurement and Control*, 2nd ed., ISA, 2004, pp. 97–116.

[4] Totten, A., "Magnetostrictive Level Sensors," *Sensors Magazine*, Vol. 19, No. 10, October 2002.

[5] Hambrice, K., and H. Hooper, "A Dozen Ways to Measure Fluid Level and How They Work," *Sensors Magazine*, Vol. 21, No. 12, December 2004.

[6] Considine, D. M., "Fluid Level Systems," *Process/Industrial Instruments and Control Handbook*, 4th ed., McGraw-Hill, 1993, pp. 4.130–4.136.

[7] Gillum, D. R., "Industrial Pressure, Level, and Density Measurement," *ISA*, 1995.

[8] Omega Engineering, Inc., "Level Measurement Systems," *Omega Complete Flow and Level Measurement Handbook and Encyclopedia*, Vol. 29, 1995.

[9] Liptak, B. E., "Level Measurement," *Instrument Engineer's Handbook Process Measurement and Analysis*, 3rd ed., Vol. 2, Chilton Book Co., pp. 269–397.

[10] Koeneman, D. W., "Evaluate the Options for Measuring Process Levels," *Chemical Engineering*, July 2000.

[11] Paul, B. O., "Seventeen Level Sensing Methods," *Chemical Processing*, February 1999.

CHAPTER 9
Flow

9.1 Introduction

The accurate measurement of fluid flow is very important in many industrial applications. Optimum performance of many processes requires specific flow rates. The cost of many liquids and gases are based on the measured flow through a pipeline, making it necessary for accounting purposes to accurately measure and control the rate of flow. This chapter discusses the basic terms, formulas, and techniques used in flow measurements and flow instrumentation. Highly accurate and rugged flow devices have now been developed and are commercially available. Developments in technology are continually improving measurement devices [1, 2]. However, one single flow device is not suitable for all applications, and careful selection is required.

9.2 Fluid Flow

At low flow rates, fluids have a laminar flow characteristic. As the flow rate increases, the laminar flow starts to break up and becomes turbulent. The speed of the liquid in a fluid flow varies across the flow. Where the fluid is in contact with the constraining walls (the boundary layer), the velocity of the liquid particles is virtually zero, while in the center of the flow, the liquid particles have the maximum velocity. Thus, the average rate of flow is used in flow calculations. The units of velocity are normally feet per second (ft/s), or meters per second (m/s). In a liquid, the fluid particles tend to move smoothly in layers with laminar flow, as shown in Figure 9.1(a). The velocity of the particles across the liquid takes a parabolic shape. With turbulent flow, the particles no longer flow smoothly in layers, and turbulence, or a rolling effect, occurs. This is shown in Figure 9.1(b). Note also the flattening of the velocity profile.

9.2.1 Flow Patterns

Flow can be considered to be laminar, turbulent, or a combination of both. Osborne Reynolds observed in 1880 that the flow pattern could be predicted from physical properties of the liquid. If the Reynolds number(R) for the flow in a pipe is equal to or less than 2,000, the flow will be laminar. If the Reynolds number ranges from 2,000 to approximately 5,000, this is the intermediate region, where the flow can be laminar, turbulent, or a mixture of both, depending upon other factors. Beyond

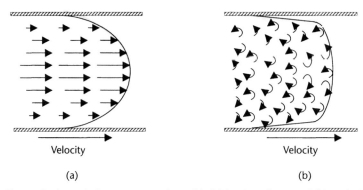

Figure 9.1 Flow velocity variations across a pipe with (a) laminar flow, and (b) turbulent flow.

approximately 5,000, the flow is always turbulent. The Reynolds number is a derived dimensionless relationship, combining the density and viscosity of a liquid with its velocity of flow and the cross-sectional dimensions of the flow, and takes the form:

$$R = \frac{VD\rho}{\mu} \tag{9.1}$$

where V is the average fluid velocity, D is the diameter of the pipe, ρ is the density of the liquid, and μ is the absolute viscosity.

Dynamic or absolute viscosity is used in the Reynolds flow equation. Table 9.1 gives a list of viscosity conversions. Typically, the viscosity of a liquid decreases as temperature increases.

Example 9.1

What is the Reynolds number for glycerin flowing at 7.5 ft/s in a 17-in diameter pipe? The viscosity of glycerin is 18×10^{-3} lb s/ft² and the density is 2.44 lb/ft³.

$$R = \frac{7.5 \times 17 \times 2.44}{12 \times 18 \times 10^{-3}} = 1,440$$

Flow rate is the volume of fluid passing a given point in a given amount of time, and is typically measured in gallons per minute (gal/min), cubic feet per minute (ft³/min), liters per minute (L/min), and so forth. Table 9.2 gives the flow rate conversion factors.

In a liquid flow, the pressures can be divided into the following: (1) static pressure, which is the pressure of fluids or gases that are stationary (see point A in

Table 9.1 Conversion Factors for Dynamic and Kinematic Viscosities

Dynamic Viscosities	Kinematic Viscosities
1 lb s/ft² = 47.9 Pa s	1 ft²/s = 9.29×10^{-2} m²/s
1 centipoise = 10 Pa s	1 stoke = 10^{-4} m²/s
1 centipoise = 2.09×10^{-5} lb s/ft²	1 m²/s = 10.76 ft²/s
1 poise = 100 centipoise	1 stoke = 1.076×10^{-3} ft²/s

9.2 Fluid Flow

Table 9.2 Flow Rate Conversion Factors

1 gal/min = 6.309 × 10^{-5} m^3/s	1 L/min = 16.67 × 10^{-6} m^3/s
1 gal/min = 3.78 L/min	1 ft^3/s = 449 gal/min
1 gal/min = 0.1337 ft^3/min	1 gal/min = 0.00223 ft^3/s
1 gal water = 231 in^3	1 ft^3 water = 7.48 gal
1 gal water = 0.1337 ft^3 = 231 in^3, 1 gal water = 8.35 lb,	
1 ft^3 water = 7.48 gal, 1,000 L water = 1 m^3, 1 L water = 1 kg	

Figure 9.2); (2) dynamic pressure, which is the pressure exerted by a fluid or gas when it impacts on a surface (point B − A); and (3) impact pressure (total pressure), which is the sum of the static and dynamic pressures on a surface, as shown by point B in Figure 9.2.

9.2.2 Continuity Equation

The continuity equation states that if the overall flow rate in a system is not changing with time [see Figure 9.3 (a)], then the flow rate in any part of the system is constant. From which:

$$Q = VA \tag{9.2}$$

where Q is the flow rate, V is the average velocity, and A is the cross-sectional area of the pipe. The units on both sides of the equation must be compatible (i.e., English units or metric units).

Example 9.2

What is the flow rate in liters per second through a pipe 32 cm in diameter, if the average velocity is 2.1 m/s?

$$Q = \frac{2.1 m/s \times \pi \times 0.32^2 \, m^2}{4} = 0.17 \, m^3/s = 0.17 \times 1,000 L/s = 170 L/s$$

If liquids are flowing in a tube with different cross-sectional areas, such as A_1 and A_2, as shown in Figure 9.3(b), then the continuity equation gives:

$$Q = V_1 A_1 = V_2 A_2 \tag{9.3}$$

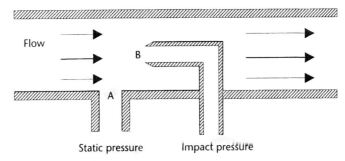

Figure 9.2 Static, dynamic, and impact pressures.

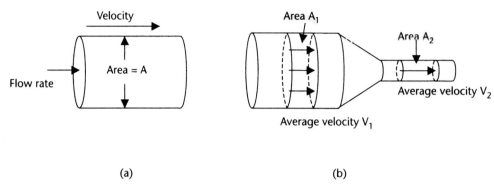

(a)   (b)

Figure 9.3 Flow diagram for use in the continuity equation with (a) constant area, and (b) differential areas.

Example 9.3

If a pipe changes from a diameter of 17 to 11 cm, and the velocity in the 17cm section is 5.4 m/s, what is the average velocity in the 11cm section?

$$Q = V_1 A_1 = V_2 A_2$$

$$V_2 = \frac{5.4 m^3/s \times \pi \times 8.5^2}{\pi \times 5.5^2 \, m^2} = 12.8 \, m/s$$

Mass flow rate (F) is related to volume flow rate (Q) by:

$$F = \rho Q \tag{9.4}$$

where F is the mass of liquid flowing, and ρ is the density of the liquid.

Since a gas is compressible, (9.3) must be modified for gas flow to:

$$\gamma_1 V_1 A_1 = \gamma_2 V_2 A_2 \tag{9.5}$$

where γ_1 and γ_2 are specific weights of the gas in the two sections of pipe.

Equation 9.3 is the rate of mass flow in the case of a gas. However, this could also apply to liquid flow, by multiplying both sides of the (9.3) by the specific weight (γ), to give the following:

$$\gamma V_1 A_1 = \gamma V_2 A_2 \tag{9.6}$$

9.2.3 Bernoulli Equation

The Bernoulli equation (1738) gives the relation between pressure, fluid velocity, and elevation in a flow system. When applied to Figure 9.4(a), the following is obtained:

$$\frac{P_A}{\gamma_A} + \frac{V_A^2}{2g} + Z_A = \frac{P_B}{\gamma_B} + \frac{V_B^2}{2g} + Z_B \tag{9.7}$$

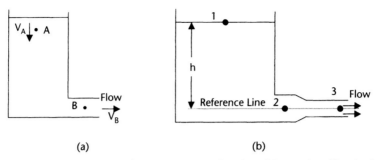

Figure 9.4 Container diagrams: (a) the pressures at points A and B are related by the Bernoulli equation, and (b) application of the Bernoulli equation to determine flow.

where P_A and P_B are absolute static pressures at points A and B, γ_A and γ_B are specific weights, V_A and V_B are average fluid velocities, g is the acceleration of gravity, and Z_A and Z_B are elevations above a given reference level (e.g., $Z_A - Z_B$ is the head of fluid).

The units in (9.6) are consistent, and reduce to units of length as follows:

$$\text{Pressure Energy} = \frac{P}{\gamma} = \frac{lb/ft^2 \left(N/m^2\right)}{lb/ft^2 \left(N/m^3\right)} = ft(m)$$

$$\text{Kinetic Energy} = \frac{V^2}{\gamma} = \frac{(ft/s)^2 (m/s)^2}{ft/s^2 (m/s^2)} ft(m)$$

$$\text{Potential Energy} = Z = ft(m)$$

This equation is a conservation of energy equation, and assumes no loss of energy between points A and B. The first term represents energy stored due to pressure; the second term represents kinetic energy, or energy due to motion; and the third term represents potential energy, or energy due to height. This energy relationship can be seen if each term is multiplied by mass per unit volume, which cancels, since the mass per unit volume is the same at points A and B. The equation can be used between any two positions in a flow system. The pressures used in the Bernoulli equation must be absolute pressures.

In the fluid system shown in Figure 9.4(b), the flow velocity V at point 3 can be derived from (9.7), and is as follows, using point 2 as the reference line:

$$\frac{P_1}{\gamma_1} + 0 + h = \frac{P_3}{\gamma_3} + \frac{V_3^2}{2g} + 0$$

$$V_3 = \sqrt{(2gh)} \tag{9.8}$$

Point 3 at the exit has dynamic pressure, but no static pressure above 1 atm. Hence, $P_3 = P_1 = 1$ atm, and $\gamma_1 = \gamma_3$. This shows that the velocity of the liquid flowing out of the system is directly proportional to the square root of the height of the liquid above the reference point.

Example 9.4

If the height of a column of water h in Figure 9.3(b) is 4.3m, what is the pressure at P_2? Assume the areas at points 2 and 3 are 29 cm^2 and 17 cm^2, respectively?

$$V_3 = \sqrt{(2 \times 9.8 \times 4.3)} = 9.18 m/s$$

Considering points 2 and 3 with the use of (9.7):

$$\frac{P_2}{9.8kN} + \frac{V_2^2}{2 \times 9.8} + 0 = \frac{101.3 kPa}{9.8kN} + \frac{V_3^2}{2 \times 9.8} + 0 \qquad (9.9)$$

Using (9.3) and knowing that the areas at point 2 and 3 are 0.0029 m^2 and 0.0017 m^2, respectively, the velocity at point 2 is given by:

$$V_2 = \left(\frac{A_3}{A_2}\right) V_3 = \left(\frac{0.0017}{0.0029}\right) 9.12 m/s = 5.35 m/s$$

Substituting the values obtained for V_2 and V_3 into (9.8) gives the following:

$$\frac{P_2}{9.8} + \frac{(5.35)^2}{2 \times 9.8} + 0 = \frac{101.3}{9.8} + \frac{(9.12)^2}{2 \times 9.8} + 0$$

$$P_2 = 128.6 \text{ kPa(a)} = 27.3 \text{ kPa(g)}$$

9.2.4 Flow Losses

The Bernoulli equation does not take into account flow losses. These losses are accounted for by pressure losses, and fall into two categories: (1) those associated with viscosity and the friction between the constriction walls and the flowing fluid; and (2) those associated with fittings, such as valves, elbows, tees, and so forth.

The flow rate Q from the continuity equation for point 3 in Figure 9.3(b), for instance, gives:

$$Q = V_3 A_3$$

However, to account for the *outlet losses*, the equation should be modified to:

$$Q = C_D V_3 A_3 \qquad (9.10)$$

where C_D is the discharge coefficient, which is dependent on the shape and size of the orifice. The discharge coefficients can be found in flow data handbooks.

Frictional losses are losses from the friction between the flowing liquid and the restraining walls of the container. These frictional losses are given by:

$$h_L = \frac{fLV^2}{2Dg} \qquad (9.11)$$

where h_L is the head loss, f is the friction factor, L is the length of pipe, D is the diameter of pipe, V is the average fluid velocity, and g is the gravitation constant.

The friction factor f depends on the Reynolds number for the flow, and the roughness of the pipe walls.

Example 9.5

What is the head loss in a 5-cm diameter pipe that is 93m long? The friction factor is 0.03, and the average velocity in the pipe is 1.03 m/s.

$$h_L = \frac{fLV^2}{2Dg} = \frac{0.03 \times 93m \times (1.03m/s)^2}{5cm \times 2 \times 9.8m/s} \frac{100}{} = 3.02m$$

This would be equivalent to 3.02m × 9.8 kN/m³ = 29.6 kPa.

Fitting losses are those losses due to couplings and fittings. Fitting losses are normally less than friction losses, and are given by:

$$h_L = \frac{KV^2}{2g} \tag{9.12}$$

where h_L is the head loss due to fittings, K is the head loss coefficient for various fittings, V is the average fluid velocity, and g is the gravitation constant.

Values for K can be found in flow handbooks. Table 9.3 gives some typical values for the head loss coefficient factor in some common fittings.

Example 9.6

Fluid is flowing at 3.7 ft/s through one inch fittings as follows: 7 × 90° ells, 5 tees, 2 gate valves, and 19 couplings. What is the head loss?

$$h_L = \frac{(7 \times 1.5 + 5 \times .08 + 2 \times 0.22 + 19 \times 0.085)3.7 \times 3.7}{2 \times 32.2}$$

$$h_L = (10.5 + 4.0 + 0.44 + 1.615)\, 0.2 = 3.3 \text{ ft}$$

To take into account losses due to friction and fittings, the Bernouilli Equation is modified as follows:

$$\frac{P_A}{\gamma_A} + \frac{V_A^2}{2g} + Z_A = \frac{P_B}{\gamma_B} + \frac{V_B^2}{2g} + Z_B + h_{Lfriction} + h_{Lfittings} \tag{9.13}$$

Table 9.3 Typical Head Loss Coefficient Factors for Fittings

Threaded ell—1 in	1.5	Flanged ell—1 in	0.43
Threaded tee—1 in inline	0.9	Branch	1.8
Globe valve (threaded)	8.5	Gauge valve (threaded)	0.22
Coupling or union—1 in	0.085	Bell mouth reducer	0.05

Form drag is the impact force exerted on devices protruding into a pipe due to fluid flow. The force depends on the shape of the insert, and can be calculated from:

$$F = C_D \gamma \frac{AV^2}{2g} \qquad (9.14)$$

where F is the force on the object, C_D is the drag coefficient, γ is the specific weight, g is the acceleration due to gravity, A is the cross-sectional area of obstruction, and V is the average fluid velocity.

Flow handbooks contain drag coefficients for various objects. Table 9.4 gives some typical drag coefficients.

Example 9.7

A 7.3-in diameter ball is traveling through the air with a velocity of 91 ft/s. If the density of the air is 0.0765 lb/ft³ and $C_D = 0.35$, what is the force acting on the ball?

$$F = C_D \gamma \frac{AV^2}{2g} = \frac{0.35 \times 0.0765 lb/ft^3 \times \pi \times 7.3^2 ft^2 \times (91 ft/s)^2}{2 \times 32.2 ft/s^2 \times 4 \times 144} = 1.0 lb$$

9.3 Flow Measuring Instruments

Flow measurements can be divided into the following groups: flow rate, total flow, and mass flow. The choice of measuring device will depend on the required accuracy, flow rate, range, and fluid characteristics (i.e., gas, liquid, suspended particulates, temperature, viscosity, and so forth).

9.3.1 Flow Rate

Many flow measurement instruments use indirect measurements, such as differential pressures, to measure the flow rate. These instruments measure the differential pressures produced when a fluid flows through a restriction. Differential pressure measuring sensors were discussed in Chapter 7. The differential pressure produced is directly proportional to flow rate. Such commonly used restrictions are the (a) orifice plate, (b) Venturi tube, (c) flow nozzle, and (d) Dall tube.

The *orifice plate* is normally a simple metal diaphragm with a constricting hole. The diaphragm is normally clamped between pipe flanges to give easy access. The differential pressure ports can be located in the flange on either side of the orifice plate, or alternatively, at specific locations in the pipe on either side of the flange, as

Table 9.4 Typical Drag Coefficient Values for Objects Immersed in Flowing Fluid

Circular cylinder with axis perpendicular to flow	0.33 to 1.2
Circular cylinder with axis parallel to flow	0.85 to 1.12
Circular disk facing flow	1.12
Flat plate facing flow	1.9
Sphere	0.1+

determined by the flow patterns (named vena contracta), as shown in Figure 9.5. Shown also is the pressure profile. A differential pressure gauge is used to measure the difference in pressure between the two ports. The differential pressure gauge can be calibrated in flow rates. The lagging edge of the hole in the diaphragm is beveled to minimize turbulence. In fluids, the hole is normally centered in the diaphragm, as shown in Figure 9.6(a). However, if the fluid contains particulates, the hole could be placed at the bottom of the pipe, as shown in Figure 9.6(b), to prevent a buildup of particulates. The hole also can be in the form of a semicircle having the same diameter as the pipe, and located at the bottom of the pipe, as shown in Figure 9.6(c).

The flow rate Q in a differential flow rate meter is given by:

$$Q = K\left(\frac{\pi}{4}\right)\left(\frac{d_S}{d_P}\right)^2 \sqrt{2gh} \qquad (9.15)$$

where K is the flow coefficient constant, d_s is the diameter of the orifice, d_p is the pipe diameter, and h is the difference in height between P_H and P_L.

Example 9.8

In a 30-in diameter pipe, a circular orifice has a diameter of 20 in, and the difference in height of the manometer levels is 2.3 ft. What is the flow rate in cubic feet per second, if K is 0.97?

$$Q = 0.97\left(\frac{3.14}{4}\right)\left(\frac{20}{30}\right)^2 \sqrt{2 \times 32.2 \times 2.3}$$

$$Q = 4.12 \text{ ft}^3/\text{s}$$

The *Venturi tube*, shown in Figure 9.7(a), uses the same differential pressure principal as the orifice plate. The Venturi tube normally uses a specific reduction in tube size, and is normally well suited for use in larger diameter pipes, but it becomes

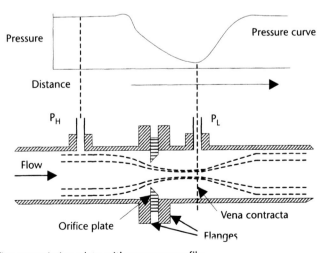

Figure 9.5 Orifice constriction plate with pressure profile.

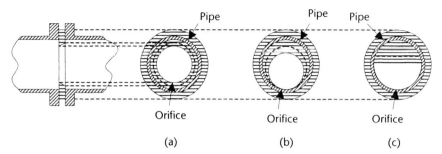

Figure 9.6 Orifice shapes and locations used (a) with fluids, and (b, c) with suspended solids.

heavy and excessively long. One advantage of the Venturi tube is its ability to handle large amounts of suspended solids. It creates less turbulence and insertion loss than the orifice plate. The differential pressure taps in the Venturi tube are located at the minimum and maximum pipe diameters. The Venturi tube has good accuracy, but is expensive.

The *flow nozzle* is a good compromise on cost and accuracy between the orifice plate and the Venturi tube for clean liquids. It is not normally used with suspended particles. Its main use is the measurement of steam flow. The flow nozzle is shown in Figure 9.7(b).

The *Dall tube*, as shown in Figure 9.7(c), has the lowest insertion loss, but is not suitable for use with slurries.

A typical ratio (i.e., a beta ratio, which is the diameter of the orifice opening (d) divided by the diameter of the pipe (D)) for the size of the constriction of pipe size in flow measurements is normally between 0.2 and 0.6. The ratios are chosen to give sufficiently high pressure drops for accurate flow measurements, but not high enough to give turbulence. A compromise is made between high beta ratios (d/D), which give low differential pressures, and low ratios, which give high differential

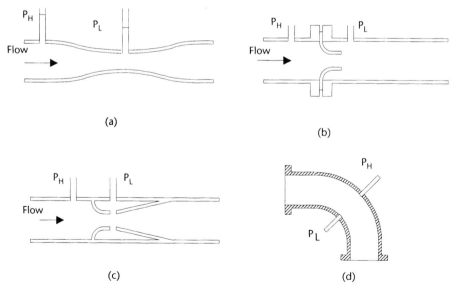

Figure 9.7 Types of constrictions used to measure flow: (a) Venturi tube, (b) flow nozzle, (c) Dall tube, and (d) elbow.

pressures but can create high losses. The Dall tube has the advantage of having the lowest insertion loss, but it cannot be used with slurries.

The *elbow* can be used as a differential flow meter. Figure 9.7(d) shows the cross section of an elbow. When a fluid is flowing, there is a differential pressure between the inside and outside of the elbow, due to the change in direction of the fluid. The pressure difference is proportional to the flow rate of the fluid. The elbow meter is good for handling particulates in solution, and has good wear and erosion resistance characteristics, but has low sensitivity.

In an elbow, the flow is given by;

$$\text{Flow} = C\sqrt{(RP_D D^3 \rho)} \tag{9.16}$$

where C is a constant, R is the center line radius of the elbow, P_D is the differential pressure, D is the diameter of the elbow, and ρ is the density of the fluid.

The *pilot static tube*, as shown in Figure 9.8, is an alternative method of measuring flow rate, but has a disadvantage, in that it really measures fluid velocity at the nozzle. Because the velocity varies over the cross section of the pipe, the pilot static tube should be moved across the pipe to establish an average velocity, or the tube should be calibrated for one area. Other disadvantages are that the tube can become clogged with particulates, and that the differential pressure between the impact and static pressures for low flow rates may not be enough to give the required accuracy. The differential pressures in any of the above devices can be measured using the pressure measuring sensors discussed in Chapter 7 (Pressure).

In a pilot static tube, the flow Q is given by:

$$Q = K\sqrt{(\rho_s - \rho_I)} \tag{9.17}$$

where K is a constant, p_s is the static pressure, and p_I is the impact pressure.

Variable-area meters, such as the *rotameter* shown in Figure 9.9(a), are often used as a direct visual indicator for flow rate measurements. The rotameter is a vertical tapered tube with a T (or similar) shaped weight, and the tube is graduated in flow rate for the characteristics of the gas or liquid flowing up the tube. The velocity of a fluid or gas flowing decreases as it goes higher up the tube, due to the increase in the bore of the tube. Hence, the buoyancy on the weight reduces as it goes higher up

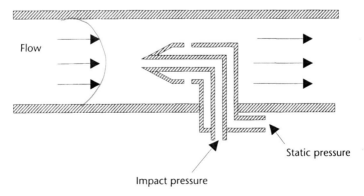

Figure 9.8 Pilot static tube.

the tube. An equilibrium point is eventually reached, where the force on the weight due to the flowing fluid is equal to that of the weight (i.e., the higher the flow rate, the higher the weight goes up the tube). The position of the weight also is dependent on its size and density, the viscosity and density of the fluid, and the bore and taper of the tube. The rotameter has low insertion loss, and has a linear relationship to flow rate. In cases where the weight is not visible, such as an opaque tube used to reduce corrosion, it can be made of a magnetic material, and tracked by a magnetic sensor on the outside of the tube. The rotameter can be used to measure differential pressures across a constriction or flow in both liquids and gases [3].

Vortex flow meters are based on the fact that an obstruction in a fluid or gas flow will cause turbulence or vortices. In the case of the vortex precession meter (for gases), the obstruction is shaped to give a rotating or swirling motion forming vortices, which can be measured ultrasonically. See Figure 9.9(b). The frequency of the vortex formation is proportional to the rate of flow. This method is good for high flow rates. At low flow rates, the vortex frequency tends to be unstable.

Rotating flow rate devices are rotating sensors. One example is the turbine flow meter, which is shown in Figure 9.10(a). The turbine rotor is mounted in the center of the pipe and rotates at a speed proportional to the rate of flow of the fluid or gas passing over the blades. The turbine blades are normally made of a magnetic material or ferrite particles in plastic, so that they are unaffected by corrosive liquids. A Hall device or an MRE sensor attached to the pipe can sense the rotating blades. The turbine should be only used with clean fluids, such as gasoline. The rotating flow devices are accurate, with good flow operating and temperature ranges, but are more expensive than most of the other devices.

Pressure flow meters use a strain gauge to measure the force on an object placed in a fluid or gas flow. The meter is shown in Figure 9.10(b). The force on the object is proportional to the rate of flow. The meter is low-cost, with medium accuracy.

A *moving vane* type of device can be used in a pipe configuration or an open channel flow. The vane can be spring loaded and have the ability to pivot. By measuring the angle of tilt, the flow rate can be determined.

Electromagnetic flow meters only can be used in conductive liquids. The device consists of two electrodes mounted in the liquid on opposite sides of the pipe. A

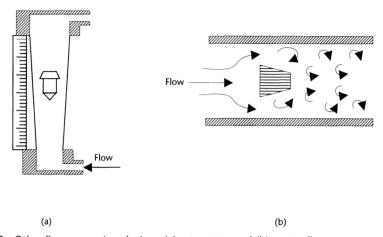

Figure 9.9 Other flow measuring devices: (a) rotameter, and (b) vortex flow meter.

9.3 Flow Measuring Instruments

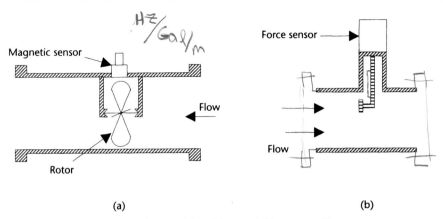

Figure 9.10 Flow rate measuring devices: (a) turbine, and (b) pressure flow meter.

magnetic field is generated across the pipe perpendicular to the electrodes, as shown in Figure 9.11. The conducting fluid flowing through the magnetic field generates a voltage between the electrodes, which can be measured to give the rate of flow. The meter gives an accurate linear output voltage with flow rate. There is no insertion loss, and the readings are independent of the fluid characteristics, but it is a relatively expensive instrument [4].

Ultrasonic flow meters can be transit-time flow meters, or can use the Doppler effect. In the transit-time flow meter, two transducers with receivers are mounted diametrically opposite to each other, but inclined at 45° to the axis of the pipe, as shown in Figure 9.12. Each transducer transmits an ultrasonic beam at a frequency of approximately 1 MHz, which is produced by a piezoelectric crystal. The transit time of each beam is different due to the liquid flow. The difference in transit time of the two beams is used to calculate the average liquid velocity. The advantage of this type of sensor is that the effects of temperature density changes cancel in the two beams. There is no obstruction to fluid flow, and corrosive or varying flow rates are not a problem, but the measurements can be affected by the Reynolds number or velocity profile. The transmitters can be in contact with the liquid, or can be clamped externally on to the pipe.

The Doppler flow meter measures the velocity of entrapped gas ($>30\ \mu$) or small particles in the liquid, as shown in Figure 9.13. A single transducer and receiver are mounted at 45° to the axis of the pipe. The receiver measures the

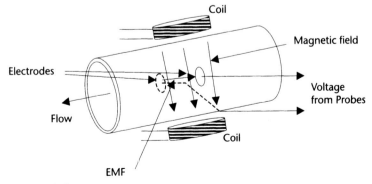

Figure 9.11 Magnetic flow meter.

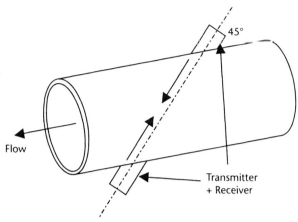

Figure 9.12 Ultrasonic transit-time flow meter.

difference in frequency of the transmitted and received signals, from which the flow velocity can be calculated. The meter can be mounted externally, and is not affected by changes in liquid viscosity.

Ultrasonic flow meters are normally used to measure flow rates in large diameter, nonporous pipes (e.g., cast iron, cement, or fiberglass), and they require periodic recalibration. Meters must not be closer than 10m to each other to prevent interference. This type of meter has a temperature operating range of $-20°$ to $+250°C$ and an accuracy of $\pm 5\%$ FSD.

9.3.2 Total Flow

Positive displacement meters are used to measure the total quantity of fluid flowing, or the volume of liquid in a flow. The meters use containers of known size, which are filled and emptied a known number of times in a given time period, to give the total flow volume. The common types of instruments for measuring total flow are:

- The Piston flow meter;
- Rotary piston;
- Rotary vane;

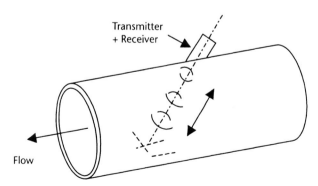

Figure 9.13 Ultrasonic Doppler flow meter.

9.3 Flow Measuring Instruments

- Nutating disk;
- Oval gear.

Piston meters consist of a piston in a cylinder. Initially, the fluid enters on one side of the piston and fills the cylinder, at which point the fluid is diverted to the other side of the piston via valves, and the outlet port of the full cylinder is opened. The redirection of fluid reverses the direction of the piston and fills the cylinder on the other side of the piston. The number of times the piston traverses the cylinder in a given time frame determines the total flow. The piston meter is shown in Figure 9.14. The meter has high accuracy, but is expensive.

Nutating disk meters are in the form of a disk that oscillates, allowing a known volume of fluid to pass with each oscillation. The meter is illustrated in Figure 9.15. Liquid enters and fills the left chamber. Because the disk is off center, the liquid pressure causes the disk to wobble. This action empties the volume of liquid from the left chamber to the right chamber, the left chamber is then refilled, and the liquid in the right chamber exits. The oscillations of the disk are counted and the total volume measured. This meter is not suitable for measuring slurries. The meter is accurate and expensive, but a low-cost version is available, which is used in domestic water metering and industrial liquid metering [5].

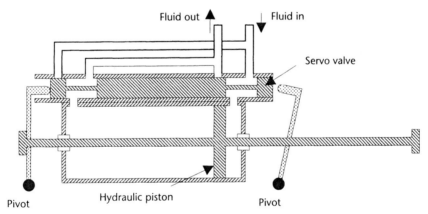

Figure 9.14 Piston meter.

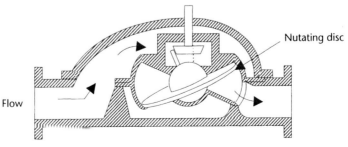

Figure 9.15 Nutating disk flow meter.

Velocity meters, normally used to measure flow rate, also can be set up to measure total flow. Multiplying the velocity by the cross-sectional area of the meter can measure total flow.

9.3.3 Mass Flow

By measuring the flow rate and knowing the density of a fluid, the mass of the flow can be measured. Mass flow instruments include constant speed impeller turbine wheel-spring combinations, which relate the spring force to mass flow, and devices that relate heat transfer to mass flow [6].

Coriolis flow meters, which can be used to measure mass flow, can be either in the form of a straight tube or a loop. In either case, the device is forced into resonance perpendicular to the flow direction. The resulting coriolis force produces a twist movement in the pipe or loop that can be measured and related to the mass flow. See Figure 9.16. The loop has a wider operating range than the straight tube, is more accurate at low flow rates, and can be used to measure both mass flow and density.

The *anemometer* is a method that can be used to measure gas flow rates. One method is to keep the temperature of a heating element in a gas flow constant and measure the power required. The higher the flow rates, the higher the amount of heat required. The alternative method (hot-wire anemometer) is to measure the incident gas temperature, and the temperature of the gas downstream from a heating element. The difference in the two temperatures can be related to the flow rate [7]. Micromachined anemometers are now widely used in automobiles for the measurement of air intake mass, as described in Chapter 6, Figure 6.13. The advantages of this type of sensor are that they are very small, have no moving parts, have minimal obstruction to flow, have a low thermal time constant, and are very cost effective with good longevity.

9.3.4 Dry Particulate Flow Rate

Dry particulate flow rate on a conveyer belt can be measured with the use of a load cell. This method is illustrated in Figure 9.17. To measure flow rate, it is only necessary to measure the weight of material on a fixed length of the conveyer belt [8].

The flow rate Q is given by:

$$Q = WR/L \qquad (9.18)$$

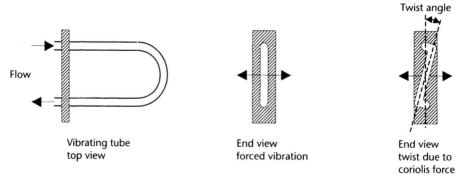

Figure 9.16 Mass flow meter using coriolis force.

9.4 Application Considerations

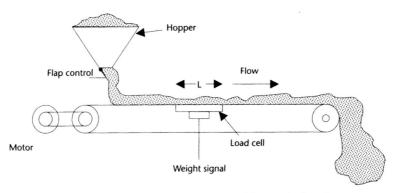

Figure 9.17 Conveyer belt system for the measurement of dry particulate flow rate.

where W is the weight of material on length L of the weighing platform, and R is the speed of the conveyer belt.

Example 9.9

A conveyer belt is traveling at 27 cm/s, and a load cell with a length of 0.72m is reading 5.4 kg. What is the flow rate of the material on the belt?

$$Q = \frac{5.4 \times 27}{100 \times 0.72} \, kg/s = 2.025 \, kg/s$$

9.3.5 Open Channel Flow

Open channel flow occurs when the fluid flowing is not contained as in a pipe, but is in an open channel. Flow rates can be measured using constrictions, as in contained flows. A Weir sensor used for open channel flow is shown in Figure 9.18(a). This device is similar in operation to an orifice plate. The flow rate is determined by measuring the differential pressures or liquid levels on either side of the constriction. A Parshall flume, which is similar in shape to a Venturi tube, is shown in Figure 9.18(b). A paddle wheel and an open flow nozzle are alternative methods of measuring open channel flow rates.

9.4 Application Considerations

Many different types of sensors can be used for flow measurements. The choice of any particular device for a specific application depends on a number of factors, such as: reliability, cost, accuracy, pressure range, size of pipe, temperature, wear and erosion, energy loss, ease of replacement, particulates, viscosity, and so forth [9].

9.4.1 Selection

The selection of a flow meter for a specific application to a large extent will depend upon the required accuracy and the presence of particulates, although the required accuracy is sometimes downgraded because of cost. One of the most accurate meters is the magnetic flow meter, which can be accurate to 1% of FSD. This meter

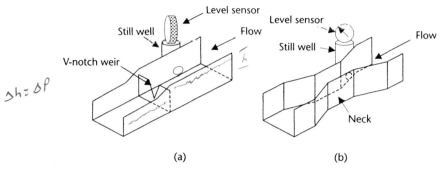

Figure 9.18 Open channel flow sensors: (a) Weir, and (b) Parshall flume.

is good for low flow rates with high viscosities, and has low energy loss, but is expensive and requires a conductive fluid.

The turbine gives high accuracies, and can be used when there is vapor present, but the turbine is better with clean, low viscosity fluids. Table 9.5 gives a comparison of flow meter characteristics [10].

The most commonly used general-purpose devices are the pressure differential sensors used with pipe constrictions. These devices will give an accuracy in the 3% range when used with solid state pressure sensors, which convert the readings directly into electrical units, or the rotameter for direct visual reading. The Venturi tube has the highest accuracy and least energy loss, followed by the flow nozzle, then the orifice plate. For cost effectiveness, the devices are in the reverse order. If large amounts of particulates are present, the Venture tube is preferred. The differential pressure devices operate best between 30% and 100% of the flow range. The elbow also should be considered in these applications.

Gas flow can be best measured with an anemometer. Solid state anemometers are now available with good accuracy and very small size, and are cost effective.

For open channel applications, the flume is the most accurate, and is preferred if particulates are present, but is the most expensive.

Table 9.5 Summary of Flow Meter Characteristics

Meter Type	Range	Accuracy	Comments
Orifice plate	3 to 1	±3% FSD	Low cost and accuracy
Venturi tube	3 to 1	±1% FSD	High cost, good accuracy, low losses
Flow nozzle	3 to 1	±2% FSD	Medium cost and accuracy
Dall tube	3 to 1	±2% FSD	Medium cost and accuracy, low losses
Elbow	3 to 1	±6%–10% FSD	Low cost, losses, and sensitivity
Pilot static tube	3 to 1	±4% FSD	Low sensitivity
Rotameter	10 to 1	±2% of rate	Low losses, line of sight
Turbine meter	10 to 1	±2% FSD	High accuracy, low losses
Moving vane	5 to 1	±10% FSD	Low cost, low accuracy
Electromagnetic	30 to 1	±0.5% of rate	Conductive fluid, low losses, high cost
Vortex meter	20 to 1	±0.5% of rate	Poor at low flow rates
Strain gauge	3 to 1	±2% FSD	Low cost, and accuracy
Ultrasonic meter	30 to1	±5% FSD	Doppler 15 to 1 range, large diameter pipe
Nutating disk	5 to 1	±3% FSD	High accuracy and cost
Anemometer	100 to 1	±2% of rate	Low losses, fast response

Particular attention also should be given to manufacturers' specifications and application notes.

9.4.2 Installation

Because of the turbulence generated by any type of obstruction in an otherwise smooth pipe, attention must be given to the placement of flow sensors. The position of the pressure taps can be critical for accurate measurements. The manufacturers' recommendations should be followed during installation. In differential pressure sensing devices, the upstream tap should be at a distance from 1 to 3 pipe diameters from the plate or constriction, and the downstream tap up to 8 pipe diameters from the constriction.

To minimize pressure fluctuations at the sensor, it is desirable to have a straight run of 10 to 15 pipe diameters on either side of the sensing device. It also may be necessary to incorporate laminar flow planes into the pipe to minimize flow disturbances, and dampening devices to reduce flow fluctuations to an absolute minimum.

Flow nozzles may require vertical installation if gases or particulates are present. To allow gases to pass through the nozzle, it should be facing upward; and for particulates, facing downward.

9.4.3 Calibration

Flow meters need periodic calibration. This can be done by using another calibrated meter as a reference, or by using a known flow rate. Accuracy can vary over the range of the instrument, and with temperature and specific weight changes in the fluid. Thus, the meter should be calibrated over temperature as well as range, so that the appropriate corrections can be made to the readings. A spot check of the readings should be made periodically to check for instrument drift, which may be caused by the instrument going out of calibration, or particulate buildup and erosion.

9.5 Summary

This chapter discussed the flow of fluids in closed and open channels, and gases in closed channels. Liquid flow can be laminar or turbulent, depending upon the flow rate and its Reynolds number. The Reynolds number is related to the viscosity, pipe diameter, and liquid density. The various continuity and flow equations are used in the development of the Bernoulli equation, which uses the concept of the conservation of energy to relate pressures to flow rates. The Bernoulli equation can be modified to allow for losses in liquids due to viscosity, friction with the constraining tube walls, and drag.

Many types of sensors are available for measuring the flow rates in gases, liquids, slurries, and free-flowing solids. The sensors vary, from tube constrictions where the differential pressure across the constriction is used to obtain the flow rate, to electromagnetic flow meters, to ultrasonic devices. Flow rates can be measured in volume, total, or mass. The choice of sensor for measuring flow rates will depend on many factors, such as accuracy, particulates, flow velocity, range, pipe size, viscosity, and so forth. Only experienced technicians should perform installation and calibration.

Definitions

Bernoulli equation   is an equation for flow based on the conservation of energy.

Flow rate   is the volume of fluid or gas passing a given point in a given amount of time.

Laminar flow   in a liquid occurs when its average velocity is comparatively low, and $R < 2,000$. The flow is streamlined and laminar without eddies.

Mass flow   is the mass of liquid or gas flowing in a given time period.

Reynolds number (R)   is a derived relationship, combining the density and viscosity of a liquid, with its velocity of flow and the cross-sectional dimensions of the flow.

Total flow   is the volume of liquid or gas flowing in a given period of time.

Turbulent flow   in a liquid occurs when the flow velocity is high, and $R > 5,000$. The flow breaks up into fluctuating velocity patterns and eddies.

Velocity   in fluids is the average rate of fluid flow across the diameter of the pipe.

Viscosity   is a property of a gas or liquid that measures its resistance to motion or flow.

References

[1] Boillat, M. A., et al., "A Differential Pressure Liquid Flow Sensor For Flow Regulation and Dosing Systems," *Proc. IEEE Micro Electro Mechanical Systems*, 1995, pp. 350–352.

[2] Konrad, B., P. Arquint, and B. van der Shoot, "A Minature Flow Sensor with Temperature Compensation," *Sensors Magazine*, Vol. 20, No. 4, April 2003.

[3] Scheer, J. E., "The Basics of Rotameters," *Sensors Magazine*, Vol. 19, No. 10, October 2002.

[4] Lynnworth, L., "Clamp-on Flowmeters for Fluids," *Sensors Magazine*, Vol. 18, No. 8, August 2001.

[5] Humphries, J. T., and L. P. Sheets, *Industrial Electronics*, 4th ed., Delmar, 1993, pp. 359–364.

[6] Jurgen, R. K., *Automotive Electronics Handbook*, 2nd ed., McGraw-Hill, 1999, pp. 4.1–4.9.

[7] Hsieh, H. Y., J. N. Zemel, and A. Spetz, "Pyroelectric Anemometers: Principles and Applications," *Proceedings Sensor Expo*, 1993, pp. 113–120.

[8] Nachtigal, C. L., "Closed-Loop Control of Flow Rate for Dry Bulk Solids," *Proceedings Sensor Expo*, 1994, pp. 49–56.

[9] Yoder, J., "Flow Meters and Their Application; an Overview," *Sensors Magazine*, Vol. 20, No. 10, October 2003.

[10] Chen, J. S. J., "Paddle Wheel Flow Sensors The Overlooked Choice," *Sensors Magazine*, Vol. 16, No. 12, December 1999.

Temperature and Heat

10.1 Introduction

Temperature is without doubt the most widely measured variable. Thermometers can be traced back to Galileo (1595). The importance of accurate temperature measurement cannot be overemphasized. In the process control of chemical reactions, temperature control is of major importance, since chemical reactions are temperature-dependent. All physical parameters are temperature-dependent, making it necessary in most cases to measure temperature along with the physical parameter, so that temperature corrections can be made to achieve accurate parameter measurements. Instrumentation also can be temperature-dependent, requiring careful design or temperature correction, which can determine the choice of measurement device. For accurate temperature control, precise measurement of temperature is required [1]. This chapter discusses the various temperature scales used, their relation to each other, methods of measuring temperature, and the relationship between temperature and heat.

10.2 Temperature and Heat

Temperature is a measure of molecular energy, or heat energy, and the potential to transfer heat energy. Four temperature scales were devised for the measurement of heat and heat transfer.

10.2.1 Temperature Units

Three temperature scales are in common use to measure the relative hotness or coldness of a material. The scales are: Fahrenheit (°F) (attributed to Daniel G. Fahrenheit, 1724); Celsius (°C) (attributed to Anders Celsius, 1742); and Kelvin (K), which is based on the Celsius scale and is mainly used for scientific work. The Rankine scale (°R), based on the Fahrenheit scale, is less commonly used, but will be encountered.

The Fahrenheit scale is based on the freezing point of a saturated salt solution at sea level (14.7 psi or 101.36 kPa) and the internal temperature of oxen, which set the 0 and 100 point markers on the scale. The Celsius scale is based on the freezing point and the boiling point of pure water at sea level. The Kelvin and Rankine scales are referenced to absolute zero, which is the temperature at which all molecular motion ceases, or the energy of a molecule is zero. The temperatures of the freezing and boiling points of water decrease as the pressure decreases, and change with the purity of the water.

Conversion between the units is shown in Table 10.1.

The need to convert from one temperature scale to another is a common everyday occurrence. The conversion factors are as follows:

To convert °F to °C

$$°C = (°F - 32)5/9 \qquad (10.1)$$

°F = 9/5 °C + 32

To convert °F to °R

$$°R = °F + 459.6 \qquad (10.2)$$

To convert °C to K

$$K = °C + 273.15 \qquad (10.3)$$

To convert K to °R

$$°R = 9/5 \times K \qquad (10.4)$$

Example 10.1

What temperature in °F corresponds to 435K?
From (10.3):

$$°C = 435 - 273.15 = 161.83$$

From (10.1):

$$°F = 161.85 \times 9/5 + 32 = 291.33 + 32 = 323.33°F$$

Example 10.2

What is the equivalent temperature of −63°F in °C?
From (10.1):

$$°C = (°F - 32)5/9$$

$$°C = (-63 - 32)5/9 = -52.2°C$$

−52.7

Table 10.1 Conversion Between Temperature Scales

Reference Point	°F	°C	°R	K
Water boiling point	212	100	671.6	373.15
Internal oxen temperature	100	37.8	559.6	310.95
Water freezing point	32	0.0	491.6	273.15
Salt solution freezing point	0.0	−17.8	459.6	255.35
Absolute zero	−459.6	−273.15	0.0	0.0

Example 10.3

Convert (a) 285K to °R and (b) 538.2°R to K.

(a) °R = 285 × 9/5 = 513°R

(b) K = 538.2 × 5/9 = 299K

10.2.2 Heat Energy

The temperature of a body is a measure of the heat energy in the body. As energy is supplied to a system, the vibration amplitude of the molecules in the system increases and its temperature increases proportionally.

Phase change is the transition between the three states that exist in matter: solid, liquid, and gas. However, for matter to make the transition from one state up to the next (i.e., solid to liquid to gas), it has to be supplied with energy. Energy has to be removed if the matter is going down from gas to liquid to solid. For example, if heat is supplied at a constant rate to ice at 32°F, then the ice will start to melt or turn to liquid, but the temperature of the ice-liquid mixture will not change until all the ice has melted. Then, as more heat is supplied, the temperature will start to rise until the boiling point of the water is reached. The water will turn to steam as more heat is supplied, but the temperature of the water and steam will remain at the boiling point until all the water has turned to steam. Then, the temperature of the steam will start to rise above the boiling point. Material also can change its volume during the change of phase. Some materials bypass the liquid stage, and transform directly from solid to gas or from gas to solid, in a transition called *sublimation*.

In a solid, the atoms can vibrate, but are strongly bonded to each other, so that the atoms or molecules are unable to move from their relative positions. As the temperature is increased, more energy is given to the molecules, and their vibration amplitude increases to a point where they can overcome the bonds between the molecules and can move relative to each other. When this point is reached, the material becomes a liquid. The speed at which the molecules move in the liquid is a measure of their thermal energy. As more energy is imparted to the molecules, their velocity in the liquid increases to a point where they can escape the bonding or attraction forces of other molecules in the material, and the gaseous state or boiling point is reached.

The *temperature and heat* relationship is given by the British thermal unit (Btu) in English units, or calories (cal) per joule in SI units. By definition, 1 Btu is the amount of energy required to raise the temperature of 1 lb of pure water 1°F, at 68°F and atmospheric pressure. It is a widely used unit for the measurement of heat energy. By definition, 1 cal is the amount of energy required to raise the temperature of 1g of pure water 1°C, at 4°C and atmospheric pressure. The joule is normally used in preference to the calorie, where 1J = 1 W×s. It is slowly becoming accepted as the unit for the measurement of heat energy in preference to the Btu. The conversion between the units is given in Table 10.2.

Thermal energy (W_{TH}), expressed in SI units, is the energy in joules in a material, and typically can be related to the absolute temperature (T) of the material, as follows:

Table 10.2 Conversions Related to Heat Energy

1 Btu = 252 cal	1 cal = 0.0039 Btu
1 Btu = 1,055J	1J = 0.000948 Btu
1 Btu = 778 ft·lb	1 ft·lb = 0.001285 Btu
1 cal = 4.19J	1J = 0.239 cal
1 ft·lb = 0.324 cal	1J = 0.738 ft·lb
1 ft·lb = 1.355J	1W = 1 J/s

$$W_{TH} = \frac{3}{2}kT \qquad (10.5)$$

where k = Boltzmann's constant = 1.38×10^{-23} J/K.

The above also can be used to determine the average velocity v_{TH} of a gas molecule from the kinetic energy equation:

$$W_{TH} = \frac{1}{2}mv_{TH}^2 = \frac{3}{2}kT$$

from which

$$v_{TH} = \sqrt{\frac{3kT}{m}} \qquad (10.6)$$

where m is the mass of the molecule in kilograms.

Example 10.4

What is the average thermal speed of an oxygen atom at 320°R? The molecular mass of oxygen is 26.7×10^{-27} kg.

$$320°R = 320 \times 5/9 K = 177.8 K$$

$$v_{TH} = \sqrt{\frac{3kT}{m}}$$

$$v_{TH} = \sqrt{\frac{3 \times 1.38 \times 10^{-23} \text{ J/K} \times 177.8 K}{26.7 \times 10^{-27} \text{ kg}}} \times \sqrt{\frac{kg \times m^2}{s^2 \times J}}$$

$$v_{TH} = 525 \text{ m/s}$$

The *specific heat* of a material is the quantity of heat energy required to raise the temperature of a given weight of the material 1°. For example, as already defined, 1 Btu is the heat required to raise 1 lb of pure water 1°F, and 1 cal is the heat required to raise 1g of pure water 1°C. Thus, if a material has a specific heat of 0.7 cal/g °C, then it would require 0.7 cal to raise the temperature of a gram of the material 1°C, or 2.93J to raise the temperature of the material 1K. Table 10.3 gives the specific heat of some common materials, in which the values are the same in either system.

10.2 Temperature and Heat

Table 10.3 Specific Heats of Some Common Materials, in Btu/lb °F or Cal/g °C

Alcohol	0.58 to 0.6	Aluminum	0.214	Brass	0.089
Glass	0.12 to 0.16	Cast iron	0.119	Copper	0.092
Gold	0.0316	Lead	0.031	Mercury	0.033
Platinum	0.032	Quartz	0.188	Silver	0.056
Steel	0.107	Tin	0.054	Water	1.0

The amount of heat needed to raise or lower the temperature of a given weight of a body can be calculated from:

$$Q = WC(T_2 - T_1) \quad (10.7)$$

where W is the weight of the material, C is the specific heat of the material, T_2 is the final temperature of the material, and T_1 is the initial temperature of the material.

Example 10.5

The heat required to raise the temperature of a 3.8 kg mass 135°C is 520 kJ. What is the specific heat of the mass in cal/g °C?

$$C = Q/W\delta T = 520 \times 1{,}000/3.8 \times 1{,}000 \times 135 \times 4.19 \text{ cal/g°C} = 0.24 \text{ cal/g°C}$$

As always, care must be taken in selecting the correct units. Negative answers indicate extraction of heat, or heat loss.

10.2.3 Heat Transfer

Heat energy is transferred from one point to another using any of three basic methods: conduction, convection, and radiation. Although these modes of transfer can be considered separately, in practice two or more of them can be present simultaneously.

Conduction is the flow of heat through a material, where the molecular vibration amplitude or energy is transferred from one molecule in a material to the next. If one end of a material is at an elevated temperature, then heat is conducted to the cooler end. The thermal conductivity of a material (k) is a measure of its efficiency in transferring heat. The units can be in British thermal units per hour per foot per degree Fahrenheit, or in watts per meter kelvin (1 Btu/ft·hr·°F = 1.73 W/m K). Table 10.4 gives typical thermal conductivities for some common materials.

Heat conduction through a material is derived from the following relationship:

$$Q = \frac{-kA(T_2 - T_1)}{L} \quad (10.8)$$

Table 10.4 Thermal Conductivity Btu/hr·ft °F (W/m K)

Air	0.016 (room temperature) (0.028)	Aluminum	119 (206)
Concrete	0.0 (1.4)	Copper	220 (381)
Water	0.36 (room temperature) (0.62)	Mercury	4.8 (8.3)
Brick	0.4 (0.7)	Steel	26 (45)
Brass	52 (90)	Silver	242 (419)

where Q is the rate of heat transfer, k is the thermal conductivity of the material, A is the cross-sectional area of the heat flow, T_2 is the temperature of the material distant from the heat source, T_1 is the temperature of the material adjacent to heat source, and L is the length of the path through the material.

Note that the negative sign in the (10.8) indicates a positive heat flow.

Example 10.6

A furnace wall 2.5m × 3m in area and 21 cm thick has a thermal conductivity of 0.35 W/m K. What is the heat loss if the furnace temperature is 1,050°C and the outside of the wall is 33°C?

$$Q = \frac{-kA(T_2 - T_1)}{L}$$

$$Q = \frac{-0.35 \times 7.5(33 - 1050)}{0.21} = 12.7 kW$$

Example 10.7

The outside wall of a room is 4.5m × 5m. If the heat loss is 2.2 kJ/hr, what is the thickness (d) of the wall? Assume the inside and outside temperatures are 23°C and −12°C, respectively, and assume the conductivity of the wall is 0.21 W/m K.

$$Q = \frac{-kA(T_2 - T_1)}{L}$$

$$2,200 kJ/hr = \frac{-0.21 W/mK \times 4.5m \times 5m \times (-12 - 23)K}{dm} \times \frac{60 \times 60 J/s}{W \times hr}$$

$$d = 0.27m = 27 \text{ cm}$$

Convection is the transfer of heat due to motion of elevated temperature particles in a liquid or a gas. Typical examples are air conditioning systems, hot water heating systems, and so forth. If the motion is due solely to the lower density of the elevated temperature material, the transfer is called free or natural convection. If blowers or pumps move the material, then the transfer is called forced convection. Heat convection calculations in practice are not as straightforward as conduction calculations. Heat convection is given by:

$$Q = hA(T_2 - T_1) \tag{10.9}$$

where Q is the convection heat transfer rate, h is the coefficient of heat transfer, A is the heat transfer area, and $T_2 - T_1$ is the difference between the source (T_2) and final temperature (T_1) of the flowing medium.

It should be noted that, in practice, the proper choice for h is difficult because of its dependence on a large number of variables (e.g., density, viscosity, and specific heat). Charts are available for h. However, experience is needed in their application.

Example 10.8

How much heat is transferred from a 12m × 15m surface by convection, if the temperature difference between the front and back surfaces is 33°C and the surface has a heat transfer rate of 1.5 W/m² K?

$$Q = 1.5 \times 12 \times 15 \times 33 = 8.91 \text{ kW}$$

Radiation is the emission of energy by electromagnetic waves, which travel at the speed of light through most materials that do not conduct electricity. For instance, radiant heat can be felt at a distance from a furnace where there is no conduction or convection. Heat radiation is dependent on factors, such as surface color, texture, shapes, and so forth. Information more than the basic relationship for the transfer of radiant heat energy given below should be factored in. The radiant heat transfer is given by:

$$Q = CA(T_2^4 - T_1^4) \tag{10.10}$$

where Q is the heat transferred, C is the radiation constant (dependent on surface color, texture, units selected, and so forth), A is the area of the radiating surface, T_2 is the absolute temperature of the radiating surface, and T_1 is the absolute temperature of the receiving surface.

Example 10.9

The radiation constant for a furnace is 0.11×10^{-8} W/m² K⁴, the radiating surface area is 3.9 m², the radiating surface temperature is 1,150K, and the room temperature is 22°C. How much heat is radiated?

$$Q = 0.11 \times 10^{-8} \times 3.9 \times [(1{,}150)^4 - (22 + 273)^4]$$

$$Q = 0.429 \times 10^{-8} \times [174.9 \times 10^{10} - 0.76 \times 10^{10}] = 7.47 \times 10^3 \text{W}$$

Example 10.10

What is the radiation constant for a wall 19 × 9 ft, if the radiated heat loss is 2.33×10^4 Btu/hr, the wall temperature is 553°R, and the ambient temperature is 12°C?

$$2.33 \times 10^4 \text{ Btu/hr} = C \times 19 \times 9 \, [(553)^4 - (53.6 + 460)^4]$$

$$C = 2.33 \times 10^4 / 171(9.35 \times 10^{10} - 6.92 \times 10^{10})$$

$$C = 3.98 \times 10^6 / 2.43 \times 10^{10} = 1.64 \times 10^4 \text{ Btu/hr ft}^2{}°\text{F}^4$$

10.2.4 Thermal Expansion

Linear thermal expansion is the change in dimensions of a material due to temperature changes. The change in dimensions of a material is due to its coefficient of thermal expansion, which is expressed as the change in linear dimension (α) per degree temperature change. The change in linear dimension due to temperature changes can be calculated from the following formula:

$$L_2 = L_1[1 + \alpha(T_2 - T_1)] \tag{10.11}$$

where L_2 is the final length, L_1 is the initial length, α is the coefficient of linear thermal expansion, T_2 is the final temperature, and T_1 is the initial temperature.

Example 10.11

What is the length of a copper rod at 420K, if the rod was 93m long at 10°F?

New length = $L_1[1 + \alpha(T_2 - T_1)]$ = 93{1 + 16.9 × 10^{-6} [147 – (–12)]}m

New Length = 93[1 + 16.9 × 10^{-6} × 159] = 93 × 1.0027m = 93.25m

Volume thermal expansion is the change in the volume (β) per degree temperature change due to the linear coefficient of expansion. The thermal expansion coefficients for linear and volume expansion for some common materials per °F (°C) are given in Table 10.5. The volume expansion in a material due to changes in temperature is given by:

$$V_2 = V_1[1 + \beta(T_2 - T_1)] \tag{10.12}$$

where V_2 is the final volume, V_1 is the initial volume, β is the coefficient of volumetric thermal expansion, T_2 is the final temperature, and T_1 the initial temperature.

Example 10.12

Calculate the new volume for a silver cube that measures 3.15 ft on a side, if the temperature is increased from 15° to 230°C.

Old Volume = 31.26 ft³

New volume = 3.15³[1 + 57.6 × 10^{-6} × (230 – 15)]

= 3.15³(1 + 0.012) = 31.63 ft³   31.255 ft³

In a gas, the relation between the pressure, volume, and temperature is given by:

$$\frac{P_1 V_1}{T_1} = \frac{P_2 V_2}{T_2} \tag{10.13}$$

Table 10.5 Thermal Coefficients of Expansion per °F (°C)

	Linear × 10^{-6}	Volume × 10^{-6}		Linear × 10^{-6}	Volume × 10^{-6}
Alcohol	—	61–66 (109.8–118.8)	Aluminum	12.8 (23.04)	—
Brass	11.3 (20.3)	—	Cast iron	5.6 (10.1)	20 (36)
Copper	9.4 (16.9)	29 (52.2)	Glass	5 (9)	14 (25.2)
Gold	7.8 (14.04)	—	Lead	16 (28.8)	—
Mercury	—	100 (180)	Platinum	5 (9)	15 (27)
Quartz	0.22 (0.4)	—	Silver	11 (19.8)	32 (57.6)
Steel	6.1 (11)	—	Tin	15 (27)	38 (68.4)
Invar	0.67 (1.2)	—	Kovar	3.28 (5.9)	—

where P_1 is the initial pressure, V_1 is the initial volume, T_1 is the initial absolute temperature, P_2 is the final pressure, V_2 is the final volume, and T_2 is the final absolute temperature.

10.3 Temperature Measuring Devices

The methods of measuring temperature can be categorized as follows:

1. Expansion of materials;
2. Electrical resistance change;
3. Thermistors;
4. Thermocouples;
5. Pyrometers;
6. Semiconductors.

Thermometer is often used as a general term for devices for measuring temperature. Examples of temperature measuring devices are described below.

10.3.1 Expansion Thermometers

Liquid in glass thermometers using mercury were, by far, the most common direct visual reading thermometer (if not the only one). Mercury also has the advantage of not wetting the glass; that is, the mercury cleanly traverses the glass tube without breaking into globules or coating the tube. The operating range of the mercury thermometer is from −30° to +800°F (−35° to +450°C). The freezing point of mercury −38°F (−38°C). The toxicity of mercury, ease of breakage, the introduction of cost-effective, accurate, and easily read digital thermometers, has brought about the demise of the mercury thermometer for room and clinical measurements. Other liquid in glass devices operate on the same principle as the mercury thermometer. These other liquids have similar properties to mercury (e.g., have a high linear coefficient of expansion, are clearly visible, are nonwetting), but are nontoxic. The liquid in glass thermometers are used to replace the mercury thermometer, and to extend its operating range. These thermometers are inexpensive, and have good accuracy (<0.1°C) and linearity. These devices are fragile, and used for local measurement. The operating range with different liquids is from −300° to +1,000°F (−170° to +530°C). Each type of liquid has a limited operating range.

A *bimetallic strip* is a relatively inaccurate, rugged temperature-measuring device, which is slow to respond and has hysteresis. The device is low cost, and therefore is used extensively in On/Off-types of applications, or for local analog applications not requiring high accuracy, but it is not normally used to give remote analog indication. These devices operate on the principle that metals are pliable, and different metals have different coefficients of expansion, as shown in Table 10.5. If two strips of dissimilar metals, such as brass and invar (copper-nickel alloy), are joined together along their length, then they will flex to form an arc as the temperature changes. This is shown in Figure 10.1(a). Bimetallic strips are usually configured as a spiral or helix for compactness, and can then be used with a pointer to make an inexpensive, compact, rugged thermometer, as shown in Figure 10.1(b).

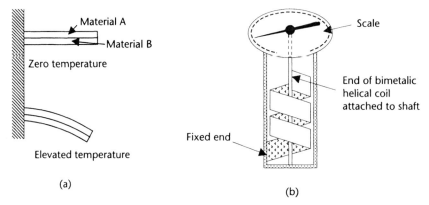

Figure 10.1 (a) Effect of temperature change on a bimetallic strip, and (b) bimetallic thermometer using a helical bimetallic coil.

When using a straight bimetallic strip, an important calculation to determine the movement of the free end of the strip is given by:

$$d = \frac{3(a_A - a_B)(T_2 - T_1)l^2}{t} \tag{10.14}$$

where d is the deflection, α_A is the coefficient of linear expansion of metal A, α_B is the coefficient of linear expansion of metal B, T_1 is the lower temperature, T_2 is the elevated temperature, l is the length of the element, and t is the thickness of the element.

The equation shows the linear relationship between deflection and temperature change. The operating range of bimetallic elements is from $-180°$ to $+430°C$, and they can be used in applications from oven thermometers to home and industrial control thermostats.

Example 10.13

A straight bimetallic strip 25 cm long and 1.4 mm thick is made of copper and invar, and has one end fixed. How much will the free end of the strip deflect if the temperature changes 18°F?

$$d = \frac{3(9.4 - 0.67)10^{-6}(18)25 \times 25}{0.14}\,cm = 2.1\,cm$$

When the bimetallic element is wound into a spiral, the deflection of the free end of the strip is given by:

$$d = \frac{9(a_A - a_B)(T_2 - T_1)rl}{4t} \tag{10.15}$$

where r is the radius of the spiral, and l is the length of the unwound element.

Example 10.14

A 42-cm length of 2.3-mm thick copper-invar bimetallic strip is wound into a spiral with a radius of 5.2 cm at 70°F. What is the deflection of the strip at 120°F?

$$d = \frac{9(9.4 - 0.67)10^{-6}(120-70)5.2 \times 42}{4 \times 0.23}$$

$$= 0.93 cm$$

Pressure-spring thermometers are used where remote indication is required, as opposed to glass and bimetallic devices, which give readings at the point of detection. The pressure-spring device has a metal bulb made with a low coefficient of expansion material along with a long metal narrow bore tube. Both contain material with a high coefficient of expansion. The bulb is at the monitoring point. The metal tube is terminated with a Bourdon spring pressure gauge (scale in degrees), as shown in Figure 10.2. The pressure system can be used to drive a chart recorder, actuator, or a potentiometer wiper to obtain an electrical signal. As the temperature in the bulb increases, the pressure in the system rises. Bourdon tubes, bellows, or diaphragms sense the change in pressure. These devices can be accurate to 0.5%, and can be used for remote indication up to a distance of 100m, but must be calibrated, since the stem and Bourdon tube are temperature-sensitive.

There are three types or classes of pressure-spring devices. These are:

- Class 1 Liquid filled;
- Class 2 Vapor pressure;
- Class 3 Gas filled.

The *liquid-filled thermometer* works on the same principle as the liquid in glass thermometer, but is used to drive a Bourdon tube. The device has good linearity and accuracy, and can be used up to 550°C.

The *vapor-pressure thermometer* system is shown in Figure 10.2. The bulb is partially filled with liquid and vapor, such as methyl chloride, ethyl alcohol, ether, or toluene. In this system, the lowest operating temperature must be above the boiling point of the liquid, and the maximum temperature is limited by the critical temperature of the liquid. The response time of the system is slow, being of the order of 20 seconds. The temperature-pressure characteristic of the vapor-pressure thermometer is nonlinear, as shown in the vapor pressure curve for methyl chloride in Figure 10.3.

A *gas thermometer* is filled with a gas, such as nitrogen, at a pressure of between 1,000 and 3,350 kPa, at room temperature. The device obeys the basic gas laws for

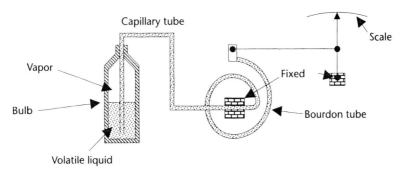

Figure 10.2 Vapor-pressure thermometer.

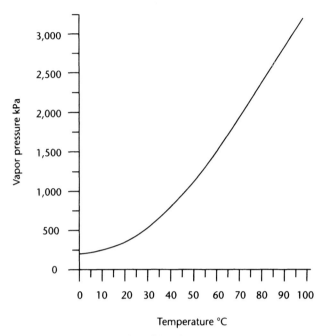

Figure 10.3 Vapor pressure curve for methyl chloride.

a constant volume system [(10.15), $V_1 = V_2$], giving a linear relationship between absolute temperature and pressure.

10.3.2 Resistance Temperature Devices

RTDs are either a metal film deposited on a form or are wire-wound resistors, which are then sealed in a glass-ceramic composite material [2]. Figure 10.4 shows a three-wire RTD encased in a stainless steel sheath for protection. The coil is wound to be noninductive. The space between the element and the case is filled with a ceramic power for good thermal conduction. The element has three leads, so that correction can be made for voltage drops in the lead wires. The electrical resistance of pure metals is positive, increasing linearly with temperature. Table 10.6 gives the temperature coefficient of resistance of some common metals used in resistance thermometers. Platinum is the first choice, followed by nickel. These devices are accurate—temperature changes of a fraction of a degree can be measured. The RTD can be used to measure temperatures from $-300°$ to $+1,400°F$ ($-170°$ to $+780°C$). The response time is typically between 0.5 and 5 seconds.

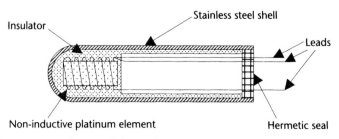

Figure 10.4 Cross section of a typical three-wire RTD.

10.3 Temperature Measuring Devices

Table 10.6 Temperature Coefficient of Resistance of Some Common Metals

Material	Coefficient per °C	Material	Coefficient per °C
Iron	0.006	Tungsten	0.0045
Nickel	0.005	Platinum	0.00385

In a resistance thermometer, the variation of resistance with temperature is given by:

$$R_{T_2} = R_{T_1}[1 + \text{Coeff.}\,(T_2 - T_1)] \tag{10.16}$$

where R_{T_2} is the resistance at temperature T_2, and R_{T_1} is the resistance at temperature T_1.

Example 10.15

What is the resistance of a platinum resistor at 480°C, if its resistance at 16°C is 110Ω?

$$\begin{aligned}\text{Resistance at } 480°C &= 110\big[1 + 0.00385(480 - 16)\big]\,\Omega \\ &= 110(1 + 1.7864)\,\Omega \\ &= 306.5\,\Omega\end{aligned}$$

Resistance devices are normally measured using a Wheatstone bridge type of system, or are supplied from a constant current source. Care should be taken to prevent the electrical current from heating the device and causing erroneous readings. One method of overcoming this problem is to use a pulse technique (i.e., by passing a current through the resistor only when a measurement is being made), keeping the average power very low so that the temperature of the resistor remains at the ambient temperature.

10.3.3 Thermistors

Thermistors are a class of metal oxide (semiconductor material) that typically has a high negative temperature coefficient of resistance. They also can be positive. Thermistors have high sensitivity, which can be up to a 10% change per degree Celsius, making it the most sensitive temperature element available, but thermistors also have very nonlinear characteristics. The typical response time is from 0.5 to 5 seconds, with an operating range typically from −50° to +300°C [3]. Devices are available with the temperature range extended to 500°C. Thermistors are low cost, and are manufactured in a wide range of shapes, sizes, and values. When in use, care has to be taken to minimize the effects of internal heating. Thermistor materials have a temperature coefficient of resistance (α) given by:

$$\alpha = \frac{\Delta R}{R_S}\left(\frac{1}{\Delta T}\right) \tag{10.17}$$

where ΔR is the change in resistance due to a temperature change ΔT, and R_s is the material resistance at the reference temperature.

The nonlinear characteristics, shown in Figure 10.5, make the device difficult to use as an accurate measuring device without compensation, but its sensitivity and low cost makes it useful in many applications. The device is normally used in a bridge circuit, and padded with a resistor to reduce its nonlinearity.

10.3.4 Thermocouples

Thermocouples (T/C) are formed when two dissimilar metals are joined together to form a junction. Joining together the other ends of the dissimilar metals to form a second junction completes an electrical circuit. A current will flow in the circuit if the two junctions are at different temperatures. The current flowing is the result of the difference in electromotive force developed at the two junctions due to their temperature difference. The voltage difference between the two junctions is measured, and this difference is proportional to the temperature difference between the two junctions. Note that the thermocouple only can be used to measure temperature differences. However, if one junction is held at a reference temperature, then the voltage between the thermocouple junctions gives a measurement of the temperature of the second junction, as shown in Figure 10.6(a). An alternative method is to measure the temperature of the reference junction and apply a correction to the output signal, as shown in Figure 10.6(b). This method eliminates the need for constant temperature enclosures.

Three effects are associated with thermocouples:

1. The *Seebeck effect* states that the voltage produced in a thermocouple is proportional to the temperature between the two junctions.

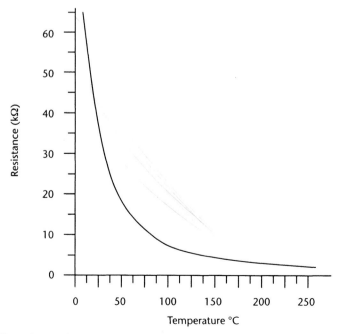

Figure 10.5 Thermistor resistance temperature curve.

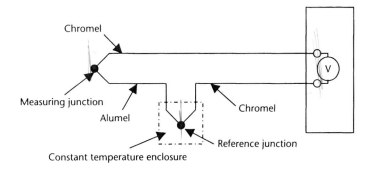

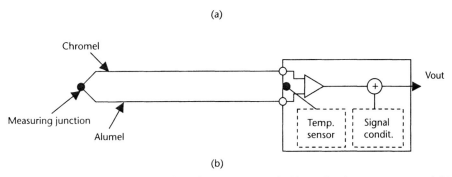

Figure 10.6 (a) Thermocouple with the reference junction held at a fixed temperature, and (b) the temperature of the reference junction measured, and a correction applied to the output signal.

2. The *Peltier effect* states that if a current flows through a thermocouple, then one junction is heated (outputs energy), and the other junction is cooled (absorbs energy).
3. The *Thompson effect* states that when a current flows in a conductor along which there is a temperature difference, heat is produced or absorbed, depending upon the direction of the current and the variation of temperature.

In practice, the Seebeck voltage is the sum of the electromotive forces generated by the Peltier and Thompson effects. There are a number of laws to be observed in thermocouple circuits. First, the law of intermediate temperatures states that the thermoelectric effect depends only on the temperatures of the junctions, and is not affected by the temperatures along the leads. Second, the law of intermediate metals states that metals other than those making up the thermocouples can be used in the circuit, as long as their junctions are at the same temperature. Other types of metals can be used for interconnections, and tag strips can be used without adversely affecting the output voltage from the thermocouple. Letters designate the various types of thermocouples. Tables of the differential output voltages for different types of thermocouples are available from manufacturers' thermocouple data sheets [4]. Table 10.7 lists some thermocouple materials and their Seebeck coefficients. The operating range of the thermocouple is reduced to the figures in parentheses if the given accuracy is required. For operation over the full temperature range, the

Table 10.7 Operating Ranges for Thermocouples and Seebeck Coefficients

Type	Approximate Range	Seebeck Coefficient μV/°C
Copper—Constantan (T)	−140° to 400°C	40 (−59 to +93) ± 1°C
Chromel—Constantan (E)	−180° to +1,000°C	62 (0 to 360) ± 2°C
Iron—Constantan (J)	30° to 900°C	51 (0 to 277) ± 2°C
Chromel—Alumel (K)	30° to 1,400°C	40 (0 to 277) ± 2°C
Nicrosil—Nisil (N)	30° to 1,400°C	38 (0 to 277) ± 2°C
Platinum (Rhodium 10 %)—Platinum (S)	30° to 1,700°C	7 (0 to 538) ± 3°C
Platinum (Rhodium 13 %)—Platinum (R)	30° to 1,700°C	7 (0 to 538) ± 3°C

accuracy would be reduced to approximately ±10% without linearization [5], or ±0.5% when compensation is used. Thermocouple tables are available from several sources such as:

- Vendors;
- NIST Monograph 125;
- ISA standards—ISA-MC96.1 − 1982.

There are three types of basic packaging for thermocouples: sealed in a ceramic bead, insulated in a plastic or glass extrusion, or metal sheathed. The metal sheath is normally stainless steel with magnesium oxide or aluminum oxide insulation. The sheath gives mechanical and chemical protection to the T/C. Sheathed T/C are available with the T/C welded to the sheath, insulated from the sheath, and exposed for high-speed response. These configurations are shown in Figure 10.7.

10.3.5 Pyrometers

Radiation can be used to sense temperature, by using devices called pyrometers, with thermocouples or thermopiles as the sensing element, or by using color

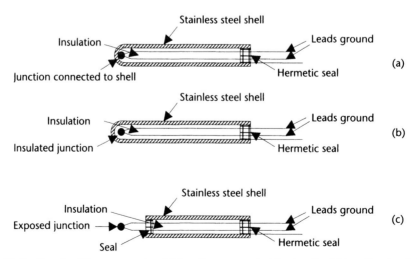

Figure 10.7 Sheathed thermocouples: (a) junction connected to sheath, (b) junction insulated from sheath, and (c) exposed junction for high-speed response.

comparison devices [6]. Pyrometers can be portable, and have a response time of a few milliseconds.

A *thermopile* is a number of thermocouples connected in series, which increases the sensitivity and accuracy by increasing the output voltage when measuring low temperature differences. Each of the reference junctions in the thermopile is returned to a common reference temperature.

Radiation pyrometers measure temperature by sensing the heat radiated from a hot body through a fixed lens, which focuses the heat energy on to a thermopile. This is a noncontact device [7, 8]. For instance, furnace temperatures are normally measured through a small hole in the furnace wall. The distance from the source to the pyrometer can be fixed, and the radiation should fill the field-of-view of the sensor. Figure 10.8(a) shows the focusing lens and thermocouple setup, using a thermopile as a detector.

Optical pyrometers compare the incident radiation to the radiation from an internal filament, as shown in Figure 10.8(b). The current through the filament is adjusted until the radiation colors match. The current then can be directly related to the temperature of the radiation source. Optical pyrometers can be used to measure temperatures from 1,100° to 2,800°C, with an accuracy of ±

10.3.6 Semiconductor Devices

Semiconductors have a number of parameters that vary linearly with temperature. Normally, the reference voltage of a zener diode or the junction voltage variations are used for temperature sensing [9]. Semiconductor temperature sensors have a limited operating range, from −50° to +150°C, but are very linear, with accuracies

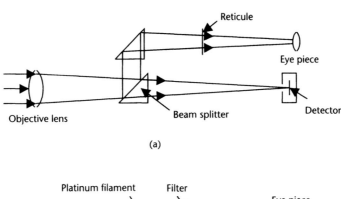

(a)

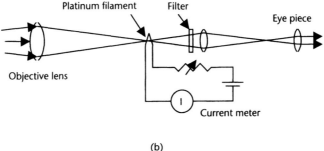

(b)

Figure 10.8 (a) Focusing lens and thermocouple set up using a thermopile as a detector. (b) Optical pyrometer comparing the incident radiation to the radiation from an internal filament.

of ±1°C or better. Other advantages are that electronics can be integrated onto the same die as the sensor, giving high sensitivity, easy interfacing to control systems, and making possible different digital output configurations [10]. The thermal time constant varies from 1 to 5 seconds, and internal dissipation can cause up to a 0.5°C offset. Semiconductor devices also are rugged, have good longevity, and are inexpensive. For the above reasons, the semiconductor sensor is used extensively in many applications, including the replacement of the mercury in glass thermometer.

10.4 Application Considerations

Many devices are available for temperature measurement. The selection of a device will be determined by the needs of the application. A comparison of the characteristics of temperature-measuring devices is given. Thermal time constant considerations, installation calibration, and protection also are discussed.

10.4.1 Selection

In process control, a wide selection of temperature sensors is available [11]. However, the required range, linearity, and accuracy can limit the selection. In the final selection of a sensor, other factors may have to be taken into consideration, such as remote indication, error correction, calibration, vibration sensitivity, size, response time, longevity, maintenance requirements, and cost [12]. The choice of sensor devices in instrumentation should not be degraded from a cost standpoint. Process control is only as good as the monitoring elements.

10.4.2 Range and Accuracy

Table 10.8 gives the temperature ranges and accuracies of temperature sensors, the accuracies shown are with minimal error correction. The ranges in some cases can be extended with the use of new materials. Table 10.9 gives a summary of temperature sensor characteristics.

Table 10.8 Temperature Range and Accuracy of Temperature Sensors

Sensor Type		Range	Accuracy (FSD)
Expansion	Mercury in glass	−35° to +430°C	±1%
	Liquid in glass	−180° to +500°C	±1%
	Bimetallic	−180° to +600°C	±20%
Pressure–spring	Liquid filled	−180° to +550°C	±0.5%
	Vapor pressure	−180° to +320°C	±2.0%
	Gas filled	−180° to +320°C	±0.5%
Resistance	Metal resistors	−200° to +800°C	±5%
	Platinum	−180° to +650°C	±0.5%
	Nickel	−180° to +320°C	±1%
	Copper	−180° to +320°C	±0.2%
Thermistor		0° to 500°C	±25%
Thermocouple		−60° to +540°C	±1%
		−180° to +2,500°C	±10%
Semiconductor IC		−40° to +150°C	±1%

10.4 Application Considerations

Table 10.9 Summary of Sensor Characteristics

Type	Linearity	Advantages	Disadvantages
Bimetallic	Good	Low cost, rugged, wide range	Local measurement, or for On/Off switching only
Pressure	Medium	Accurate, wide range	Needs temperature compensation, vapor is nonlinear
Resistance	Very good	Stable, wide range, accurate	Slow response, low sensitivity, expensive, self-heating, range
Thermistor	Poor	Low cost, small, high sensitivity, fast response	Nonlinear, range, self-heating
Thermocouple	Good	Low cost, rugged, very wide range	Low sensitivity, reference needed
Semiconductor	Excellent	Low cost, sensitive, easy to interface	Self-heating, slow response, range, power source

10.4.3 Thermal Time Constant

A temperature detector does not react immediately to a change in temperature. The reaction time of the sensor, or thermal time constant, is a measure of the time it takes for the sensor to stabilize internally to the external temperature change, and is determined by the thermal mass and thermal conduction resistance of the device [13]. Thermometer bulb size, probe size, or protection well can affect the response time of the reading. For example, a large bulb contains more liquid for better sensitivity, but this will increase the time constant to fully respond to a temperature change.

The thermal time constant is related to the thermal parameters by the following equation:

$$t_c = \frac{mc}{kA} \quad (10.18)$$

where t_c is the thermal time constant, m is the mass, c is the specific heat, k is the heat transfer coefficient, and A is the area of thermal contact.

When temperature changes rapidly, the temperature output reading of a thermal sensor is given by:

$$T - T_2 = (T_1 - T_2)e^{-t/t_c} \quad (10.19)$$

where T is the temperature reading, T_1 is the initial temperature, T_2 is the true system temperature, and t is the time from when the change occurred.

The time constant of a system t_c is defined as the time it takes for the system to reach 63.2% of its final temperature value after a temperature change. As an example, a copper block is held in an ice-water bath until its temperature has stabilized at 0°C. It is then removed and placed in a 100°C steam bath. The temperature of the copper block will not immediately go to 100°C, but its temperature will rise on an exponential curve as it absorbs energy from the steam, until after some time period (i.e., its time constant), it will reach 63.2°C, eventually reaching 100°C. During the second time constant, the copper will rise another 63.2% of the remaining temperature to get to equilibrium. That is, (100 − 63.2) × 63.2% = 23.3°C, or at the end of two time constant periods, the temperature of the copper will be 86.5°C. At the end of three periods, the temperature will be 95°C, and so on. Where a fast response time is required, thermal time constants can be a serious problem. In some cases, the constants can be several seconds in duration, and a correction may have to be electronically applied to the output reading to obtain a faster response. Measuring the rate of rise of the temperature indicated by the sensor and extrapolating the actual aiming temperature can do this. The thermal time constant of a body is similar to an electrical time constant, which is discussed in the chapter on electricity under electrical time constants.

10.4.4 Installation

Care must be taken in locating the sensing portion of the temperature sensor. It should be fully encompassed by the medium whose temperature is being measured, and not be in contact with the walls of the container. The sensor should be screened from reflected and radiant heat, if necessary. The sensor also should be placed downstream from fluids being mixed, to ensure that the temperature has stabilized, but as close as possible to the point of mixing, to give as fast as possible temperature measurement for good control. A low thermal time constant in the sensor is necessary for a quick response [14].

Compensation and calibration may be necessary when using pressure-spring devices with long tubes, especially when accurate readings are required.

10.4.5 Calibration

Temperature calibration can be performed on most temperature sensing devices by immersing them in known temperature standards, which are the equilibrium points of solid/liquid or liquid/gas mixtures (also known as the triple point). Some of these standards are given in Table 10.10. Most temperature-sensing devices are rugged and reliable, but they can go out of calibration, due to leakage during use or contamination during manufacture, and therefore should be checked on a regular basis.

10.4.6 Protection

In some applications, temperature-sensing devices are placed in thermowells or enclosures, to prevent mechanical damage or for ease of replacement. This kind of protection can greatly increase the system response time, which in some circumstances may be unacceptable. Sensors also may need to be protected from overtemperature, so that a second, more rugged device may be needed to protect the main sensing device. Semiconductor devices may have built-in overtemperature protection. A fail-safe mechanism also may be incorporated for system shutdown, when processing volatile or corrosive materials.

Table 10.10 Temperature Scale Calibration Points

Calibration Material	Temperature			
	K	°R	°F	°C
Zero thermal energy	0	0	−459.6	−273.15
Oxygen: liquid-gas	90.18	162.3	−297.3	−182.97
Water: solid-liquid	273.15	491.6	32	0
Water: liquid-gas	373.15	671.6	212	100
Gold: solid-liquid	1,336.15	2,405	1,945.5	1,063

10.5 Summary

Temperature is the most important physical parameter, because all other physical parameters are temperature-dependent. Temperature can be measured using Celsius or Kelvin, in the SI system of units, and Fahrenheit or Rankine in the English system of units. This chapter described the relations between the units and introduced the basic concept that temperature is a measure of the heat energy contained in a body or the amplitude of molecular vibration. The vibration amplitude and molecular attraction determines the phase of the material. Heat energy can be measured in British thermal units or joules, and can be transferred between bodies either by conduction, convection, or radiation. The mechanics of heat transfer was described.

A large number of instruments is available for temperature measurement. The choice of instrument is determined by the requirements of the application. New innovations, such as the semiconductor digital thermometer, have brought about the demise of the mercury thermometer. Inexpensive On/Off applications used bimetallic devices, but these are also being replaced by semiconductor devices. Wide temperature range applications typically use thermocouples, and high accuracy devices are RTDs. Some devices have long settling times due to their thermal time constant, which can be compensated for electronically [15].

Definitions

Absolute zero is the temperature at which all molecular motion ceases, or the energy of a molecule is zero.

British thermal unit (Btu) is defined as the amount of energy required to raise the temperature of 1 lb of pure water 1°F, at 68°F and 1 atm.

Calorie unit (SI) is defined as the amount of energy required to raise the temperature of 1g of pure water 1°C, at 4°C at 1 atm.

Celsius or Centigrade scale (°C) uses 0° and 100° (100° range) for the freezing and boiling points, respectively, of pure water at 1 atm.

Fahrenheit scale (°F) uses 32° and 212° (180° range) as the freezing and boiling points, respectively, of pure water at 1 atm.

Heat is a form of energy, and is a measure of the vibration amplitude of its molecules, which is indicated by its temperature.

Joules (SI) are units of heat energy.

Kelvin scale (K) is referenced to absolute zero, and based on the Celsius scale.

Rankine scale (°R) is a temperature scale referenced to absolute zero, and based on the Fahrenheit scale.

Sublimation is the direct transition from the gas state to the solid state without going into the liquid state, or the direct transition from the solid state to the gas state.

References

[1] Mathews, D., "Choosing and Using a Temperature Sensor," *Sensors Magazine*, Vol. 17, No. 1, January 2000.

[2] Krysciar, T., "Low Temperature Measurements with Thin Film Platinum Resistance Elements," *Sensors Magazine*, Vol. 16, No. 1, January 1999.

[3] Lavenuth, G., "Negative Temperature Coefficient Thermistors," *Sensors Magazine*, Vol. 14, No. 5, May 1997.

[4] Humphries, J. T., and L. P. Sheets, *Industrial Electronics*, 4th ed., Delmar, 1993, pp. 320–323.

[5] Klopfenstien, Jr., R., "Linearization of a Thermocouple," *Sensors Magazine*, Vol. 14, No. 12, December 1997.

[6] Bedrossian, Jr., J., "Infrared Noncontact Temperature Measurement," *Proceedings Sensor Expo*, 1994, pp. 597–602.

[7] Barron, W. R., "Infrared Thermometry—Today," *Proceedings Sensor Expo*, 1993, pp. 1–12.

[8] DeWitt, D. P., "Infrared Thermometry—Tomorrow," *Proceedings Sensor Expo*, 1993, pp. 13–16.

[9] Ristic, L., *Sensor Technology and Devices*, Artech House 1994, pp. 287–315.

[10] Scolio, J., "Temperature Sensor IC's Simplify Designs," *Sensors Magazine*, Vol. 17, No. 1, January 2000.

[11] Maxwell, D., and R. Williamson, "Wireless Temperature Monitoring in Remote Systems," *Sensors Magazine*, Vol. 19, No. 10, October 2002.

[12] Desmarais, R., and J. Breuer, "How to Select the Right Temperature Sensor," *Sensors Magazine*, Vol. 18, No. 1, January 2001.

[13] Johnson, C. D., *Process Control Instrumentation Technology*, 7th ed., Prentice Hall, 2003, pp. 36–38.

[14] Paluch, R., "Field Installation of Thermocouple & RTD Temperature Sensor Assemblies," *Sensors Magazine*, Vol. 19, No. 8, August 2002.

[15] Stokes J., and G. Palmer, "A Fiber Optic Temperature Sensor," *Sensors Magazine*, Vol. 19, No. 8, August 2002.

CHAPTER 11
Position, Force, and Light

11.1 Introduction

There are many sensors other than those used for measuring fluid characteristics required in process control, and which are equally important. This chapter covers many of the varied sensors used to monitor position, motion, force, and light. The sensors discussed in this chapter include magnetic sensors, light sensors, accelerometers, ultrasonic sensors, load cells, and torque sensors.

11.2 Position and Motion Sensing

Many industrial processes require information on and control of both linear and angular position, and rate of motion. These are required in robotics, rolling mills, machining operations, numerically controlled tool applications, and conveyers. In some applications, it is also necessary to measure and control acceleration and vibration. Some transducers use position-sensing devices to convert temperature, level, and/or pressure into electrical units, and controllers can use position-sensing devices to monitor the position of an adjustable valve for feedback control.

Position measurements can be absolute or incremental, and can apply to linear motion or angular motion. A body can have a constant velocity, or the velocity can be changing, in which case the motion will have an acceleration.

11.2.1 Position and Motion Measuring Devices

Potentiometers are a convenient cost-effective method for converting the displacement in a sensor to an electrical variable. The wiper or slider arm of a linear potentiometer can be mechanically connected to the moving section of a sensor. Where angular displacement is involved, a single or multiturn (up to 10 turns) rotational type of potentiometer can be used. For stability, wire-wound devices should be used, but in environmentally unfriendly conditions, the lifetime of the potentiometer may be limited by dirt, contamination, and wear.

Linear variable differential transformers (LVDTs) are devices used for measuring small distances, and are an alternative to the potentiometer. The device consists of a primary coil with two secondary windings, one on either side of the primary, as shown in Figure 11.1(a). A movable core, when centrally placed in the primary, will give equal coupling to each of the secondary coils. When an ac voltage is applied to the primary, equal voltages will be obtained from the secondary windings, which

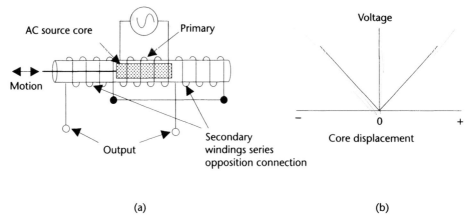

Figure 11.1 (a) The LVDT with a movable core and three windings, and (b) the secondary voltage V_s core displacement for the connections.

are shown wired in series opposition. This gives zero output voltage, as shown in Figure 11.1(b), when the core is centrally positioned. An output voltage proportional to displacement is obtained for limited travel. These devices are not as cost-effective as potentiometers, but have the advantage of being noncontact. The outputs are electrically isolated, accurate, and have better longevity than potentiometers.

Figure 11.2(a) shows an alternative method of connecting the output windings of the LVDT using a special interfacing IC. A phase-sensitive detector is used to give a linear output with displacement, as shown in Figure 11.2(b). A variety of LVDTs is available, with linear ranges from ±1 mm to ±25 cm. The transfer function is normally expressed in mV/mm.

Capacitive devices can be used to measure angular or linear displacement. The capacitance between two parallel plates is given by:

$$C = 8.85KA/d \text{ pF}$$

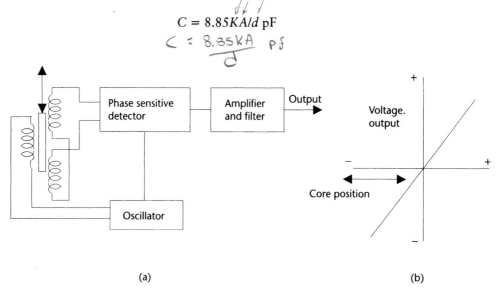

Figure 11.2 LVDT with (a) phase sensitive detector, and (b) plot of output voltage versus position with phase-sensitive detector.

where K is the dielectric of the material between the plates, A is the area of the plates in square meters, and d is the distance between the plates in meters.

There are three methods of changing the capacitance, as shown in Figure 11.3: (1) changing the distance between the plates, (2) moving one plate with respect to the other plate to reduce the overlapping area between the plates, and (3) moving a dielectric material between fixed plates. Typically, differential sensing is used, as shown in Figure 7.5(a). Capacitance variation is a very accurate method of measuring displacement, and is used extensively in micromechanical devices where the distance is small, giving high capacitance per unit area. Capacitive devices are used in the measurement of pressure, acceleration, and level.

Light interference lasers are used for very accurate incremental position measurements. Monochromatic (i.e., single frequency) light can be generated with a laser and collimated into a narrow beam. The light from the laser beam is split and a percentage goes to a detector, but the main beam goes to a mirror attached to an object whose change in distance is being measured, as shown in Figure 11.4(a). The reflected beam is then directed to the detector via the beamsplitter and a mirror. A change in the position of the object of one-quarter wavelength increases both the incident and reflected beam length one-quarter wavelength, giving a change at the detector of one-half wavelength. When the reflected beam is in phase with the incident beam, (d) is N times an even number of quarter-wavelengths of the laser beam, the light amplitudes add, and an output is obtained from the detector. When the distance (d) is N times an odd number of quarter-wavelengths, the beam amplitudes subtract, and the output from the detector is zero, as shown in Figure 11.4(b). The movement of the object generates interference fringes between the incident light and the reflected light. These fringes can be counted to give the distance the object moves. The wavelength of the light generated by a laser is about 5×10^{-7}m, so that relative positioning to one-quarter wavelength ($0.125\,\mu$m) over a distance of 50 cm to 1m is achievable.

Ultrasonic, Infrared, Laser, and Microwave Devices can be used for distance measurement. The time for a pulse of energy to travel to an object and be reflected back to a receiver is measured, from which the distance can be calculated. The speed

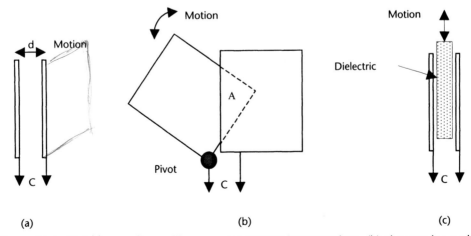

Figure 11.3 Variable capacitors, with varying (a) distance between plates, (b) plate overlap, and (c) dielectric.

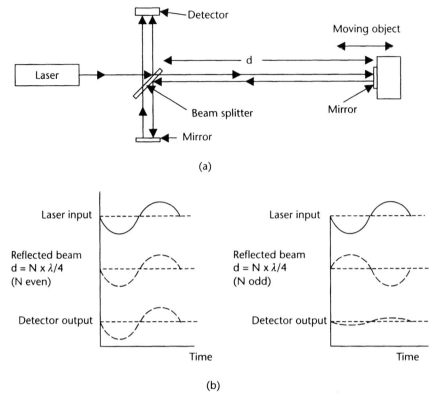

Figure 11.4 Laser (a) setup to measure distance, and (b) associated waveforms.

of ultrasonic waves is 340 m/s, and the speed of light and microwaves is 3×10^8 m/s. Ultrasonic waves can be used to measure distances from 1m to approximately 50m, while light and microwaves are used to measure larger distances.

If an object is in motion, the Doppler effect can be used to determine its speed. The Doppler effect is the change in frequency of the reflected waves caused by the motion of the object. The difference in frequency between the transmitted signal and the reflected signal can be used to calculate the velocity of the object.

Magnetic sensors that use either the Hall effect or magnetoresistive devices are commercially available. Other devices, such as the magnetotransistor, are also commercially available. These devices were discussed in Section 6.2.5.

These devices are used as proximity detectors in applications where ferrous material is used, such as the detection of the rotation of a toothed ferrous wheel. The device is placed between a small magnet and the toothed wheel. As the teeth move past the magnetic field sensor, the strength of the magnetic field is greatly enhanced, giving an output signal. The device can be used to measure linear, as well as rotational, position or speed. Magnetic devices are often preferred in dirty environments rather than optical devices, which cease to work due to dirt covering the lenses. Magnetic devices are also used in applications where opaque materials are used, such as the walls of plastic pipes.

Optical devices are used in On/Off applications to detect motion and position by sensing the presence or absence of light. A light sensor, such as a photodiode or transistor, is used to detect light from a source, such as a light emitting diode (LED).

As an example, Figure 11.5 shows the use of optical sensors to detect the position of an empty container, so that the conveyor can be stopped, and the feed for filling the container can be started. Shown also is the use of optical sensing to detect when can is full, to turn off the feed, and to restart the conveyor belt for the next container.

Accelerometers sense speed changes by measuring the force produced by the change in the velocity of a known mass (seismic mass). See (11.1). These devices can be made with a cantilevered mass and a strain gauge for force measurement, or can use capacitive measurement techniques. Accelerometers made using micromachining techniques are now commercially available [1]. The devices can be as small as 500×500 μm, so that the effective loading by the accelerometer on a measurement is very small. The device is a small cantilevered seismic mass that uses capacitive changes to monitor the position of the mass, as shown in Chapter 6, Figure 6.15. Piezoelectric devices, similar to the one shown in Figure 11.6(a), also are used to measure acceleration. The piezoelectric effect is discussed in Chapter 6. The seismic mass produces a force on the piezoelectric element during acceleration, which causes the lattice structure to be strained [2]. The strain produces an electric charge on the edges of the crystal, as shown in Figure 11.6(b). An amplifier can be integrated into the package to buffer the high output impedance of the crystal. Accelerometers are used in industry for the measurement of changes in velocity of moving equipment, in the automotive industry as crash sensors for air bag deployment, and in shipping crates to measure shock during the shipment of expensive and fragile equipment.

Vibration is a measure of the periodic motion about a fixed reference point, or the shaking that can occur in a process, due to sudden pressure changes, shock, or unbalanced loading in rotational equipment. Peak accelerations of 100g can occur during vibrations, which can lead to fracture or self-destruction. Vibration sensors are used to monitor the bearings in heavy rollers, such as those used in rolling mills. Excessive vibration indicates failure in the bearings or damage to rotating parts, which can be replaced before serious damage occurs.

Vibration sensors typically use acceleration devices to measure vibration. Micromachined accelerometers make good vibration sensors for frequencies up to approximately 1 kHz. Piezoelectric devices make good vibration sensors, with an

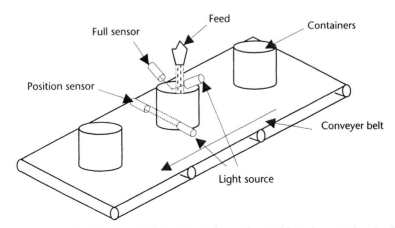

Figure 11.5 Conveyor belt with optical sensors to detect the position of a container for filling, and to detect when full.

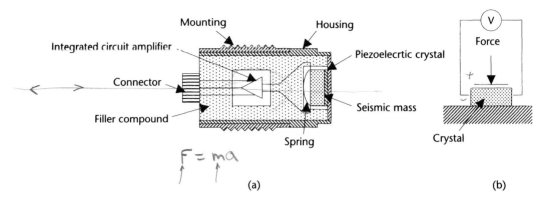

Figure 11.6 (a) Piezoelectric accelerometer. (b) Electric charge produced on the edges of a crystal, due to the piezoelectric effect.

excellent high frequency response for frequencies up to 100 kHz. These devices have very low mass, so the damping effect is minimal.

11.2.2 Position Application Considerations

Hall or MRE devices are replacing optical position sensors in dirty or environmentally unfriendly applications, since optical position sensors require clean operating conditions. These devices are small, sealed, and rugged, with very high longevity, and will operate correctly in fluids, in a dirty environment, or in contaminated areas, for both rotational and linear applications.

Optical devices can be used for reading bar codes on containers, and for imaging. Sensors in remote locations can be powered by solar cells, which fall into the light sensor category.

The *LVDT transducer application*, shown in Figure 11.7, gives a method of converting the linear motion output from a bellows into an electrical signal using an LVDT. The bellows converts the differential pressure between P_1 and P_2 into linear motion, which changes the position of the core in the LVDT. The device can be used as a gauge sensor when P_2 is open to the atmosphere.

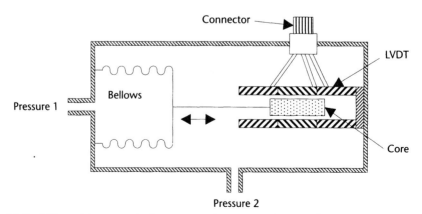

Figure 11.7 Differential pressure bellows converting pressure into an electrical signal using an LVDT.

11.3 Force, Torque, and Load Cells

Figure 11.8 shows an incremental optical shaft encoder, which is an example of an *optical sensor application.* Light from the LED shines through windows in the disk onto an array of photodiodes. As the shaft turns, the position of the image moves along the array of diodes. At the end of the array, the image of the next slot is at the start of the array. The position of the wheel with respect to its previous location can be obtained by counting the number of photodiodes traversed, and multiplying them by the number of slots monitored. The diode array enhances the accuracy of the position of the slots. The resolution of the sensor is 360°, divided by the number of slots in the disk, divided by the number of diodes in the array. Reflective strips also can replace the slots, in which case the light from the LED is reflected back to a photodiode array.

Only one slot in the disk would be required to measure revolutions per minute. Absolute position encoders are made using patterned disks. An array of LEDs with a corresponding photodetector for each LED can give the position of the wheel at any time. Disk encoders are typically designed with the binary code, or Gray code pattern [3].

Optical devices have many uses in industry, other than for the measurement of the position and speed of rotating equipment. Optical devices are used for counting objects on conveyor belts on a production line, measuring and controlling the speed of a conveyor belt [4], locating the position of objects on a conveyor [5], locating registration marks for alignment, reading bar codes, measuring and controlling thickness, and detecting breaks in filaments.

Power lasers also can be included with optical devices, since they are used for scribing and machining of materials, such as metals and laminates.

11.3 Force, Torque, and Load Cells

Many applications in industry require the measurement of force or load. Force is a vector and acts in a straight line. The force can be through the center of a mass, or be offset from the center of the mass to produce a torque. Two parallel forces acting to produce rotation are termed a couple. In other applications where a load or weight is required to be measured, the sensor can be a load cell, using devices such as strain gauges.

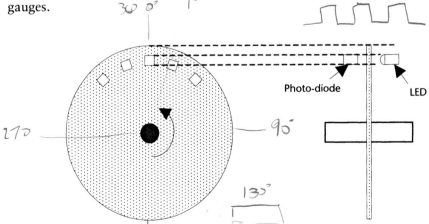

Figure 11.8 Incremental optical disk shaft encoder.

11.3.1 Force and Torque Introduction

Mass is a measure of the quantity of material in a given volume of an object. *Force* (F) is a term that relates the mass (m) of an object to its acceleration (a), and is given by:

$$F = m \times a \tag{11.1}$$

Example 11.1

What force is required to accelerate a mass of 75 kg at 8.7 m/s²?

$$F = (75 \times 8.7) \text{N} = 652.5 \text{N}$$

(kg m/s²)

Weight (w) of an object is the force on a mass due to the pull of gravity (g), which gives:

$$w = m \times g \tag{11.2}$$

Example 11.2

What is the mass of a block of metal that weighs 17N?

$$m = 17/9.8 = 1.7 \text{ kg}$$

(g = 9.81 m/s²)

Torque (t) occurs when a force acting on a body tends to cause the body to rotate, and is given by:

$$t = F \times d \tag{11.3}$$

where d is the perpendicular distance from the line of the force to the fulcrum.
A *Couple* (c) occurs when two parallel forces of equal amplitude, but in opposite directions, are acting on an object to cause rotation, and is given by:

$$c = F \times d \tag{11.4}$$

where d is the perpendicular distance between the forces.

11.3.2 Stress and Strain

Stress is the force acting on a unit area of a solid. The three types of stress most commonly encountered are tensile, compressive, and shear. See Figure 11.9.

Strain is the change or deformation in shape of a solid resulting from the applied force.

Tensile forces are forces that are trying to elongate a material, as shown in Figure 11.9(a). In this case, tensile stress (σt) is given by:

$$\sigma t = F/A \tag{11.5}$$

$$\sigma_t = F/A$$

where σt is in lb/in² (N/m²), F is in lb (N), and A is in² (m²).

The tensile strain (εt) in this case is given by:

11.3 Force, Torque, and Load Cells

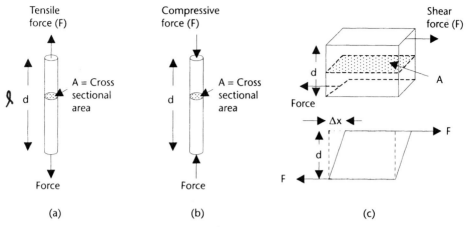

Figure 11.9 Types of stress forces: (a) tensile, (b) compressive, and (c) shear.

$$\varepsilon t = \Delta d/d \tag{11.6}$$

where Δd is the increase in length of the material, and d is the original length of the material. Strain is dimensionless.

Compressive forces are similar to tensile forces, but in the opposite direction, as shown in Figure 11.9(b). Compressive stress (σc) is given by:

$$\sigma c = F/A \tag{11.7}$$

where σc is in lb/in² (N/m²), F is in lb (N), and A is in² (m²).
The compressive strain (εc) in this case is given by:

$$\varepsilon c = \Delta d/d \tag{11.8}$$

where Δd is the decrease in length of the material, and d is the original length of the material. Strain is dimensionless, and is usually a very small number that is often expressed in micros (μ). For instance, a strain of 0.0004 would be expressed as 400 micros or 400 μm/m.

Shear forces are forces that are acting as a couple or tending to shear a material, as shown in Figure 11.9(c). In this case, shear stress (σs) is given by:

$$\sigma s = F/A \tag{11.9}$$

where σs is in lb/in² (N/m²), F is in lb (N), and A is in² (m²).
The shear strain (εs) in this case is given by:

$$\varepsilon s = \Delta x/d \tag{11.10}$$

where Δx is the bending of the material, and d is the original height of the material. Strain is dimensionless.

The relationship between stress and strain for most materials is linear within the *elastic limit* of the material (*Hooke's Law*) [6]. Beyond this limit, the material will have a permanent deformation or change in its shape and will not recover. The ratio of stress divided by strain (within the elastic limit) is called the *modulus of elasticity* or *Young's Modulus* (E). In the case of tensile and compressive stress, the modulus of elasticity is given by:

$$E = \frac{\sigma}{\varepsilon} = \frac{Fd}{A\Delta d} \qquad (11.11)$$

where E is in psi (kPa).

In the case of shear forces, the modulus of elasticity is given by:

$$E = \frac{\sigma}{\varepsilon} = \frac{Fd}{A\Delta x} \qquad (11.12)$$

The Young's Modulus of some common materials is given in Table 11.1. To convert Young's Modulus from SI units to English units, use the relationship 1 kPa (N/m^2) = 0.145 psi.

Gauge factor (G) is a ratio of the change in length to the change in area of a material under stress. For instance, a rod under tension lengthens, and this increase in length is accompanied a decrease in area (A) [7]. The effect of the change in d and A defines the sensitivity of the gauge for measuring strain, and the changes can be applied to any measurand. The gauge factor is given by:

$$G = \frac{\Delta M}{M_0 \varepsilon} \qquad (11.13)$$

where M is the measured parameter, and ε is the strain ($\Delta d/d_0$).

Equation (11.13) is the basic strain gauge equation, and can be rewritten as:

$$G = \frac{\Delta M/M_0}{\varepsilon} = \frac{\Delta M d_0}{M_0 \Delta d} \qquad (11.14)$$

Considering the electrical resistance of a metal is a good example of this effect. The gauge factor is given by:

$$G = \frac{\Delta R}{R_0 \varepsilon} \qquad (11.15)$$

Example 11.3

What is the change in resistance of a copper wire when the strain is 5,500 microstrains Assume the initial resistance of the wire is 275Ω, and the gauge factor is 2.7.

Table 11.1 Modulus of Elasticity or Some Common Materials

Material	Modulus (N/m^2)
Aluminum	6.89 × 10^{10}
Copper	11.73 × 10^{10}
Steel	20.7 × 10^{10}
Polyethylene	3.45 × 10^8

11.3 Force, Torque, and Load Cells

$$\Delta R = 275 \times 2.7 \times 5{,}500 \times 10^{-6} \Omega = 4.1 \Omega$$

The resistance of a metal is given by;

$$R_0 = \rho d_0 / A_0 \tag{11.16}$$

where R_0 is the resistance in Ω, ρ is the resistivity of the metal, d_0 is the length, and A_0 is the cross-sectional area.

For a metal under stress, the value of R_0 (ΔR) increases not only due to the increase Δd in d_0, but also due to the decrease ΔA in A_0. Assuming the volume remains constant,

$$V = d_0 A_0 = (d_0 + \Delta d)(A_0 - \Delta A) \tag{11.17}$$

from which we get

$$R_0 + \Delta R = \frac{\rho(d_0 + \Delta d)}{A_0 - \Delta A} \tag{11.18}$$

this gives the approximation

$$R_0 + \Delta R \approx \frac{\rho d_0}{A_0}\left(1 + 2\frac{\Delta d}{d_0}\right) \tag{11.19}$$

From which the change in resistance is given by:

$$\Delta R \approx 2 R_0 \frac{\Delta d}{d_0} \tag{11.20}$$

Combining (11.15) and (11.20), we get:

$$G \approx 2 \tag{11.21}$$

This shows that, if the volume of a metal remained constant under stress, then G would be 2. However, due to impurities in the metal, G can vary from 1.8 to 5.0 for some alloys, and carbon devices can have a G as high as 10. Semiconductor devices are now being used in strain gauges, and the piezoresistive effect in silicon causes the resistivity of the resistors to change with strain, giving a gauge factor as high as 200. Semiconductor devices are very small, can have a wide resistance range, do not suffer from fatigue, can have a positive or negative gauge factor, and have a temperature range from $-50°$ to $+150°C$. The linearity is poor, but variations within $\pm 1\%$ can be achieved in an integrated device.

11.3.3 Force and Torque Measuring Devices

Force and weight can be measured by comparison, as in a lever-type balance, which is an On/Off system. A spring balance or load cell can be used to generate an electrical signal, which is required in most industrial applications.

An *analytical* or *lever balance* is a device that is simple and accurate, and operates on the principle of torque comparison. When in balance, the torque on one side of the fulcrum is equal to the torque on the other side of the fulcrum, from which we get:

$$W_1 \times L = W_2 \times R \tag{11.22}$$

where W_1 is a weight at a distance L from the fulcrum, and W_2 the counterbalancing weight at a distance R from the fulcrum.

Example 11.4

If 8.3 kg of oranges are being weighed with a balance, the counterweight on the balance is 1.2 kg, and the length of the balance arm from the oranges to the fulcrum is 51 cm, then how far from the fulcrum must the counterbalance be placed?

$$8.3 \text{ kg} \times 51 \text{ cm} = 1.2 \text{ kg} \times d \text{ cm}$$

$$d = 8.3 \times 51/1.2 = 353 \text{ cm} = 3.53 \text{m}$$

A *spring transducer* is a device that measures weight by measuring the deflection of a spring when a weight is applied, as shown in Figure 11.10(a). The deflection of the spring is proportional to the weight applied (provided the spring is not stressed), according to the following equation:

$$F = Kd \tag{11.23}$$

where F is the force in lb (N), K is the spring constant in lb/in (N/m), and d is the spring deflection in inches (m).

Example 11.5

When a bag of apples is placed on a spring balance with an elongation constant of 9.8 kg/cm, the spring stretches 4.7 cm. What is the weight of the container in pounds?

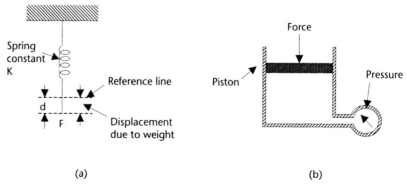

Figure 11.10 (a) Spring balance used as a force measuring device, and (b) using pressure to measure force.

11.3 Force, Torque, and Load Cells

$W = 9.8 \text{ kg/cm} \times 4.7 \text{ cm} = 46.1 \text{ kg } (451\text{N})$

$= 46.1 \text{ kg} \div 0.454 \text{ kg/lb} = 101.5 \text{ lb}$

Hydraulic and *pneumatic devices* can be used to measure force. Monitoring the pressure in a cylinder when the force is applied to a piston, as shown in Figure 11.10(b), can do this. The relation between force (F) and pressure (p) is given by:

$$F = pA \tag{11.24}$$

where A is the area of the piston.

Example 11.6

What will the pressure gauge read, if the force acting on a 23.2-cm diameter piston is 203N?

$$P = \frac{203 \times 4}{0.232 \times 0.232 \times 3.14} = 4.8 kPa \quad 4.3 \text{ N/m}^2$$

Piezoelectric devices use the piezoelectric effect, which is the coupling between the electrical and mechanical properties of certain materials to measure force. The piezoelectric effect was discussed in Chapter 6. PZT material has high sensitivity when measuring dynamic forces, but are not suitable for static forces.

11.3.4 Strain Gauge Sensors

Strain gauges are resistive sensors, which can be deposited resistors or piezo-electricresistors (Figure 11.11(a)). The resistive conducting path in the deposited gauge is copper or nickel particles deposited onto a flexible substrate in a serpentine form. If the substrate is bent in a concave shape along the axis perpendicular to the direction of the deposited resistor, or if the substrate is compressed in the direction of the resistor, then the particles themselves are forced together and the resistance decreases. If the substrate is bent in a convex shape along this axis, or if the substrate is under tension in the direction of the resistor, then the particles tend to separate and the resistance increases. Bending along an axis parallel to the resistor, compressing, or placing tension in a direction perpendicular to the direction of the resistor does not compress or separate the particles in the strain gauge, so the resistance does not change. Piezoresistor devices are often used as strain gauge elements. The mechanism is completely different from that of the deposited resistor. In this case, the resistance change is due to the change in electron and hole mobility in a crystal structure when strained. These devices can be very small with high sensitivity. Four elements are normally configured as a Wheatstone Bridge and integrated with the conditioning electronics. The resistance change in a strain gauge element is proportional to the degree of bending, compression, or tension. For instance, if the gauge were attached to a metal pillar, as shown in Figure 11.11(b), and a load or compressive force applied to the pillar, then the change in resistance of the strain gauge attached to the pillar is then proportional to the force applied. Because the resistance of the strain gauge element is temperature-sensitive, a reference or dummy

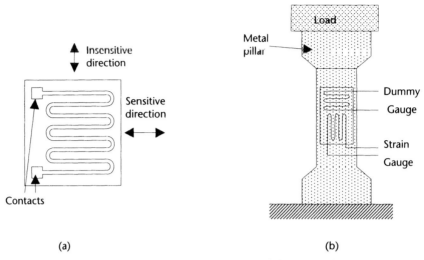

Figure 11.11 Strain gauge: (a) as a serpentine structure, and (b) as a load sensor.

strain gauge element is also added. Thus, when a bridge circuit is used to measure the change in the resistance of the strain gauge, the dummy gauge can be placed in the adjacent arm of the bridge to compensate for the temperature changes. This second strain gauge is positioned adjacent to the first, so that it is at the same temperature. It is rotated 90°, so that it is at right angles to the pressure-sensing strain gauge element, and therefore will not sense the deformation as seen by the pressure-sensing element. This structure is also used in load cells [8].

Figure 11.12 shows an alternative use of a strain gauge for measuring the force applied to a cantilever beam. The force on the beam causes the beam to bend, producing a sheer stress. This would be the type of strain encountered in a diaphragm pressure sensor.

The resistance change in strain gauges is small and requires the use of a bridge circuit for measurement, as shown in Figure 11.13. The strain gauge elements are mounted in two arms of the bridge, and two resistors, R_1 and R_2, form the other two arms. The output signal from the bridge is amplified and impedance matched. The strain gauge elements are in opposing arms of the bridge, so that any change in the resistance of the elements due to temperature changes will not affect the balance of the bridge, giving temperature compensation. More gain and impedance matching

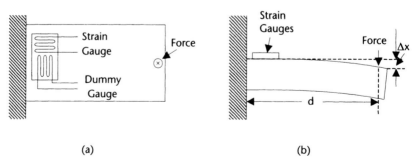

Figure 11.12 Alternative use of a strain gauge for measuring the force applied to a cantilever beam. (a) Top view; (b) side view.

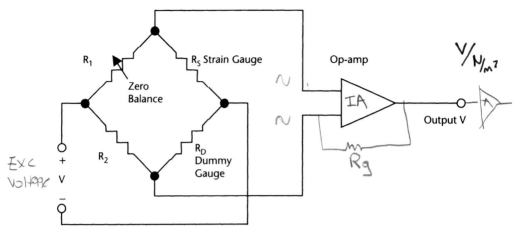

Figure 11.13 (a) Configuration for strain gauge elements, and (b) resistive bridge for signal conditioning of a strain gauge.

stages than what is shown may be required, or an ADC will be required to make the signal suitable for transmission. Additional linearization also may be needed, and this information can be obtained from the manufacturers' device specifications [9].

Example 11.7

A force of 6.5 kN is applied to the end of an aluminum cantilever beam 0.35m from its fixed end. If the beam has a cross-sectional area of 8 cm², what will be the deflection of the end of the beam?

From (11.12):

$$\Delta x = \frac{Fd}{EA} = \frac{6{,}500 \times 0.35}{6.89 \times 10^{10} \times 8 \times 10^{-4}}$$

$$\Delta x = 0.4 \times 10^{-4} \, m$$

Example 11.8

Using the information from Example 11.7, a strain gauge with a gauge factor of 4.3 is attached to the beam. If the resistance of the gauge is 1,200Ω, and the gauge is placed in a bridge circuit with 1,200Ω resistors in the other arms, what would be the change in resistance of the gauge, and the output voltage from the bridge, if the supply to the bridge were 16V?

$$\Delta R = 1{,}200 \times 4.3 \times 0.4 \times 10^{-4}/0.35 = 0.6\Omega$$

$$\Delta V = 16/2 - \frac{16 \times 1{,}200.6}{1{,}200 + 1{,}200.6} = 8 - 8.002V = 2mV$$

The *load cells* previously discussed use strain gauges or piezoelectric devices, but they also can use capacitive or electromagnetic sensing devices [10].

A *dynamometer* is a device that uses the twisting or bending in a shaft due to torque to measure force. One such device is the torque wrench used to tighten bolts to a set torque level, which can be required in some valve housings to prevent seal leakage or valve distortion. The allowable torque for correct assembly will be given in the manufacturer's specification. The twist in a shaft from a motor can be used to measure the torque output from a drive motor [11].

11.3.5 Force and Torque Application Considerations

In most applications, compensation must be made for temperature effects, which can be larger than the measurement signal. Electrical transducers can be compensated by using them in a bridge circuit with a compensating device in the adjacent arm of the bridge. Changes in material characteristics due to temperature changes also can be compensated using temperature sensors and applying a correction factor to the measurement. This method is difficult with the low-level signals obtained from strain gauges. Vibration also can be a problem when measuring force, but this usually can be corrected by damping the movement of the measuring system. The choice of measuring device will depend upon the application. A far more sensitive strain gauge is the semiconductor strain gauge using the piezoresistive effect. These devices have gauge factors up to 90 times greater than those of the deposited resistive gauges, are very small (0.5×0.3 mm), have wide resistance ranges, and unlike deposited resistive devices, do not suffer from fatigue.

11.4 Light

11.4.1 Light Introduction

Light and its measurement is important as it also relates to the sense of sight, as well as to many industrial applications, such as location, proximity, and linear distance measurement. Light and its measurement is used in many industrial applications for high accuracy linear measurements, location of overheating, temperature measurement (e.g., infrared), object location and position measurements, photoprocessing, scanning, readers (e.g., bar codes), and so forth [12].

11.4.2 EM Radiation

Light is an ultrahigh frequency EM wave that travels at 2.998×10^8 m/s. Light amplitude is measured in foot-candles or lux. The wavelength of visible light is from 4×10^{-7} m to 7×10^{-7} m. Longer wavelengths of EM waves are termed infrared, and

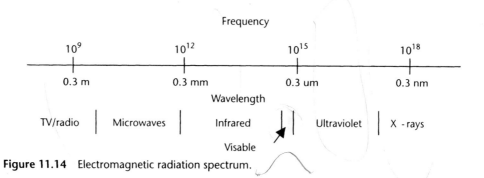

Figure 11.14 Electromagnetic radiation spectrum.

shorter wavelengths are called ultraviolet. Light wavelengths are sometimes expressed in terms of angstroms (Å), where $1\text{Å} = 1 \times 10^{-10}$ m. The electromagnetic radiation spectrum is shown in Figure 11.14. The relation between the frequency (f) and the wavelength (λ) of electromagnetic waves in free space is given by:

$$c = f\lambda \tag{11.25}$$

where c is the speed of light.

When traveling through any medium, the propagation velocity of EM waves is reduced by the refractive index of the material. The refractive index (n) is defined as:

$$n = c/v \tag{11.26}$$

where v is the velocity of EM radiation through the material.

Example 11.9

What is the wavelength of light in Å, if the wavelength is 493 nm?

$$1\text{Å} = 10^{-10}\text{m}$$

$$493 \text{ nm} = 493 \times 10^{-9} \div 10^{-10} \text{Å} = 4{,}930\text{Å}$$

Intensity is the brightness of light. The unit of measurement of light intensity in the English system is the foot-candle (fc), which is one lumen per square foot (lm/ft^2). In the SI system, the unit is the lux (lx), which is one lumen per square meter (lm/m^2). The phot (ph) is also used, which is defined as one lumen per square centimeter (lm/cm^2). The lumen replaces the candela (cd) in the SI system. A 1 cd or 1 lm source is defined as a source that emits radiation at such an intensity that there is 1/683 W passing through 1 sr of solid angle by monochromatic light with a frequency of 340×10^{12} Hz. Because a point source emits radiation in all directions, the intensity (I) on the surface of a sphere radius (R) is given by:

$$I = \frac{P}{4\pi R^2} \tag{11.27}$$

where P is the power of the source in watts.

This also shows that the intensity of a point source decreases as the inverse square of the distance from the source.

The *decibel* (dB) is a logarithmic measure used to measure and compare amplitudes and power levels in electrical units, sound, and light. When used for the comparison of light intensity, the equation is as follows:

$$\text{Light level ratio in } dB = 10\log_{10}\left(\frac{\Phi_1}{\Phi_2}\right) \tag{11.28}$$

where Φ_1 and Φ_2 are the light intensities at two different points.

The change in intensity levels at varying distances from a source is given by:

$$\text{Change in levels} = 10\log_{10}\left(\frac{d_1}{d_2}\right) \qquad (11.29)$$

where d_1 and d_2 are the distances from the source to the points being considered.

Example 11.10

Two points are 43 and 107 ft from a light bulb. What is the difference in light intensity at the two points?

$$\text{Difference} = 10\log_{10}\left(\frac{43}{107}\right) = -3.96 dB$$

X-rays should be mentioned at this point, since they are used in the process industry, and are electromagnetic waves. X-rays are used primarily as inspection tools. The X-rays can be sensed by some light sensing cells, and can be very hazardous if proper precautions are not taken.

11.4.3 Light Measuring Devices

Photocells are used for the detection and conversion of light intensity into electrical signals. Photocells can be classified as photovoltaic, photoconductive, photoemissive, and semiconductor.

Photovoltaic cells develop an EMF in the presence of light. Copper oxide and selenium are examples of photovoltaic materials. A microammeter calibrated in lux is connected across the cells and measures the current output.

Photoconductive devices change their resistance with light intensity. Examples of these materials are selenium, zirconium oxide, aluminum oxide, and cadmium sulfide.

Photoemissive materials, such as mixtures of rare Earth elements (e.g., cesium oxide), liberate electrons in the presence of light.

Semiconductors are photosensitive, and are commercially available as photodiodes and phototransistors. Light generates hole-electron pairs, which cause leakage in reversed biased diodes, and base current in phototransistors. Commercial high resolution optical sensors are available with the electronics integrated onto a single die to give temperature compensation, and a linear voltage output with incident light intensity. Visible light intensity to voltage converters (TSL 250), Infrared (IR) light to voltage converters (TSL 260), light to frequency converters (TSL 230), and light intensity to digital converters are commercially available. Note that Texas Instruments manufactures the TSL family.

11.4.4 Light Sources

Incandescent light is produced by electrically heating a resistive filament, or by the burning of certain combustible materials. A large portion of the energy emitted is in the infrared spectrum, as well as the visible spectrum.

Atomic type sources cover gas discharge devices, such as neon and fluorescent lights.

Laser emissions are obtained by excitation of the atoms of certain elements.

Semiconductor diodes, such as LEDs, are the most common commercially available light sources used in industry. When forward biased, the diodes emit light in the visible or IR region. Certain semiconductor diodes emit a narrow wavelength of visible rays, where the color is determined by material and doping. A list of LEDs and their color is given in Table 11.2.

11.4.5 Light Application Considerations

Selection of sensors for the measurement of light intensity will depend upon the application. In instrumentation, requirements include: a uniform sensitivity over a wide frequency range, low inherent noise levels, consistent sensitivity with life, and a means of screening out unwanted noise and light.

In some applications, such as the sensing of an optical disk, it is only necessary to detect absence or presence of a signal, which enables the use of simple and inexpensive sensors. For light detection, the phototransistor is widely used, because of the ability with integrated circuits to put temperature correction and amplification in the same package, for high sensitivity. The device is cost-effective and has good longevity.

Figure 11.15 shows the schematic symbols used for optoelectronic sensors, and Table 11.3 gives a comparison of photosensor characteristics.

Table 11.2 LED Characteristics

Material	Dopant	Wavelength (nm)	Color
GaAs	Zn	900	IR
GaP	Zn	700	Red
GaAsP	N	632	Orange
GaAsP	N	589	Yellow
GaP	N	570	Green
SiC	—	490	Blue

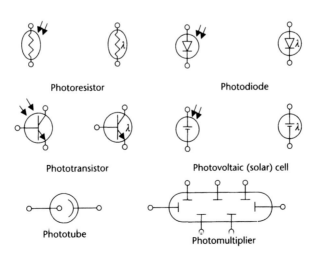

Figure 11.15 Schematic symbols for optosensors.

Table 11.3 Summary of Optosensor Characteristics

Type	Device	Response (μm)	Response Time	Advantages	Disadvantages
Photoconductive	Photoresistor CdS	0.6–0.9	100 ms	Small, high sensitivity, low cost	Slow, hysteresis, limited temperature
	Photoresistor CdSe	0.6–0.9	10 ms	Small, high sensitivity, low cost	Slow, hysteresis, limited temperature
Semiconductor	Photodiode	0.4–0.9	1 ns	Very fast, good linearity, low noise	Low-level output
	Phototransistor	0.25–1.1	1 μs		Low frequency response, nonlinear
Photovoltaic	Solar cell	0.35–0.75	20 μs	Linear, self-powered	Slow, low-level output

11.5 Summary

A number of important sensors used in process control, for applications other than for fluid characteristics, were discussed in this chapter. Sensors for measuring linear and rotational position, speed, and acceleration were introduced. These devices have many applications in process control. They can be used as transducers, to convert linear motion into electrical signals, using potentiometers, magnetic coupling, or capacitive changes. Other devices are used for absolute or incremental position control and measurement, velocity, vibration, and acceleration. Optical sensors and magnetic devices are used as proximity detectors for positioning, and the reading of bar codes and magnetic strips.

The measurement of force, torque, and load measurements are important in process control. Stress, strain, and Young's Modulus play an important part in the measurement of loads and torque. The sensitivity of strain devices and load cells is enhanced by the gauge factor, but is much higher using piezoelectric elements. Extra elements are incorporated in strain gauges and load cells to compensate for temperature effects, when used in a bridge circuit.

Light can be produced from several sources. Many materials are light sensitive, and can be used to measure light intensity. Semiconductor devices are produced to emit specific frequencies or different colors of light, and integrated semiconductor devices with temperature compensation and conditioning also can be used to convert light into voltage, frequency, or digital signals.

Definitions

Absolute position is the distance measured with respect to a fixed reference point, and can be measured whenever power is applied.

Acceleration is the rate of change of speed, for linear motion or for rotational motion.

Angular motion is a measure of the rate of rotation. Angular velocity is a measure of the rate of rotation when rotating at a constant speed about a fixed point, and angular acceleration is measured when the rotational speed is changing.

Angular position   is a measurement of the change in position of a point about a fixed axis, measured in degrees or radians.

Arc-minute   is an angular displacement of 1/60 of a degree.

Couple   occurs when two parallel forces of equal amplitude, but in opposite directions, are acting on an object to cause rotation.

Force   is a term that relates the mass of an object to its acceleration.

Incremental position   is a measure of the change in position, and is not referenced to a fixed point. If power is interrupted, the incremental position is lost.

Mass   is a measure of the quantity of material in a given volume of an object.

Rectilinear motion   is measured by the distance traversed in a given time, velocity when moving at a constant speed, or acceleration when moving is changing in a straight line.

Torque   occurs when a force that is acting on a body tends to cause the body to rotate.

Velocity or Speed   is the rate of change of position. This can be a linear measurement or an angular measurement.

Vibration   is a measure of the periodic motion about a fixed reference point.

Weight   of an object is the force on a mass due to the pull of gravity.

References

[1] Clifford, M. A., "Accelerometers Jump into the Consumer Goods Market," *Sensors Magazine*, Vol. 21, No. 8, August 2004.

[2] Kulwanosli, G., and J. Schnellinger, "The Principles of Piezoelectric Accelerometers," *Sensors Magazine*, Vol. 21, No. 2, February 2004.

[3] Hoffman, F.J., "A New Dimension in Encoder Technology," *Sensors Magazine*, Vol. 19, No. 5, May 2002.

[4] Salt, H., "A New Linear Optical Encoder," *Sensors Magazine*, Vol. 16, No. 11, November 1999.

[5] Massa, D. P., "Choosing an Ultrasonic Sensor for Proximity or Distance Measurement," *Sensors Magazine*, Vol. 16, No. 2, February 1999.

[6] Young, W. C., *Roark's Formulas for Stress and Strain*, 6th ed., McGraw-Hill.

[7] Shigley, J., *Mechanical Engineering Design*, McGraw-Hill, 1963, pp. 284–289.

[8] Nagy, M. L., C. Spanius, and J. W. Siekkinen, "A User Friendly High Sensitivity Strain Gauge," *Sensors Magazine*, Vol. 18, No. 6, June 2001.

[9] Emery, J.C., "Simplifying the Electrolytic Balance Load Cell," *Sensors Magazine*, Vol. 19, No. 6, June 2002.

[10] Bruns, R. W., "The Helix Load Cell," *Sensors Magazine*, Vol. 15, No. 5, May 1998.

[11] Andreescu, R., M. Gupta, and B. Speelman, "A Magnetostrictive Torque Sensor," *Sensors Magazine*, Vol. 21, No. 11, November 2004.

[12] Johnson, C. D., *Process Control Instrumentation Technology*, 7th ed., Prentice Hall, 2003, pp. 273–315.

Chapter 12
Humidity and Other Sensors

There are many sensors, other than level, pressure, flow, and temperature sensors, which may not be encountered on a daily basis, but still play an equally important part in process control in high technology industries and in operator protection. This chapter discusses several sensors that are very important in modern processing. They are:

1. Humidity;
2. Density, specific weight, and specific gravity;
3. Viscosity;
4. Sound;
5. pH;
6. Chemical.

12.1 Humidity

It is necessary to control the amount of water vapor present in many industrial processes. Textile, wood, and chemical processing is very sensitive to humidity.

12.1.1 Humidity Introduction

Humidity is a measure of the relative amount of water vapor present in the air or a gas.

Relative humidity (Φ) is given by:

$$\% \text{ Relative humidity} = \frac{\text{Amount of water vapor present in a given volume of air or gas}}{\text{Maximum amount of water vapor soluble in the same volume of air or gas } (P \text{ and } T \text{ constant})} \times 100 \quad (12.1)$$

where P is the pressure and T is the temperature. [Isobaric / Isothermal]

An alternative definition using vapor pressures is:

$$\% \text{ Relative humidity} = \frac{\text{water vapor pressure in air or gas}}{\text{water vapor pressure in saturated air or gas} (T \text{ constant})} \times 100 \quad (12.2)$$

The term *saturated* means the maximum amount of water vapor that can be dissolved or absorbed by a gas or air at a given pressure and temperature.

Specific humidity, *humidity ratio*, or *absolute humidity* can be defined as the mass of water vapor in a mixture in grains (where 7,000 gr = 1 lb), divided by the mass of dry air or gas in the mixture. The measurement units could also be defined as a ratio (pounds of water vapor per pound of dry air), or be defined in SI units (grams of vapor per kilogram of dry air).

$$\text{Humidity ratio} = \frac{\text{mass of water vapor in a mixture}}{\text{mass of dry air or gas in the mixture}} \tag{12.3}$$

$$= \frac{\text{mass(water vapor)}}{\text{mass(air or gas)}} = \frac{0.622\, P(\text{water vapor})}{P(\text{mixture}) - P(\text{water vapor})} \tag{12.4}$$

$$= \frac{0.622\, P(\text{water vapor})}{P(\text{air or gas})} \tag{12.5}$$

where P (water vapor) is pressure, and P (air or gas) is a partial pressure. The number 0.622 is a conversion factor between mass and pressure.

Example 12.1

Examples of water vapor in the atmosphere are as follows:

- Dark storm clouds (cumulonimbus) can contain 10 g/m^3 of water vapor.
- Medium density clouds (cumulus congestus) can contain 0.8 g/m^3 of water vapor.
- Light rain clouds (cumulus) can contain 0.2 g/m^3 of water vapor.
- Wispy clouds (cirrus) can contain 0.1 g/m^3 of water vapor.

In the case of the dark storm clouds, this equates to 100,000 tons of water vapor per square mile for a 10,000-ft-tall cloud.

12.1.2 Humidity Measuring Devices

Humidity can be measured using materials that absorb water vapor, giving a change in their characteristics that can be measured; by measuring the latent heat of evaporation; by dew point measurement; or by microwave absorption.

12.1.2.1 Psychrometers

A psychrometer uses the latent heat of vaporization to determine relative humidity. If the temperature of air is measured with a dry bulb thermometer and a wet bulb thermometer, then the two temperatures can be used with a psychrometric chart to obtain the relative humidity, water vapor pressure, heat content, and weight of water vapor in the air. Water evaporates from the wet bulb, trying to saturate the surrounding air. The energy needed for the water to evaporate cools the thermometer, so that the drier the day, the more water evaporates and the lower the temperature of the wet bulb.

To prevent the air surrounding the wet bulb from saturating, there should be some air movement around the wet bulb. This can be achieved with a small fan, or by using a sling psychrometer, which is a frame holding both the dry and wet thermometers that can rotate about a handle. The thermometers are rotated for 15 to 20 seconds. The wet bulb temperature is taken as soon as rotation stops, before it can change, and then the dry bulb temperature (which does not change) is taken. Figure 12.1 shows two temperature sensors, one dry and one in a moist wick (the end of the wick is dipped in a water reservoir). Air is moved over the sensors with a small fan. This setup gives a continuous electrical signal that can be correlated to humidity.

12.1.2.2 Hygrometers

Devices that indirectly measure humidity by sensing changes in physical or electrical properties in materials due to their moisture content are called hygrometers. Some materials, such as hair, skin, membranes, and thin strips of wood, change their length as they absorb water. The change in length is directly related to the humidity. Such devices are used to measure relative humidity from 20% to 90%, giving an accuracy of approximately ±5%. Their operating temperature range is limited to less than 70°C. Other devices use hydroscopic materials that change their electrical properties with humidity [1].

A *Laminate hygrometer* is made by attaching thin strips of wood to thin metal strips, forming a laminate. The laminate is formed into a spiral, as shown in Figure 12.2(a), or a helix. As the humidity changes, the spiral flexes, due to the change in the length of the wood. One end of the spiral is anchored, and the other is attached to a pointer (similar to a bimetallic strip used in temperature measurements), and the scale is graduated in percentage of humidity.

The *hair hygrometer* is the simplest and oldest hygrometer, and is made using hair, as shown in Figure 12.2(b). Human hair lengthens by 3% when the humidity changes from 0% to 100%. The change in length can be used to control a pointer for visual readings, or to control a transducer, such as an LVDT, for an electrical output. The hair hygrometer has an accuracy of approximately 5% for the humidity range 20% to 90%, over the temperature range −15° to 70°C.

Resistive hygrometer humidity sensors use the change in resistance of a hydroscopic material between two electrodes on an insulating substrate, as shown

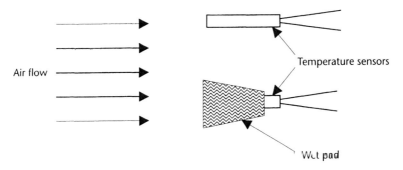

Figure 12.1 Wet and dry bulb psychrometer.

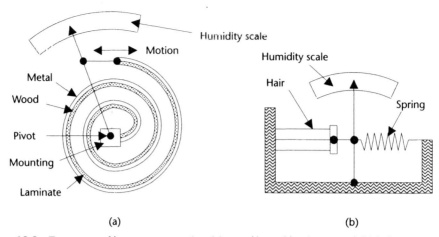

Figure 12.2 Two types of hygrometers, using (a) metal/wood laminate, and (b) hair.

in Figure 12.3(a). The electrodes are coated with a hydroscopic material (i.e., one that absorbs water, such as lithium chloride) [2]. The hydroscopic material provides a conductive path between the electrodes, and the coefficient of resistance of the path is inversely proportional to humidity. Alternatively, the electrodes can be coated with a bulk polymer film that releases ions in proportion to the relative humidity. Temperature correction can be applied again for an accuracy of 2% over the operating temperature range −40° to 70°C, and relative humidity range 2% to 98%. An ac voltage is normally used with this type of device. At 1 kHz, a relative humidity change from 2% to 98% typically will give a resistance change from 10 MΩ to 1 kΩ. Variations of this device are the *electrolytic*, and the *resistance-capacitance hygrometer*.

A *capacitive hygrometer* uses the change of the dielectric constant of certain thin polymer films that change linearly with humidity, so that the capacitance between two plates using the polymer as the dielectric is directly proportional to humidity. The device is shown in Figure 12.3(b). The capacitive device has good longevity, a working temperature range of 0° to 100°C, and a fast response time, and it can be temperature compensated to give an accuracy of ±0.5% over the full humidity range.

Piezoelectric or *sorption hygrometers* use two piezoelectric crystal oscillators. One is used as a reference and is enclosed in a dry atmosphere, and the other is

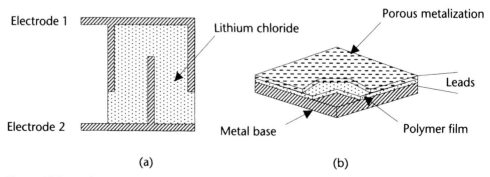

Figure 12.3 Hydrometers: (a) resistive, and (b) capacitive.

exposed to the humidity to be measured. Moisture increases the mass of the crystal, which decreases its resonant frequency. By comparing the frequencies of the two oscillators, the humidity can be calculated. Moisture content of gases from 1 to 25,000 p/m can be measured.

12.1.2.3 Dew Point Measuring Devices

A simple method of measuring the humidity is to obtain the dew point, which is the temperature at which the air becomes saturated with water vapor. Cooling the air or gas until water condenses on an object, and then measuring that temperature, achieves this. Typically, a mirrored surface, polished stainless steel, or silvered surface is cooled from the backside, by cold water, refrigeration, or Peltier cooling. As the temperature drops, a point is reached where dew from the air or gas starts to form on the mirror surface. The condensation is detected by the reflection of a beam of light from the mirror to a photocell. The intensity of the reflected light reduces as condensation takes place, and the temperature of the mirror at that point can be measured [3].

12.1.2.4 Moisture Content Measuring Devices

The moisture content of materials is very important in some processes. There are two methods commonly used to measure the moisture content: by using microwaves, or by measuring the reflectance of the material to infrared rays.

Microwave absorption by water vapor is a method used to measure the humidity in a material. Microwaves (1 to 100 GHz) are absorbed by the water vapor in the material, and the relative amplitudes of the transmitted microwaves and of the microwaves that passed through a material are measured. The ratio of these amplitudes is a measure of the humidity content of the material.

Infrared absorption, an alternative to microwave absorption, uses infrared rays. In the case of infrared absorption, the measurements are based on the ability of materials to absorb and scatter infrared radiation (i.e., reflectance). Reflectance depends on chemical composition and moisture content. An infrared beam is directed onto the material, and the energy of the reflected rays is measured. The measured wavelength and amplitude of the reflected rays are compared to the incident wavelength and amplitude, and the difference between the two is related to the moisture content.

Other methods of measuring moisture content are by observations of: color changes; neutron reflection; nuclear magnetic resonance; or the absorption of moisture by certain chemicals and subsequent measurement of the change in mass.

12.1.3 Humidity Application Considerations

Although wet and dry bulbs formerly were the standard for making relative humidity measurements, more modern and simpler electrical methods, such as capacitance and resistive devices, are now available. These devices are small, rugged, reliable, and accurate with high longevity, and if necessary, can be calibrated by the NIST against accepted gravimetric hygrometer methods. Using these methods, the water vapor in a gas is absorbed by chemicals, which are weighed before and after to determine the amount of water vapor absorbed from a given volume of gas, from which the relative humidity can be calculated. Table 12.1 gives a comparison of humidity sensor characteristics. In a production environment, the device of choice would depend upon its application. For atmospheric humidity control in most facilities, the capacitive or resistive device would be used, but for moisture content in a material, a radiation absorption technique would be used.

12.2 Density and Specific Gravity

12.2.1 Density and Specific Gravity Introduction

The density, specific weight, and specific gravity were defined in Chapter 7. The relation between density ρ and specific weight γ [4] is given by:

$$\gamma = \rho g \qquad (12.6)$$

where g is the acceleration of gravity is 32.2 ft/s² or 9.8 m/s², depending upon the units being used.

Example 12.2

What is the specific weight of a material whose density is 3.45 Mg/m³?

$$\gamma = \rho g = 3.45 \times 9.8 \text{ kN/m}^3 = 33.81 \text{ kN/m}^3$$

Table 12.2 gives a list of the density and specific weight of some common materials.

Table 12.1 Humidity Sensor Characteristics

Type	Humidity Range	Temperature Range	Accuracy
Psychrometer	5% to 95%	0° to 100°C	±5%
Hair hydrometer	20% to 90%	−15° to +70°C	±5%
Resistance hydrometer	2% to 98%	−40° to +70°C	±2%
Resistance-capacitance hydrometer	0% to 100%	0° to 150°C	±2%
Capacitive hydrometer	0% to 100%	0° to 100°C	±0.5%
Dew point	5% to 95%	5° to 95°C	±5%

Table 12.2 Density and Specific Weights

	Specific Weight		Density		SG
	lb/ft³	kN/m³	Slug/ft³	Mg/m³	
Acetone	49.4	7.74	1.53	0.79	0.79
Ammonia	40.9	6.42	1.27	0.655	0.655
Benzene	56.1	8.82	1.75	0.9	0.9
Gasoline	46.82	7.35	3.4	0.75	0.75
Glycerin	78.6	12.4	2.44	1.26	1.26
Mercury	847	133	26.29	13.55	13.55
Water	62.43	9.8	1.94	1.0	1.0

12.2.2 Density Measuring Devices

The density of liquids can be measured by measuring the buoyancy of a known mass, by vibration techniques, by measuring pressure at known depths, or by measuring radiation absorption [5].

Hydrometers are the simplest device for measuring the specific weight or density of a liquid. The device consists of a graduated glass tube with a weight at one end, which causes the device to float in an upright position. The device sinks in a liquid until an equilibrium point between its weight and buoyancy is reached. The specific weight or density then can be read directly from the graduations on the tube.

A *thermohydrometer* is a combination hydrometer and thermometer, so that both the specific weight/density and temperature can be recorded, and the specific weight/density can be corrected from look-up tables for temperature variations to improve the accuracy of the readings.

Induction hydrometers are used to convert the specific weight or density of a liquid into an electrical signal. In this case, a fixed volume of liquid set by the overflow tube is used in the type of setup shown in Figure 12.4(a). The displacement device, or hydrometer, has an attached core made of soft iron or a similar metal. The core is positioned in a coil, which forms part of a bridge circuit. As the density/specific weight of the liquid changes, the buoyant force on the displacement device changes. This movement can be measured by the coil and converted into a density reading.

Vibration sensors are an alternate method of measuring the density of a fluid, as shown in Figure 12.4(b). Fluid is passed through a U-tube that has a flexible mount, which can vibrate when driven from an outside source, such as a piezoelectric vibrator. The frequency of the vibration decreases as the specific weight or density of the fluid increases, so that by measuring the vibration frequency, the specific weight/density can be calculated. The temperature is also monitored, so that density changes with temperature can be taken into account [6]. The sensor can be used at temperatures up to 200°C, and with tube pressures up to 14 MPa(g). The device is accurate and reliable, but is limited to flow rates from 2 to 25 L/min, and is expensive.

Pressure can be used to determine liquid density, if the pressure at the base of a column of liquid of known height (h) can be measured to determine the density and specific gravity of a liquid. The density of the liquid is given by:

$$\rho = \frac{p}{gh} \tag{12.7}$$

The specific weight is given by:

$$\gamma \times \rho = \frac{p}{h} \tag{12.8}$$

Example 12.3

The pressure at the base of a column of liquid is 532 Pa. If the density of the liquid is 2.1 Mg/m^3, what is the height of the column?

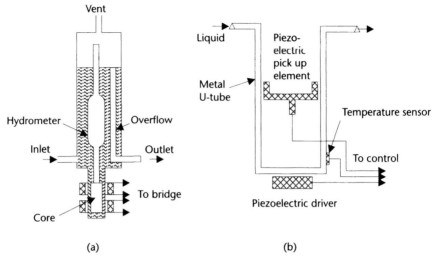

Figure 12.4 Density measuring devices: (a) induction hydrometer, and (b) U-tube liquid density meter.

$$h = \frac{p}{\rho g} = \frac{532}{2.1 \times 9.8} = 25.85 m$$

Differential bubblers can be used to measure liquid density or specific weight, as shown in Figure 12.5. Two air supplies are used to supply two tubes whose ends are at different depths in a liquid. The difference in air pressures between the two air supplies is directly related to the density of the liquid, by the following equation:

$$\rho = \frac{\Delta p}{g \Delta h} \qquad (12.9)$$

where Δp is the difference in the pressures, and Δh the difference in the height of the bottoms of the two tubes.

Example 12.4

What is the density of a liquid in a bubbler system, if pressures of 420 Pa and 38 kPa are measured at depths of 21 cm and 5.4m, respectively?

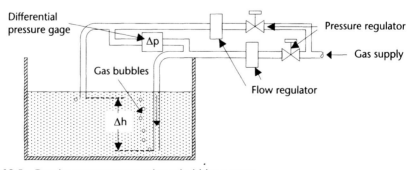

Figure 12.5 Density measurement using a bubbler system.

$$\rho = \frac{38 - \dfrac{420}{1000}}{\left(5.4 - \dfrac{21}{100}\right) \times 9.8} = \frac{38 - 0.42}{(5.4 - 0.21) \times 9.8} = 0.74\,Mg/m^3$$

The *weight* of a known volume of the liquid can be used to determine density (ρ), if a container of known volume can be filled with a liquid, and weighed both full and empty. The difference in weight gives the weight of liquid, from which the density can be calculated using:

$$\rho = \frac{W_f - W_c}{g \times Vol} \qquad (12.10)$$

where W_f is the weight of container plus liquid, W_c is the weight of container, and *Vol* is the volume of the container.

This was discussed in Chapter 8, under load cells; see also Example 8.5.

Radiation density sensors, as shown in Figure 12.6, consist of a clamped-on radiation source located on one side of a pipe or container and a sensing device on the other. The sensor is calibrated with the pipe or container empty, and then with the pipe or contained filled. The density of the liquid causes a difference in the measured radiation, the difference in the measured radiation can then be used to calculate the density of the liquid. The sensor can be used for continuous flow measurement. The device has an accuracy of ±1% of span, with a response time of about 10 seconds. The disadvantages of the device are the 30-min warm-up time, the high cost, and the use of hazardous materials.

Gas densities are normally measured by sensing the frequency of vibration of a vane in the gas, or by weighing a volume of the gas and comparing it to the weight of the same volume of air.

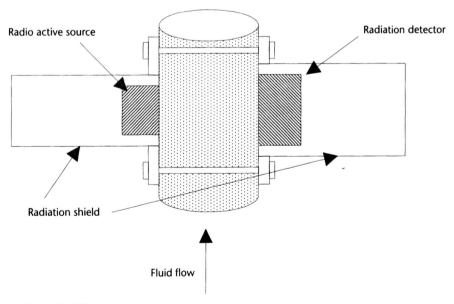

Figure 12.6 Radiation density sensor.

12.2.3 Density Application Considerations

Ideally, when measuring the density of a liquid, there should be some agitation to ensure uniform density throughout the liquid (e.g., to eliminate temperature gradients). Excessive agitation should be avoided [7].

Density measuring equipment is available for extreme temperatures, from −150° to 600°F, and pressures, in excess of 1,000 psi. When measuring corrosive, abrasive, and volatile liquids, radiation devices should be considered. In a production environment, the vibrating U-tube or the radiation sensor are normally the devices of choice.

12.3 Viscosity

Viscosity was discussed in Chapter 9; it will be discussed in this chapter in more detail.

12.3.1 Viscosity Introduction

Viscosity (μ) in a fluid is the resistance to its change of shape. Viscosity is related to the attraction between the molecules in a liquid, which resists any change due to flow or motion. When a force is applied to a fluid at rest, the molecular layers in the fluid tend to slide on top of each other, exhibiting a laminar flow [8]. These fluids are called *Newtonian fluids*, and the flow is consistent over temperature. Non-Newtonian fluid dynamics is very complex. The force (F) resisting motion in a Newtonian fluid is given by:

$$F = \frac{\mu A V}{y} \qquad (12.11)$$

where A is the boundary area being moved, V is the velocity of the moving boundaries, y is the distance between boundaries, and μ is the coefficient of viscosity, or dynamic viscosity. The units of measurement must be consistent.

Sheer stress (τ) is the force per unit area, and is given in the following formula:

$$\mu = \frac{\tau y}{V} \qquad (12.12)$$

where τ is the shear stress, or force per unit area.

If F is in lb, A in ft^2, V is in ft/s, and y is in ft, then μ is in lb s/ft^2. If F is in N, A is in m^2, V is in m/s, and y is in m, then μ is in N s/m^2. A sample list of fluid viscosities is given in Table 12.3.

The standard unit of viscosity is the poise, where a centipoise (poise/100) is the viscosity of water at 68.4°F. Conversions are given in Table 9.1. (1 centipoise = 2.09×10^{-5} lb s/ft^2.)

When the temperature of a body increases, more energy is imparted to the atoms, making them more active, and thus effectively reducing the molecular attraction. This in turn reduces the attraction between the fluid layers, lowering the viscosity. Therefore, viscosity decreases as temperature increases.

Table 12.3 Dynamic Viscosities, at 68°F and Standard Atmospheric Pressure

Fluid	μ (lb s/ft^2)	Fluid	μ (lb s/ft^2)
Air	38×10^{-8}	Carbon dioxide	31×10^{-8}
Hydrogen	19×10^{-8}	Nitrogen	37×10^{-8}
Oxygen	42×10^{-8}	Carbon tetrachloride	20×10^{-6}
Ethyl alcohol	25×10^{-6}	Glycerin	18×10^{-3}
Mercury	32×10^{-6}	Water	21×10^{-6}
Water	1×10^{-2} poise		

12.3.2 Viscosity Measuring Instruments

Viscometers or *viscosimeters* are used to measure the resistance to motion of liquids and gases. Several different types of instruments have been designed to measure viscosity, such as the inline falling-cylinder viscometer, the drag-type viscometer, and the Saybolt universal viscometer. The rate of rise of bubbles in a liquid also can be used to give a measure of the viscosity of a liquid.

The *falling-cylinder viscometer* uses the principle that an object, when dropped into a liquid, will descend to the bottom of the vessel at a fixed rate. The rate of descent is determined by the size, shape, and density of the object, and the density and viscosity of the liquid. The higher the viscosity, the longer the object will take to reach the bottom of the vessel. The falling-cylinder device measures the rate of descent of a cylinder in a liquid, and correlates the rate of descent to the viscosity of the liquid.

A *rotating disc viscometer* is a drag-type device. The device consists of two concentric cylinders, with the space between the two cylinders filled with the liquid being measured, as shown in Figure 12.7. The outer cylinder is driven by an electric motor at a constant speed using a synchronous motor, and the force on the inner cylinder is measured using a torque sensor. The viscosity of the liquid then can be

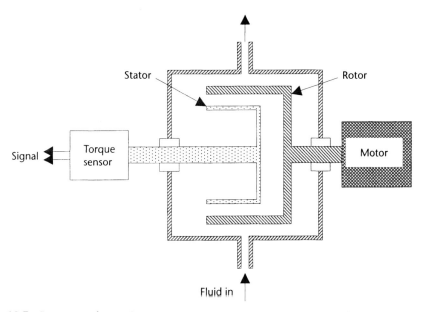

Figure 12.7 Drag-type viscometer.

determined. This type of viscometer can be used for viscosities from 50 to 50,000 centipoises, with an accuracy of ±1.0% and repeatability of 0.5% of span. The device can be used for viscosity measurements from −40° to +150°C, and pressures up to 28 MPa(g). In a production environment, the rotating disk viscometer is normally the device chosen.

The *Saybolt instrument* measures the time for a given amount of fluid to flow through a standard size orifice, or through a capillary tube with an accurate bore. The time is measured in Saybolt seconds, which is directly related to, and can be easily converted to, other viscosity units.

Example 12.5

Two parallel plates separated by 1.45 cm are filled with a liquid with a viscosity of 3.6×10^{-2} Pa·s. What is the force acting on 1m² of the plate, if the other plate is given a velocity of 2.3 m/s?

$$F = \frac{3.6 \times 10^{-2}\, Pa \cdot s \times 1m^2 \times 2.3m \times 100}{1.45 m \cdot s} = 5.71 N$$

12.4 Sound

Sound and its measurement is important, since it relates to the sense of hearing, as well as many industrial applications, such as for the detection of flaws in solids, and for location and linear distance measurement. Sound pressure waves can induce mechanical vibration and hence failure.

12.4.1 Sound Measurements

Sounds are pressure waves that travel through air, gas, solids, and liquids, but cannot travel through space or a vacuum, unlike radio (electromagnetic) waves. Pressure waves can have frequencies up to approximately 50 kHz. Sound waves start at 16 Hz and go up to 20 kHz; above 30 kHz, sonic waves become ultrasonic. Sound waves travel through air at approximately 340 m/sec, depending on factors such as temperature and pressure. The amplitude or loudness of sound is measured in phons.

The *Sound pressure level* (SPL) units are often used in the measurement of sound levels, and are defined as the difference in pressure between the maximum pressure at a point and the average pressure at that point. The units of pressure are normally expressed as follows:

$$1 \text{ dyne/cm}^2 = 1\, \mu\text{bar} = 1.45 \times 10^{-5} \text{ psi} \qquad (12.13)$$

where $1N = 10^5$ dyn and 1μbar = 0.1 Pa (See Section 7.2.3).

The *decibel* (dB) is a logarithmic measure used to measure and compare amplitudes and power levels in electrical units, sound, light, and so forth. The sensitivity of the ears and eyes are logarithmic. To compare different sound intensities, the following applies:

$$\text{Sound level ratio in dB} = 10\log_{10}\left(\frac{I_1}{I_2}\right) \quad (12.14)$$

where I_1 and I_2 are the sound intensities at two different locations, and are scalar units. A reference level (for I_2) is 10^{-16} W/cm^2 (the average level of sound that can be detected by the human ear at 1 kHz) to measure sound levels.

When comparing different pressure levels, the following is used:

$$\text{Pressure level ratio in dB} = 20\log_{10}\left(\frac{P_1}{P_2}\right) \quad (12.15)$$

where P_1 and P_2 are the pressures at two different locations. (Pressure is a measure of sound power, hence the use of 20 log.) A value of 20 μN/m^2 for P_2 is accepted as the average pressure level of sound that can be detected by the human ear at 1 kHz, and is therefore the reference level for measuring sound pressures.

Typical figures for SPL are:

- Threshold of pain: 140 to 150 dB;
- Rocket engines: 170 to 180 dB;
- Factory: 80 to 100 dB.

12.4.2 Sound Measuring Devices

Microphones are pressure transducers, and are used to convert sound pressures into electrical signals. The following types of microphones can be used to convert sound pressure waves into electrical signals: electromagnetic, capacitance, ribbon, crystal, carbon, and piezoelectric. Figure 12.8(a) shows the cross section of a dynamic microphone, which consists of a coil in a magnetic field driven by sound waves impinging on a diaphragm. An EMF is induced in the coil by the movement of the diaphragm. Figure 12.8(b) shows the cross section of capacitive microphone, which is an accepted standard for accurate acoustical measurements. Sound pressure

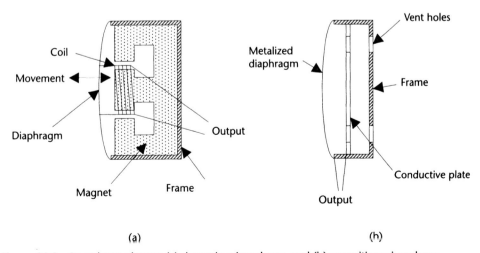

Figure 12.8 Sound transducers: (a) dynamic microphone, and (b) capacitive microphone.

waves on the diaphragm cause variations in the capacitance between the diaphragm and the rigid plate. The electrical signals then can then be analyzed in a spectrum analyzer for the various frequencies contained in the sounds, or just to measure amplitude.

Sound level meter is the term given to any of a variety of meters for measuring and analyzing sounds.

12.4.3 Sound Application Considerations

Selection of sensors for the measurement of sound intensity will depend upon the application. In instrumentation, requirements include: a uniform sensitivity over a wide frequency range, low inherent noise levels, consistent sensitivity with life, and a means of screening out unwanted noise from other sources.

12.5 pH Measurements

In many process operations, pure and neutral water (i.e., not acidic or alkaline) is required for cleaning or diluting other chemicals. Water contains both hydrogen ions and hydroxyl ions. When these ions are in the correct ratio, the water is neutral, but an excess of hydrogen ions causes the water to be acidic, and an excess of hydroxyl ions causes the water to be alkaline [9].

12.5.1 pH Introduction

The pH (i.e., power of hydrogen) of the water is a measure of its acidity or alkalinity. Neutral water has a pH value of 7 at 77°F (25°C). When water becomes acidic, the pH value decreases. Conversely, when the water becomes alkaline, the pH value increases. pH values use a base 10 log scale. That is, a change of 1 pH unit means that the concentration of hydrogen ions has increased (or decreased) by a factor of 10, and a change of 2 pH units means the concentration has changed by a factor of 100. The pH value is given by:

$$pH = \log_{10} [1/\text{hydrogen ion concentration}] \qquad (12.16)$$

The pH value of a liquid can range from 0 to 14. The hydrogen ion concentration is in grams per liter. That is, a pH of 4 means that the hydrogen ion concentration is 0.0001 g/L at 25°C.

Strong hydrochloric or sulfuric acids will have a pH of 0 to 1.

- 4% caustic soda: pH =14;
- Lemon and orange juice: pH = 2 to 3;
- Ammonia: pH is approximately 11.

Example 12.6

The hydrogen ion content in water goes from 0.203 g/L to 0.0032 g/L. How much does the pH change?

$$pH_1 = \log\left(\frac{1}{0.203}\right) = 0.69$$

$$pH_2 = \log\left(\frac{1}{0.0032}\right) = 2.495$$

Change in pH = 0.69 − 2.495 = −1.805

12.5.2 pH Measuring Devices

The pH is normally measured by chemical indicators or by pH meters. The final color of chemical indicators depends on the hydrogen ion concentration, and their accuracy is only from 0.1 to 0.2 pH units. For indication of acid, alkali, or neutral water, litmus paper is used, which turns pink if acidic, turns blue if alkaline, and remains white if neutral.

A *pH sensor* normally consists of a sensing electrode and a reference electrode immersed in the test solution, which forms an electrolytic cell, as shown in Figure 12.9. One electrode contains a saturated potassium chloride (alkaline) solution to act as a reference. The electrode is electrically connected to the test solution via the liquid junction. The other electrode contains a buffer, which sets the electrode in contact with the liquid sample. The electrodes are connected to a differential amplifier, which amplifies the voltage difference between the electrodes, giving an output voltage that is proportional to the pH of the solution. A temperature sensor in the liquid is used by the signal conditioning electronics to correct the output signal for changes in pH caused by changes in temperature.

12.5.3 pH Application Considerations

The pH of neutral water varies with temperature. Neutral water has a pH of approximately 7.5 at 32°F, and approximately 6 at 212°F. pH systems are normally automatically temperature compensated. pH test equipment must be kept clean and free from contamination. Calibration of test equipment is done with commercially available buffer solutions with known pH values. Cleaning between each reading is

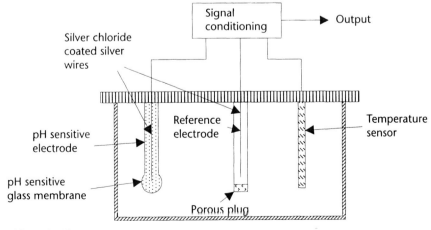

Figure 12.9 A pH sensor.

essential to prevent contamination. For continuous monitoring of pH in a production environment, a conductivity method is normally used.

12.6 Smoke and Chemical Sensors

The detection of smoke, radiation, and chemicals is of great importance in industrial processing, not only as it relates to the safety of humans, and to the control of atmospheric and ground environment pollution, but also is used in process control applications to detect the presence, absence, or levels of impurities in processing chemicals.

Smoke detectors and heat sensors (e.g., automatic sprinklers) are now commonplace in industry for the protection of people and equipment, and for the monitoring and detection of hazardous chemicals. Low-cost smoke detectors using infrared sensing or ionization chambers are commercially available. Many industrial processes use a variety of gases in processing, such as inert gases (e.g., nitrogen), to prevent contamination from oxygen in the air. Conversely, gases or chemicals can be introduced to give a desired reaction. It is necessary to be able to monitor, measure, and control a wide variety of gases and chemicals. A wide variety of gas and chemical sensors are available, and the Taguchi type of sensor is one of the more common.

12.6.1 Smoke and Chemical Measuring Devices

Infrared sensors detect changes in the signal received from an LED due to the presence of smoke, or some other object, in the light path.

Ionization chambers are devices that detect the leakage current between two plates that have a voltage between them. The leakage occurs when carbon particles from smoke are present and provide a conductive path between the plates.

Taguchi-type sensors are used for the detection of hydrocarbon gases, such as carbon monoxide, carbon dioxide, methane, and propane. The Taguchi sensor has an element coated with an oxide of tin, which combines with the hydrocarbon to give a change in electrical resistance that can be detected. To prevent depletion of the tin oxide, the element is periodically heated and the chemical reaction is reversed, in order to reduce the coating back to tin oxide.. The tin oxide can be made sensitive to different hydrocarbons by using different oxides of tin and different deposition techniques.

12.6.2 Smoke and Chemical Application Consideration

Many hazardous, corrosive, toxic, and environmentally unfriendly chemicals are used in the processing industry. These chemicals require careful monitoring during use, transportation, and handling. Analysis labs and control rooms must meet safety codes. Further information can be obtained from the ISA series RP 60 practices. All processing plants and labs must have an alarm system, which can shut down certain operations if a problem occurs. These systems are regularly tested, and are often duplicated to provide built-in fail-safe features, such as redundancy as protection against sensor failure. Table 12.4 gives some of the chemicals used in industry and the type of sensor used for measurement.

Table 12.4 Industrial Chemicals and Sensors

Chemical	Sensor	Alternate Sensor
Ammonia	Ultraviolet	Catalytic
Carbon dioxide	Mass spectrometer	Thermal conductivity detector
Carbon monoxide	Electrochemical	Infrared absorption
Chlorine	Thermal conductivity detector	Gas chromatograph
Hydrocarbons	Catalytic	Flame ionization detector
Hydrogen	Mass spectrometer	Thermal conductivity
Nitric oxide	Ultraviolet	Chemiluminescence
Nitrogen	Mass spectrometer	Gas chromatograph
Nitrogen dioxide	Ultraviolet	Amperometric
Oxygen	Paramagnetic	Zirconia oxide
Ozone	Polarographic	Gas chromatograph
Sulphur dioxide	Gas chromatograph	Ultraviolet

12.7 Summary

A number of different types of sensors were introduced in this chapter. These are not the main sensors used in process control, but are very important in many industries. This chapter introduced humidity, the definition of water vapor and its relation to a saturated gas using both volume and pressure definitions, and its relation to dew point. Humidity measuring devices, such as psychrometers, hydrometers, and dew point measuring devices, were described, as well as methods for measuring moisture content in materials.

Density, specific weight, and specific gravity were defined for both liquids and gases. Some of the various methods and instruments for measuring these quantities are described.

Viscosity was introduced, along with the formulas used in its measurement, the various types of viscometers used, and its effect on motion within a fluid.

An introduction to sound intensity and pressure waves has been provided, as well as the use of sonic and ultrasonic waves for distance measurement. Sound reference levels were discussed with the formulas used to measure sound levels.

The need for measuring pH is given, and its relation to acidity and alkalinity is discussed. The types of instruments used in its measurement were given.

Smoke and chemical sensors were introduced, and the various types of sensors used in their detection and measurement listed.

Definitions

Dew point is the temperature of a saturated mixture of water vapor in air or in a gas.

Dry-bulb temperature is the temperature of a mixture of water vapor and air (gas), as measured by a dry thermometer element.

Humidity is a measure of the relative amount of water vapor present in the air or in a gas.

Psychrometric chart   is a combined graph showing the relation between dry-bulb temperatures, wet-bulb temperatures, relative humidity, water vapor pressure, weight of water vapor per weight of dry air, and enthalpy (Btus per pound of dry air).

Relative humidity (Φ)   is the percentage of water vapor by weight that is present in a given volume of air or gas, compared to the weight of water vapor that is present in the same volume of air or gas saturated with water vapor, at the same temperature and pressure.

Specific humidity, humidity ratio, or absolute humidity   is the mass of water vapor in a mixture, divided by the mass of dry air or gas in the mixture.

Wet-bulb temperature   is the temperature of the air (gas), as measured by a moist thermometer element.

References

[1] Roveti, D. K., "Choosing a Humidity Sensor; A Review of Three Technologies," *Sensors Magazine*, Vol. 18, No. 7, July 2001.

[2] Lauffer, C., "Trace Moisture Measurement with Aluminum Oxide Sensors," *Sensors Magazine*, Vol. 20, No. 5, May 2003.

[3] Wiederhold, P. R., "The Principles of Chilled Mirror Hygrometry," *Sensors Magazine*, Vol. 17, No. 7, July 2000.

[4] Sparks, D., et al., "A Density/Specific Gravity Meter Based on Silicon Microtube Technology," *Proceedings Sensors Expo*, September 2002.

[5] Zang, Y., S. Tadigadara, and N. Najafi, "A Micromachined Coriolis-force Based Mass Flowmeter for Direct Mass Flow and Fluid Density Measurements," *Proceedings Tranducers*, 2001.

[6] Gillum, D., "Industrial Pressure, Level, and Density Measurement," ISA, 1995.

[7] Sparks, D., and N. Najafi, "A New Densitometer," *Sensors Magazine*, Vol. 21, No. 2, February 2004.

[8] *CRC Handbook of Chemistry and Physics*, 62nd ed., Table F-12, CRC Press Inc., 1981–1982.

[9] Walsh, K., "Simplified Electrochemical Diagnostics and Asset Management," *Sensors Magazine*, Vol. 17, No. 5, May 2000.

Regulators, Valves, and Motors

13.1 Introduction

Regulators and valves are the last, most expensive, and least understood element in a process control loop. They are used to control the process variable by regulating gas flow, liquid flow, and pressure. In many processes, this involves control of many thousands of cubic meters of a liquid, using low-level analog, digital, or pneumatic signals. Regulating gas and/or liquid flow also can be used to control temperature. Control loops can be local self-regulating loops under pneumatic, hydraulic, or electrical control; or the loops can be processor controlled, with additional position feedback loops. Electrical signals from a controller are either low-level signals, which require the use of relays for power control, or amplification and power switching devices, and possibly optoisolators for signal isolation. These power control devices are normally at the point of use, so that electrically controlled actuators and motors can be supplied directly from the power lines.

13.2 Pressure Controllers

Gases used in industrial processing, such as oxygen, nitrogen, hydrogen, and propane, are stored in high-pressure containers in liquid form. The high-pressure gases from above the liquid are reduced in pressure and regulated with gas regulators. Before they can be distributed through the facility, the gas lines may have additional regulators at the point of use. Other types of regulators are used for release of excessive pressures and control of liquid levels.

13.2.1 Pressure Regulators

A *spring controlled regulator* is an internally controlled pressure regulator, as shown in Figure 13.1(a). Initially, the spring holds the inlet valve open, gas under pressure flows into the main cylinder, and expands at a rate higher than the gas can exit the cylinder. As the pressure in the cylinder increases, a predetermined pressure is reached at which the spring loaded diaphragm starts to move up, causing the valve to partially close. That is, the pressure on the diaphragm controls the flow of gas into the cylinder, in order to ideally maintain a constant pressure in the main cylinder and at the output, regardless of flow rate. The output pressure can be adjusted by the spring screw adjustment. A double-seated valve as shown is normally used. This type of valve is not loaded by the pressure of the incoming gases, as

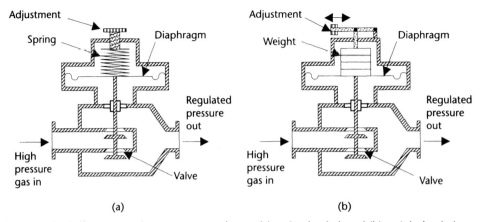

Figure 13.1 Self-compensating pressure regulators: (a) spring loaded, and (b) weight loaded.

would be the case with a single-seated valve. The pressure on one face of the valve is balanced by the pressure on the face of the other valve, so that the diaphragm is not loaded by the incoming gas pressure acting on the valve.

A *weight-controlled regulator* is shown in Figure 13.1(b). The internally controlled regulator has a weight loaded diaphragm. The operation is the same as the spring-loaded diaphragm, except the spring is replaced with a weight. The pressure can be adjusted by the position of a sliding weight on a cantilever arm.

A *pressure controlled diaphragm regulator* is shown in Figure 13.2. The internally controlled regulator has a pressure loaded diaphragm. Pressure from a regulated external air or gas supply is used to load the diaphragm via a restriction. The pressure to the regulator then can be adjusted by a bleed valve, which in turn is used to set the output pressure of the regulator.

An alternative to the internal pressure diaphragm, as in the regulators shown above, is to apply the pressure to the top of the diaphragm, as shown in Figure 13.3. The cross section shows the output pressure being fed externally to a spring-loaded pressure regulator. The spring holds the valve open until the output pressure, which is fed to the upper surface of the diaphragm, overcomes the force of the spring on the diaphragm and starts to close the valve, thus regulating the output pressure. The

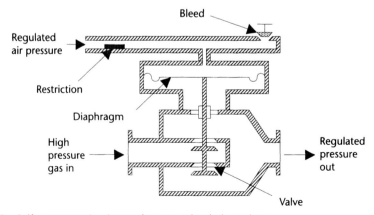

Figure 13.2 Self-compensating internal pressure loaded regulator.

13.2 Pressure Controllers

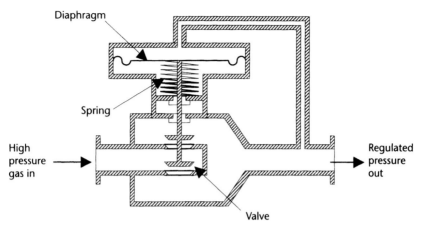

Figure 13.3 External connection to a spring loaded regulator.

valve is inverted from the internal regulator connection, and the internal pressure is isolated from the lower side of the diaphragm. Weight loaded and air loaded diaphragms also are available for externally connected regulators.

Pilot-operated pressure regulators use an internal or external pilot for the feedback signal amplification and control. In this case, the pilot is a small regulator that is positioned between the pressure connection to the regulator and the loading pressure on the diaphragm. Figure 13.4 shows such an externally connected pilot regulator. The pressure from the output of the regulator is used to control the pilot, which in turn amplifies the signal and controls the pressure from the air supply to the diaphragm, giving greater control than that available with the internal pressure control diaphragm. A small change in the output pressure is required to produce a full pressure range change of the regulator giving a high gain system for good output pressure regulation.

13.2.2 Safety Valves

Safety valves are fitted to all high-pressure containers, from steam generators to domestic water heaters. Figure 13.5 shows the cross section of a safety valve. The

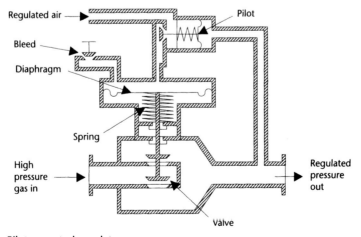

Figure 13.4 Pilot-operated regulator.

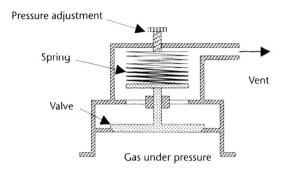

Figure 13.5 Automatic pressure safety valve.

valve is closed until the pressure on the lower face of the valve reaches a predetermined level set by the spring. When this level is reached, the valve moves up, allowing the excess pressure to escape through the vent.

13.2.3 Level Regulators

Level regulators are in common use in industry to maintain a constant fluid pressure, or a constant fluid supply to a process. Level regulators can be a simple float and valve arrangement, as shown in Figure 13.6(a), or an arrangement using capacitive sensors, as given in Chapter 6, to control a remote pump. The arrangement shown in Figure 13.6(a) is a simple, cost-effective method used to control water levels in many applications; two common uses of this device are in swimming pools and toilet cisterns. When the fluid level drops, the float moves downward, opening the inlet valve and allowing fluid to flow into the tank. As the tank fills, the float rises, causing the inlet valve to close, maintaining a constant level and preventing the tank from overflowing.

The float controls the position of the weight in Figure 13.6(b). The position of the weight is monitored by position sensors A and B. When the weight is in position A (container empty), the sensor can be used to turn on a pump to fill the tank, and when sensor B (container full) senses the weight, it can be used to turn the pump off. The weight can be made of a magnetic material, and the level sensors would then be Hall effect or MRE devices.

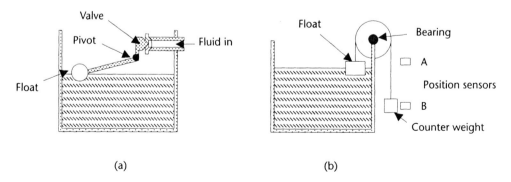

Figure 13.6 (a) Automatic fluid level controller, and (b) means of detecting full level or empty level in a fluid reservoir.

13.3 Flow Control Valves

When a change in a measured variable with respect to a reference has been sensed, it is necessary to apply a control signal to an actuator to make corrections to an input controlled variable, via a valve bringing the measured variable back to its preset value. In most cases, any change in the variables (e.g., temperature, pressure, mixing ingredients, and level) can be corrected by controlling flow rates. In general, the actuators used to control valves for flow rate control, and can be electrically, pneumatically, or hydraulically controlled. Actuators can be self-operating in local feedback loops, in such applications as temperature sensing with direct hydraulic or pneumatic valve control, pressure regulators, and float level controllers. The two most common types of variable aperture devices used for flow control are the globe valve and the butterfly valve.

13.3.1 Globe Valve

The globe valve's cross section is shown in Figure 13.7(a). The actuator controlling the valve can be driven electrically (using a solenoid or motor), pneumatically, or hydraulically. The actuator determines the speed of travel and the distance that the valve shaft travels. The globe valve can be designed for quick opening operation, for equal percentage operation, or with a linear relationship between flow and lift, or any combination of these. In equal percentage operation, the flow is proportional to the percentage the valve is open, or there is a logarithmic relationship between the flow and valve travel. The shape of the plug determines the flow characteristics of the actuator, and is normally described in terms of percentage of flow versus percentage of lift or travel. Various valve configurations for quick opening operation, for equal percentage operation, or with a linear relationship between flow and lift, are shown in Figure 13.8.

The valve plug shown in Figure 13.7(a) gives a linear relationship between flow and lift. The valve characteristic is given in Figure 13.7(b). Shown in the graph are the characteristics for a quick opening plug and an equal percentage plug, illustrating some of the characteristics that can be obtained from the large number of plugs that are available. The selection of the type of control plug should be carefully

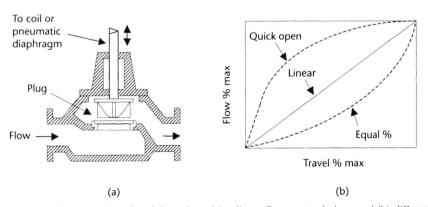

Figure 13.7 (a) Cross section of a globe valve with a linear flow control plug, and (b) different flow patterns for various plugs versus plug travel.

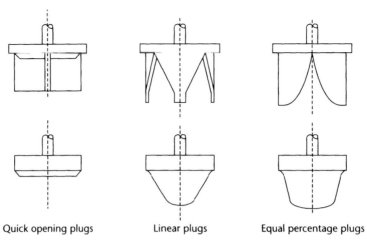

Figure 13.8 Examples of types of quick opening, linear, and equal percentage plugs.

chosen for any particular application. The type will depend on a careful analysis of the process characteristics. If the load changes are linear, then a linear plug should be used; conversely, if the load changes are nonlinear, then a plug with the appropriate nonlinear characteristics should be used.

The globe valve can be straight through with single seating, as illustrated in Figure 13.7(a), or can be configured with double seating, which is used to reduce the actuator operating force, but is expensive, difficult to adjust and maintain, and does not have a tight seal when shut off. Angle valves also are available, in which the output port is at right angles or 45° to the input port.

Many other configurations are available in the globe valve family. Figure 13.9(a) shows a two-way valve (diverging type) that is used to switch the incoming flow from one exit to another. When the valve stem is up, the lower port is closed and the incoming liquid exits to the right; and when the valve stem is down, the upper port is closed and the liquid exits from the bottom. A converging type is also available,

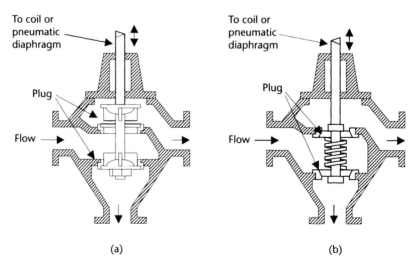

Figure 13.9 Cross sections of globe valve configurations: (a) two-way valve, and (b) three-way valve.

which is used to switch either of two incoming flows to a single output. Figure 13.9(b) illustrates a three-way valve. In the neutral position, both exit ports are held closed by the spring; when the valve stem moves down, the top port is opened; and when the valve stem moves up from the neutral position, the lower port is opened.

Other types of globe valves are the needle valve, which have a diameter from 1/8 to 1 in; the balanced cage-guided valve; and the split body valve. In the cage-guided valve, the plug is grooved to balance the pressure in the valve body, and the valve has good sealing when shut off. The split body valve is designed for ease of maintenance, and can be more cost-effective than the standard globe valve, but pipe stresses can be transmitted to the valve and cause it to leak. Globe valves are not well-suited for use with slurries [1].

13.3.2 Butterfly Valve

The butterfly valve consists of a cylindrical body with a disk the same size as the internal diameter of the valve body, mounted on a shaft that rotates perpendicular to the axis of the body. The action is similar to a louver damper. The disk pivots to the vertical position to shut off any flow, and to the horizontal position when fully open. The valve is shown in Figure 13.10(a), and its flow versus travel characteristics are shown in Figure 13.10(b). The relation between flow and lift is

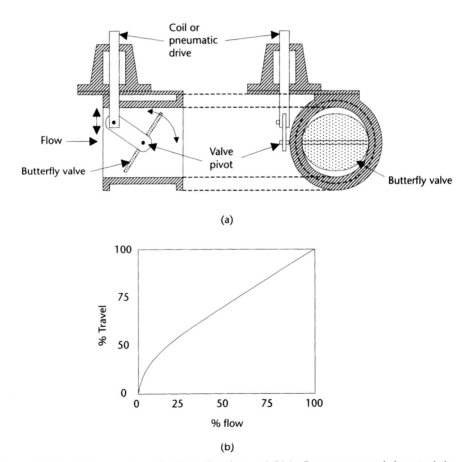

Figure 13.10 (a) Cross section of a butterfly valve, and (b) its flow versus travel characteristics.

approximately equal up to approximately 50% open, after which it is linear. Butterfly valves offer high capacity at low cost, are simple in design, are easy to install, and have tight closure. The torsion force on the shaft increases until the valve is open 70°, and then reverses. Butterfly valves have a limited pressure range, and are not used for slurries. The pressure produces a strong thrust on the valve bearings.

13.3.3 Other Valve Types

A number of other types of valves are in common use, including the weir-type diaphragm, the ball valve, and the rotary plug valve. The cross sections of these valves are shown in Figure 13.11.

A *weir-type diaphragm* valve is shown in Figure 13.11(a). The valve is shown open, and can be closed by forcing a flexible membrane down onto a lower flexible membrane, using a pincer action. A flexible membrane also can be forced down onto a weir. Diaphragm valves are good low-cost choices for slurries and liquids with suspended solids, but tend to require high maintenance and have poor flow characteristics.

A one-piece *ball valve* is shown in Figure 13.11(b). The valve is a partial sphere that rotates, and the valve tends to be slow to open. The ball valve is available in other configurations with various shaped spheres for different flow characteristics. The valve is good for slurries and liquids with solid matter because of its self-cleaning operation. Ball valves have tight turnoff characteristics, are simple in design, and have greater capacity than similar-sized globe valves.

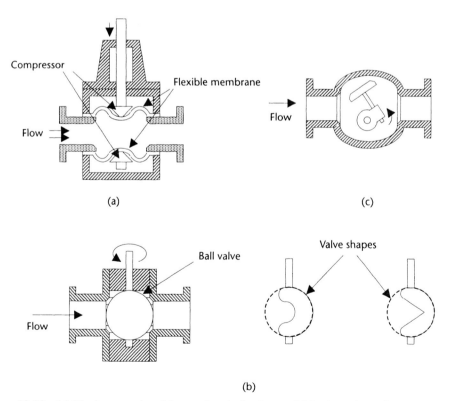

Figure 13.11 (a) Diaphragm valve, (b) one-piece ball valve, and (c) rotary plug valve.

An *eccentric rotary plug valve* is shown in Figure 13.11(c). The valve is medium cost, requires less closing force than many other types of valves, and can be used for forward or reverse flow. The valve has tight shutoff characteristics with a positive metal-to-metal seating action without a rubbing action in the seal ring, and has a high capacity. The good shutoff characteristics, low wear, and few moving parts make it a good valve for use with corrosive liquids.

13.3.4 Valve Characteristics

Other factors that determine the choice of valve type are corrosion resistance, operating temperature ranges, high and low pressures, velocities, pipe size, and fluids containing solids. Correct valve installation is essential, and vendor recommendations must be carefully followed. In situations where sludge or solid particulates can be trapped upstream of a valve, a means of purging the pipe must be available. To minimize disturbances and obtain good flow characteristics, a clear run from one to five pipe diameters upstream and downstream should be allowed.

Valve sizing is based on pressure loss. Valves are given a C_V capacity number that is based on test results, and indicates the number of gallons per minute of water at 60°F (15.5°C), which, when flowing through the fully opened valve, will have a pressure drop of 1 psi (6.9 kPa). That is, a valve with a capacity of 25 C_V means that the valve will have a pressure drop of 1 psi when 25 gal/min of water are flowing. For liquids, the relation between pressure drop P_d (psi), flow rate Q (gal/min), and capacity C_V is given by:

$$C_V = Q\sqrt{(SG/P_d)} \qquad (13.1)$$

where SG is the specific gravity of the liquid.

Example 13.1

What is the capacity of a valve, if there is a pressure drop of 3.5 psi when 2.3 gal/s of a liquid with an SG of 60 lb/ft³ are flowing?

$$C_V = 2.3 \times 60\sqrt{\frac{60}{62.4 \times 3.5}} = 138 \times 0.52 = 72.3$$

Table 13.1 gives a comparison of some of the valve characteristics. The values shown are typical of the devices available and may be exceeded by some manufacturers with new designs and materials.

13.3.5 Valve Fail Safe

An important consideration in many systems is the position of the actuators when there is a loss of power (i.e., if chemicals or the fuel to the heaters continue to flow, or if total system shutdown occurs). Figure 13.12 shows an example of a pneumatically or hydraulically operated globe valve design that can be configured to open or close during a system failure. The modes of failure are determined by simply changing the spring position and the pressure port.

Table 13.1 Valve Characteristics

Parameter	Globe	Diaphragm	Ball	Butterfly	Rotary Plug
Size	1 to 36 in	1 to 20 in	1 to 24 in	2 to 36 in	1 to 12 in
Slurries	No	Yes	Yes	No	Yes
Temperature Range	−200° to +540°C	−40° to +150°C	−200° to +400°C	−50° to +250°C	−200° to +400°C
Quick-opening	Yes	Yes	No	No	No
Linear	Yes	No	Yes	No	Yes
Equal percentage	Yes	No	Yes	Yes	Yes
Control range	20:1 to 100:1	3:1 to 15:1	50:1 to 350:1	15:1 to 50:1	30:1 to 100:1
Capacity (C_V) (d = diameter)	10 to 12 × d^2	14 to 22 × d^2	14 to 24 × d^2	12 to 35 × d^2	12 to 14 × d^2

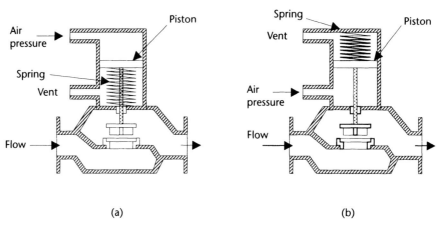

Figure 13.12 Fail-safe pneumatic or hydraulic operated valves. If there is a loss of operating pressure, the valve (a) opens, and (b) closes.

In Figure 13.12(a), applying pressure to the pressure port to oppose the spring action will close the globe valve. If the system fails (i.e., if there is a loss of pneumatic pressure), then the spring acting on the piston will force the valve to its open position. In Figure 13.12(b), the spring is removed from below the piston to a position above the piston, and the inlet and exhaust ports are reversed. In this case, applied pressure working against the spring action will open the valve. If the system fails and there is a loss of control pressure, the spring action will force the piston down and close the valve. Similar fail-safe electrically and hydraulically operated valves are available. Two-way and three-way fail-safe valves also are available, which can be configured to be in a specific position when the operating system fails.

13.3.6 Actuators

Actuators are used to control various types of valves. Shown in Figure 13.13 are two types of pneumatic diaphragm actuators. Figure 13.13(a) shows a reverse action for lifting a valve against the pressure of the liquid on the valve, and Figure 13.13(b) shows direct action for moving a valve downwards against the pressure on the valve. Depending upon the valve, the pressure can occur on closing or opening. The

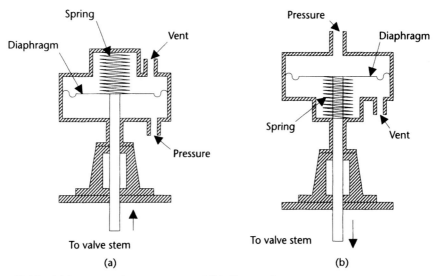

Figure 13.13 (a) Reverse acting actuator, and (b) direct action actuator.

actuator must be able to operate the valve against the pressure acting on the valve plus the spring, and must be able to tightly close the valve.

Example 13.2

A valve is used to turn off the water at the base of a 122m tall water column. (a) If the valve is 45 cm in diameter, what is the force required by the actuator to turn off the water, assuming the water pressure is acting on the face of the valve? (b) If the pneumatic actuator pressure line has a maximum pressure of 100 psi and 12% of the actuator pressure is required to overcome the spring and tightly close the valve, what is the diameter of the diaphragm in the actuator?

(a) Required actuator force = $9.8 \times 122 \times 45^2 \times 3.14/4 \times 10^4$ kN = 190 kN

(b) 190 kN $\times$ 1.12 = $100 \times 249.1 \times 3.14 \times d^2/4 \times 10^3$

$d^2 = 212.8 \times 4/78.2 \text{m}^2 = 10.88 \text{m}^2$

$d = 3.3$m

This is an excessively large diaphragm, but is used to illustrate a point on required actuator forces.

13.4 Power Control

Electrical power for actuator operation can be controlled from low-level analog and digital signals, using electronic power devices, relays, or magnetic contactors. Relays and magnetic contactors have a lower On resistance than electronic devices have, but they require higher drive power. Relays and contactors provide voltage isolation between the control signals and output circuits, but are slow to switch, have lower current handling capability than electronic power devices, and have a limited switching life. Relays typically are used to switch low-power signal lines, whereas contactors are used to switch higher powers, such as power to motors. In electronic devices, the problem of electrical isolation between drive circuits and output power circuits can be easily overcome by design or with the use of optoisolators. Electronic power devices have excellent longevity and many advantages due to their high switching speeds in variable power control circuits.

13.4.1 Electronic Devices

A number of electronic devices, such as silicon controlled rectifiers (SCR), TRIAC, and MOS devices, can be used to control several hundred kilowatts of power from low-level electrical signals. Electronic power control devices fall into two categories. First, a triggered devices such as the SCR and TRIAC, which are triggered by a pulse on the gate into the conduction state. Once triggered, these devices can be turned off only by reducing the anode/cathode current to below their sustaining current (i.e., when the supply voltage/current drops close to zero). These devices can block high reverse voltages. They are used extensively in ac circuits, where the supply regularly transcends through zero, automatically turning the device Off. The second group of devices include: Darlington Bipolar Junction Transistors (BJT), Power MOSFET, Insulated Gate Bipolar Transistors (IGBT), and MOS-Controlled Thyristors (MCT). These devices are turned On and Off by an input control signal, but do not have the capability of high reverse voltage blocking. This group of devices is more commonly used with dc power supplies, or is biased to prevent a reverse voltage across the device.

The SCR is a current-operated device, and only can be triggered to conduct in one direction. When used with an ac supply, it blocks the negative half cycle, and only will conduct on the positive half-cycle when triggered. Once triggered, the SCR remains On for the remaining portion of the half-cycle. Figure 13.14(a) shows the circuit of an SCR with a load [2]. Figure 13.14(b) shows the effects of triggering on the load voltage (V_L). By varying the triggering in relation to the positive half-cycle, the power in the load can be controlled from 0% to 50% of the total available power. Power can be controlled from 50% to 100% by putting a diode in parallel with the SCR to conduct current on the negative half-cycles. Light-activated SCRs also are available.

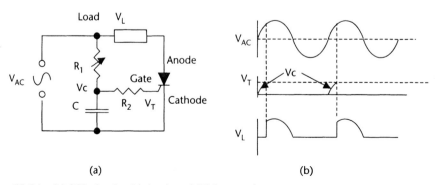

Figure 13.14 (a) SCR circuit with load, and (b) its waveforms.

One method of triggering the SCR is shown in Figure 13.15(a), with the corresponding circuit waveforms shown in Figure 13.15(b). During the positive half-cycle, the capacitor C is charged via R_1 and R_2 until the triggering point of the SCR is reached. The diode can be connected on either side of the load. The advantage of connecting the diode to the SCR side of the load is to turn Off the voltage to the gate when the SCR is fired, reducing dissipation. The diode is used to block the negative half-cycle from putting a high negative voltage on the gate and damaging the SCR. The zener diode is used to clamp the positive half-cycle at a fixed voltage (V_z), so that the capacitor (V_c) has a fixed aiming voltage, giving a linear relation between triggering time and potentiometer setting. V_z and V_c in Figure 13.15(b) show this.

Example 13.3

In Figure 13.13, an SCR with a 5V gate trigger level is used with a 12V zener diode, and the capacitor is 0.15 µF. What value of R_2 will give full control of the power to the load down to zero?

Time duration of half-sine wave at 60 Hz = 1/60 × 2 = 8.3 ms

Charging time can be found from capacitor charging equation $V_C = V_0 (1 - e^{-t/RC})$

$5 = 12(1 - e^{-t/RC})$

From which

$t = 0.54RC = 8.3$ ms

$R = 8.3 \times 10^6 / 0.54 \times 0.15 \times 10^3 = 102.5$ kΩ

Control from 0% to 100% can be obtained with a single SCR in a bridge circuit, as shown in Figure 13.16(a). The waveforms are shown in Figure 13.16(b). The bridge circuit changes the negative half-cycles into positive half-cycles, so that

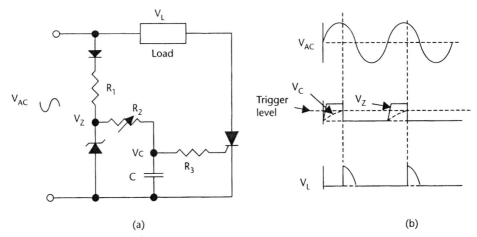

(a)   (b)

Figure 13.15 (a) Typical SCR triggering circuit with trigger point control, and (b) corresponding triggering waveforms.

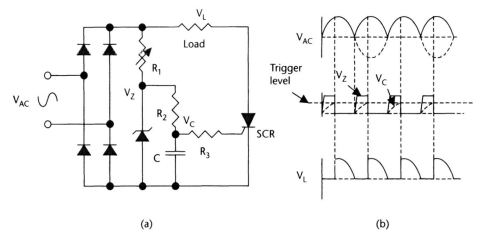

Figure 13.16 Bridge circuit for SCR control (a) using full wave rectification, and (b) corresponding waveforms.

the SCR only sees positive half-cycles. The SCR is triggered during every half-cycle, and is turned Off every half-cycle when the supply voltage goes to zero. As shown, the system is controlled by a low-level signal coupled by an optoisolator. The potentiometer R and capacitor C set the triggering point, and since the SCR only sees positive voltages, the diode is not required. For cost savings, the zener diode is omitted. As in the previous figure, resistor R can be connected to either side of the load.

The *DIAC* is a semiconductor device developed for trigger control, primarily for use with TRIACs. Figure 13.17(a) shows the symbol for the device, and Figure 13.17(b) shows the device's characteristics. The DIAC is a two-terminal symmetrical switching device. As the voltage increases across the device, little current flows until the breakdown voltage V_L is reached, at which point the device breaks down and conducts as shown. The breakdown occurs with both positive and negative voltages. The breakdown voltage of the DIAC is used to set the trigger voltage for the TRIAC. When the device breaks down, the TRIAC triggers.

TRIACs can be considered as two reversed SCRs connected in parallel. They can be triggered on both the positive and negative half-cycles of the ac waveform. A circuit for triggering a TRIAC is shown in Figure 13.18(a), with the associated waveforms shown in Figure 13.18(b). The TRIAC can be used to control power to the

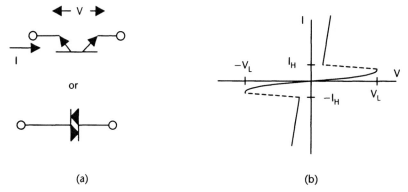

Figure 13.17 DIAC used in SCR and TRIAC triggering circuits: (a) symbol, and (b) characteristic.

13.4 Power Control

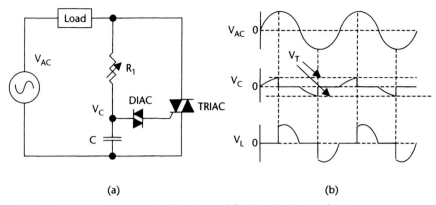

Figure 13.18 (a) TRIAC power control circuit, and (b) the circuit waveforms.

load from 0% to 100%, by controlling the trigger points with respect to the ac sine wave as the ac voltage increases from zero. The capacitor C is then charged via R_1 until the breakdown voltage of the DIAC is reached, and the TRIAC is triggered on both the positive and negative half-cycles, as shown by the waveforms in Figure 13.18(b).

Example 13.4

A TRIAC is used to supply 750A to a load from a 120V supply. What is the maximum power that can be supplied to the load, and the power loss in the TRIAC? Assume the voltage drop across the TRIAC is 2.1V.

Power loss in TRIAC = 2.1 × 750W = 1.575 kW

Power from supply = 750 × 120W = 90 kW

Power to load = 90 − 1.575 kW = 88.425 kW

This example illustrates that the efficiency of the switch is greater than 98%, and the high dissipation that can occur in the switch and the need for cooling fins with low thermal resistance. Precautions in the design of power switching circuits, choices of devices for specific applications, and thermal limitations are outside the scope of this book. Device data sheets must be consulted and advice obtained from device manufacturers before designing power controllers [3].

When opening and closing switches with voltage applied in power control circuits, problems can occur, such as power surges that cause large current transients in the supply line. These transients produce unwanted RF interference and potentially damaging high voltage inductive transients. A solution to this problem is to fire the thyristor when the supply voltage is at or near zero. Several commercial devices are available for this function. These devices are called *zero-voltage switches* (ZVS). These devices control power by eliminating cycles. Figure 13.19(a) shows a zero-voltage crossover switch driving a TRIAC. The waveforms are shown in Figure 13.19(b). Power is supplied in complete cycles, and one power cycle for three line cycles, or 33% power to the load, is shown in the figure.

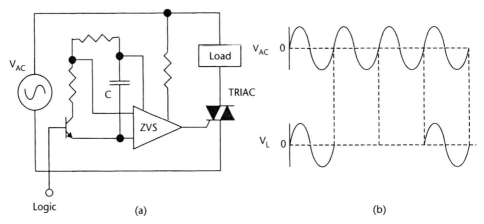

Figure 13.19 Zero-voltage crossover switch (a) driving a TRIAC, and (b) associated waveforms.

Zero-voltage crossover switches are suitable for controlling heating elements, valve control, and lighting, but are not ideally suited for motor speed control, since motors under heavy load tend to slow down, and the missing power cycles tend to cause vibration.

Power devices that are turned On and Off by the input are:

1. BJTs, which are current-controlled devices. Power bipolar devices have low gain, so are normally used in a Darling configurations to give high current gain and the ability to control high currents with low drive currents [4].
2. Power MOSFETs, which are voltage-controlled devices designed for high-speed operation, but their high saturation voltage and temperature sensitivity limit their application in power circuits.
3. IGBTs [5]. An MOS transistor, as opposed to the Darlington bipolar configuration, controls the power bipolar output device, making it a voltage-controlled device. The IGBT has fast switching times. Older devices had a high saturation voltage, and newer devices have a saturation voltage about the same as a BJT.
4. MCTs are voltage-controlled devices with a low saturation voltage and medium speed switching characteristics.

A comparison of the characteristics of power devices is given in Table 13.2. These devices are used for power and motor control. Applications include rectification of multiphase ac power to give a variable voltage dc power level output, or the

Table 13.2 Comparison of Power Device Characteristics

Device	Power handling	Saturation voltage	Turn-on time	Turn-off time
SCR	2 kV 1.5 kA	1.6V	20 μs	N/A
TRIAC	2 kV 1 kA	2.1V	20 μs	N/A
BJT	1.2 kV 800A	1.9V	2 μs	5 μs
MOSFET	500V 50A	3.2V	90 ns	140 ns
IGBT	1.2 kV 800A	1.9V	0.9 μs	200 ns
MCT	600V 60A	1.1V	1.0 μs	2.1 μs

control of dc motors from an ac power source. Other applications include the control of multiphase motors from a dc power source, or the conversion of dc power to multiphase ac power [6].

13.4.2 Magnetic Control Devices

A signal from a controller is a low-level signal, but can be amplified to control an actuator or small motor. Power for actuators are normally generated close to the point of use, to prevent energy loss in the leads and to prevent large currents from flowing in the ground return lines to the controller, minimizing offset and ground line noise. Because of the isolation that the relay gives between the driving circuit and the motor circuit, the motor and power supply can be either dc or ac. Such a relay can have multiple contacts to control three-phase ac motors.

Contactors are designed for switching high currents and voltages, such as used in motor control applications. A single-pole-single-throw double-break contactor is shown in Figure 13.20. In Figure 13.20(a), the contactor is shown de-energized and with the contacts open. When a current is passed through the coil, the magnetic field in the core attracts and pulls in the soft iron keeper, which closes the contacts, as shown in Figure 13.20(b). Contactors can have multiple contacts for multiphase motors. Contact material is critical, since chemical and metallurgical actions occur during switching, causing wear, high contact resistance, and welding. Gold or rhodium can be used for currents below 1A. Silver is used for currents in the range from 1A to 10A when the supply voltage is above 6V. Silver cadmium is sometimes used for currents in the range from 5A to 25A when the supply voltage is above 12V. Mercury wetted contacts are available for currents up to 100A. The contact life in relays typically is limited to between 100,000 and 500,000 operations.

13.5 Motors

The student needs to be aware of the types of functions that motors perform in industrial applications, but details of motors and control circuits are outside the scope of this text. Motors are used for pumping fluids, compressors, driving

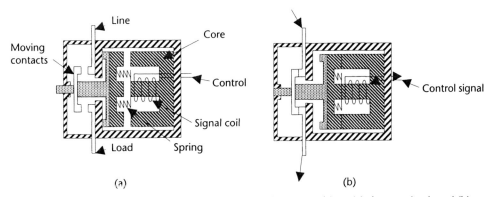

Figure 13.20 Contactor used for high current and voltage switching: (a) de-energized, and (b) energized.

conveyor belts, and any form of positioning required in industry. For control applications or positioning, servos or stepper motors are used.

13.5.1 Servo Motors

Servo motors can rotate to a given position, and can be stopped and reversed. In the case of a servo motor, the angular position and speed can be precisely controlled by a servo loop, which uses feedback from the output to the input. The position of the output shaft is monitored by a potentiometer, which provides an analog feedback voltage to the control electronics. The control electronics can use this information to power the output motor and stop it in any desired position, or reverse the motor to stop at any desired position [7].

In Figure 13.21, the actuator for a globe valve operated by an electric motor is shown. The screw driven by the motor can move the plug in the valve up or down. A potentiometer wiper is attached to the valve stem, and gives a feedback voltage that is directly proportional to the amount the valve is open. This value is fed back to the input node of the amplifier controlling the motor and is compared to a reference voltage, so that the reference voltage sets the position of the valve. The system also could be digital, in which case a digital encoding technique would be used for the feedback and a digital comparator used to compare the reference data to the feedback data.

13.5.2 Stepper Motors

Stepper motors rotate a fixed angle with each input pulse. The rotor is normally a fixed magnet with several poles and a stator with several windings. A single magnet rotor and a four-section stator are shown in Figure 13.22(a), which gives a 90° rotation for each input phase. Stepper motors are the only motor that is digital, in that they step one position for each input pulse. The driving waveform is shown in Figure 13.22(b). Stepper motors are available in many different designs, with a wide selection of the number of poles and drive requirements, all of which define the stepper motor characteristics and rotation angle for each input phase. Changing the

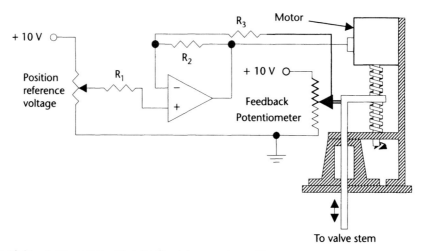

Figure 13.21 Servo motor with a feedback loop and amplifier.

13.5 Motors

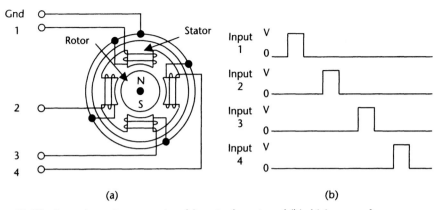

Figure 13.22 Four-phase stepper motor: (a) motor layout, and (b) driving waveforms.

sequence of the driving phases can reverse the stepper motors. Stepper motors are available with stepping angles of 0.9°, 1.8°, 3.6°, 7.5°, 15°, and 18°, up to 90°. Because the motor steps by a fixed angle, a known angle with each input pulse feedback is not required. However, since only the relative position is known, loss of power can cause loss of position information, so that in a system using stepper motors, a position reference is usually required.

13.5.3 Synchronous Motors

Synchronous motors operate on a principle similar to that of the transformer, because the rotor induces voltages in the stator windings. Synchronous motors (synchros) are small motors working as a master and slave, in that when the position of the rotor in the master is changed, the rotor in the slave synchronous motor will follow. Figure 13.23 shows the diagram of the master and slave synchronous motors. An ac voltage is fed to the rotors of both motors. The magnetic field produced by the rotor in the local synchronous motor induces a voltage into the three field windings (positioned 120° apart). The value of the induced voltages in each field winding will be different, and is directly related to the position of the rotor.

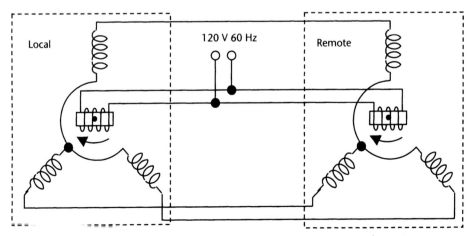

Figure 13.23 Synchronous motor operation of local transmitter (master) and remote receiver (slave).

These field windings are connected directly or amplified, and then fed to three identical field windings in the remote synchronous motor, which will replicate the magnetic field in the local motor. Because the two rotors are fed from the same ac supply, the rotor in the slave will seek the same position as the rotor in the master. Typically, synchronous motors have many applications when duplicating position in formation to remote locations. The output from the master synchronous motor can be digitized in a synchro to digital converter (SDC) for transmission and processing. The signal can be converted back to its original format using a digital to synchro converter (DSC), for use by the slave synchronous motor.

13.6 Application Considerations

13.6.1 Valves

The selection of control valves for a particular application depends on many variables, such as the corrosive nature of the fluid, temperature of operation, pressure of the fluid, velocity of the flow, volume of the flow, and the amount of suspended solids.

Valves are the final element in a control loop, and are critical in providing the correct flow for process control. The valve is subject to operation in very harsh conditions, and is one of the most costly elements in the process control system. Their choice and correct installation require both knowledge and experience. Careful attention must be made to the system requirements and manufacturers specifications, before a careful valve selection can be made. Additional information can be obtained from the ISA 75 series of standards.

Some of the factors affecting the choice of valves are:

1. Fail-safe considerations for two-way and three-way types of valves;
2. Valve size from flow requirements, avoiding both oversizing and undersizing;
3. Materials used in the valve construction, ranging from PVC to brass to steel, based on considerations of pressure, size, and corrosion;
4. Tightness of shutoff, as classified by quality of shutoff by leakage at maximum pressure (Valves are classified into six classes depending on leakage, from 0.5% of rated capacity to 0.15 mL/min for a 1-in diameter valve);
5. Level of acceptable pressure drop across the valve;
6. Linear or rotary motion of type of valve (e.g., globe, diaphragm, ball, or butterfly valves).

The type of valve or plug depends on the nature of the process reaction. In the case of a fast reaction with small load changes, control is only slightly affected by valve characteristics. When the process is slow with large load changes, valve characteristics are important. If the load change is linear, then a valve with a linear characteristic should be used. In the case of a nonlinear load change, a valve with an equal percentage change may be required. In some applications, valves are required to be completely closed when Off. Other considerations are: maintenance; serviceability; fail-safe features; pneumatic, hydraulic, solenoid, or motor control; and the need for feedback. The above is a limited review of actuator valves, and, as previously noted, the manufacturers' data sheets should be consulted when choosing a valve for a particular application.

Electrically operated servo or stepper motors normally control position and speed. In applications such as pumping, compressors, or conveyer belts, three-phase motors are normally used.

13.6.2 Power Devices

Power switching devices, from contactors to solid state devices, will be chosen from considerations of power handling, switching speed, isolation, and cost. Some of the considerations are:

1. For low-speed operation, mechanical relay devices give isolation, relatively low dissipation, and are low cost.
2. Light control and ac motor control can use SCRs and TRIACs, which are available in a wide range of packages, depending on current handling and heat dissipation requirements [8].
3. For power control, multiphase motor control, and high-speed switching applications, BJT or IGBTs can be used. These devices also come in a variety of low thermal resistance packages.
4. MOSFET devices can be used in medium power applications, since they the advantage that control circuits can be integrated on to the same die as the power device [9].

13.7 Summary

Regulators and valves are available in many shapes and sizes, and since they are one of the most expensive and most important components in a process control system, great care has to be taken in their selection. The various types of regulators, including internal and external connected regulators, were discussed. Regulators can be loaded using spring, weight, or pressure. More expensive devices use pilot devices in the feedback loop for higher system feedback gain, which gives better regulation, control, and flexibility. The most common valve is the globe valve. This device is available in many configurations, having many types of plugs to give fast opening, linear, or equal percentage characteristics. Valve sizes depend on rates of flow and acceptable losses. Materials used depend on pressure, temperature, and resistance to corrosion. Globe valves can be configured as two-way or three-way fail-safe modes, split body for ease of maintenance, and so forth. Other types of valves are the butterfly, diaphragm, ball, and rotary plug valves. Actuators can be controlled pneumatically or electronically. The more common electronic power handling device types are the SCR, TRIAC, and IGBT. Electronic control devices have fast operation, are robust, and can handle large amounts of power for control. Devices can operate from an ac or dc supply. Actuator positions can be controlled by stepper motors or motors using feedback. Other types of motors for controlling position are synchronous motors.

References

[1] Battikha, N. E., *The Condensed Handbook of Measurement and Control*, 2nd ed., ISA, 2004, pp. 223–238.

[2] Johnson, C. D., *Process Control Instrumentation Technology*, 7th ed., Prentice Hall, 2003, pp. 331–340.

[3] Chan, C. C., "An Overview of Electric Vehicle Technology," *Proceedings of the IEEE*, Vol. 81, No. 9, September 1993, pp. 1302–1313.

[4] van de Wouw, T., "Darlingtons for High Power Systems," *P. C. I 88 Conference Proceedings*, Vol. 15, June 1988, pp. 204–213.

[5] Yilmaz, H., et al., "50A 1,200V N-channel IGT," *IEE Proceedings*, Vol. 132, Part 1, No. 6, December 1985.

[6] Jurgen, R. K., *Automotive Electronics Handbook*, 2nd ed., McGraw-Hill, 1999, Chapter 33.

[7] Humphries, J. T., and L. P. Sheets, *Industrial Electronics*, 4th ed., Delmar, 1993, pp. 464–474.

[8] Schuster, D., "Know Your Power," *Sensors Magazine*, Vol. 16, No. 8, August 1999.

[9] Dunn, B., and R. Frank, "Guidelines for Choosing a Smart Power Technology," *P. C. I 88 Conference Proceedings*, Vol. 15, June 1988, pp. 143–157.

CHAPTER 14
Programmable Logic Controllers

14.1 Introduction

Modern industrial control systems are microprocessor-based programmable systems containing hardware and software for direct digital control, distributed control, programmable control, and PID action. The systems are designed not only for continuous monitoring and adjustment of process variables, but also for sequential control, which is an event-based process, and alarm functions. This chapter discusses the Programmable Logic Controller (PLC), its operation, and the use of the PLC for sequential, continuous control, and alarm functions.

14.2 Programmable Controller System

Prior to the PLC, standalone devices, such as indicators, controllers, and recorders, were used for monitoring and control. These devices are still in use in small operations, but are not cost-effective, and are unsuitable for modern control requirements [1].

The processor in a PLC system has software that is easily programmable and flexible, making the initial program, updates, modifications, and changes easy to implement. Because of the complexity and large number of variables in many process control systems, microprocessor-based PLCs are used for decision making. The PLC can be configured to receive a small number of inputs (both analog and digital), and control a small number of outputs. The system also can be expanded with plug-in modules to receive a large number of signals, and simultaneously control a large number of actuators, displays or other types of devices. PLCs are categorized into low-end, midrange, and high-end, where low-end is from 64 expandable up to 256 I/Os, midrange is expandable up to 2,048 I/Os, and high-end is expandable up to 8,192 I/Os. PLCs have the ability to communicate with each other on a local area network (LAN) or a wide area network (WAN), and to send operational data to, and be controlled from, a central computer terminal. Figure 14.1 shows a typical controller setup for monitoring sequential logic. The input module of the controller senses the condition of the sensors. A decision then can be made by the PLC and the appropriate control signal sent via the output module to the actuators or motors in the process control system.

Figure 14.2 shows the block diagram of the basic controller. A variety of input modules are available for interfacing between the digital and analog signals, PID functions, and the processor's input bus. Output modules are used for actuator

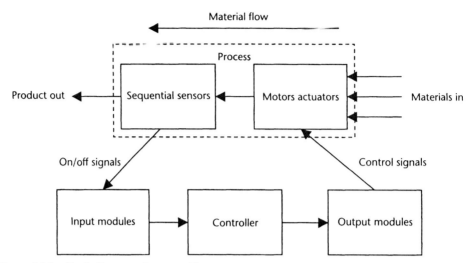

Figure 14.1 Block diagram of a control loop.

control, indicators, alarm outputs, and timing functions, and to interface between the processor's output bus and peripheral units. The modules are rack-mounted, so that only the required modules can be used, leaving rack space for expansion. The memory can be divided into RAM for system operation, ROM, and EEPROM (nonvolatile memory) for storing set point information, and look-up tables [2].

The processor not only controls the process but must be able to communicate to the outside world, as well as to the Foundation Fieldbus (see Section 15.7.2) for communication to smart sensors [3]. All of these control functions may not be required in a small process facility, but are necessary in large facilities. The individual control loops are not independent in a process but are interrelated, and many measured variables may be monitored and manipulated variables controlled simultaneously. Several processors also may be connected to a mainframe computer for complex control functions. Figure 14.3 shows the block diagram of a processor controlling two analog loops. The analog output from the monitors is converted to a digital signal in an ADC. The digital signal is selected in a multiplexer and put into memory by the process, awaiting evaluation and further action. After processing,

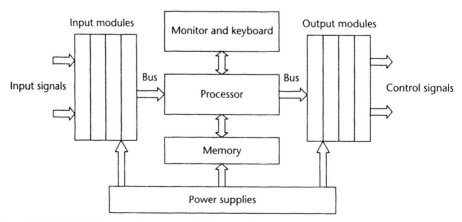

Figure 14.2 Block diagram of programmable controller.

14.3 Controller Operation

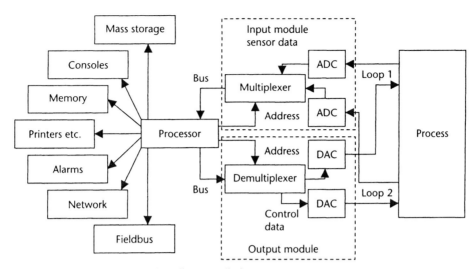

Figure 14.3 Computer-based digitally controlled process.

the digital output signal is fed to the actuator through a demultiplexer and converted to an analog signal by a DAC. The processor also will have mass storage for storing process data for later use or for making charts and graphs. The process also will be able to control a number of peripheral units and sensors as shown, and perform alarm functions [4].

14.3 Controller Operation

The central processing unit can be divided into the processor, memory, and input and output units or modules, as shown in Figure 14.4. The units are interconnected by a two-way 16-bit data bus, a one-way address bus, and a one-way enable bus. The address and data buses are common to all units, and the enable bus will select the individual units being addressed and connect the unit's data bus drivers or receivers to the data bus. The enable and address buses are controlled by the processor, which uses software instructions for its direction. When addressing an input module, the module is selected with its enable bus code. The address bus then can be used to select which external input data is to be put onto the two-way data bus. This data is then transferred to memory to wait for the next step from the software instructions. The output modules are addressed and selected in the same way as the

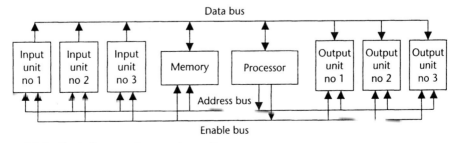

Figure 14.4 Block diagram of processing unit.

input modules. The unit is selected by the enable code, and the address bus directs the data placed on the data bus from the memory by the processor to its output.

The operation cycle in the PLC is made up of two separate modes; these are the I/O scan mode, followed by the execution mode [5].

I/O scan mode is the period when the processor updates the output control signals, based on the information received from the previous I/O scan cycle after its evaluation of the signals. The processor then scans the inputs in a serial mode and updates its internal memory as to the status of the inputs.

Execution mode follows the I/O scan mode. In this mode, the processor evaluates the input data stored in memory against the data programmed into the CPU. The programs usually are set up using ladder networks, where each rung of the ladder is an instruction for the action to be taken for each given input data level. The rung instructions are sequentially scanned, and the input data evaluated. The processor then can determine the actions to be taken by the output modules, and puts the data into memory for transfer to the output modules during the next I/O scan mode.

Scan time is the time required for the PLC to complete one I/O scan plus the execution cycle. This time depends on the number of input and output channels, the length of the ladder instruction sets, and the speed of the processor. A typical scan time is between 5 and 20 ms. As well as evaluating data, the PLC also can generate accurate time delays, store and record data for future use, and produce data in chart or graph form.

14.4 Input/Output Modules

Input/output modules act as the signal interface between the monitoring sensors and actuators, and the controller. I/O modules also provide electrical isolation, if necessary, to convert the input signals into an electronic format suitable for evaluation by the controller; provide memory storage; and format the output signals for displays and control functions. Modules fall into three categories: (1) those for use with discrete I/O levels, (2) those with analog signal levels, and (3) those that have intelligence to evaluate and modify the input signals before they can be used by the controller. Some modules are configured for local signals up to 500 ft, and some are for remote signals from 500 to 10,000 ft. Input/output modules will typically have 16 inputs or outputs, but can be as high as 32, or as low as 4. Modules that have both input and output ports are also available.

14.4.1 Discrete Input Modules

Discrete input modules serve as On/Off signal receivers for the processor. The basic function of the input module is to determine the presence or absence of a signal. The inputs from peripheral devices to the input modules can be ac or dc signals. The voltage ratings for input modules can vary from 24V to 240 V, ac or dc, as well as 5V and 12V TTL levels. The various types of applications that can be used with the discrete input modules are given in Table 14.1.

Figure 14.5 shows examples of the input module tag strips that are normally on the front of the modules. A discrete module typically will have 16 inputs, which can

14.4 Input/Output Modules

Table 14.1 Discrete Input Applications

Type of Input	Application
Discrete input	Push buttons, switch, relay contacts, starter contacts, proximity switch, photoelectric device, float switch
TTL Input	CMOS logic or TTL level
dc or ac Inputs	General purpose high-, medium-, or low-level inputs
Discrete parallel inputs	Thumbwheel switches, bar code readers, weigh scales, position encoders, ADC, BCD/parallel data devices

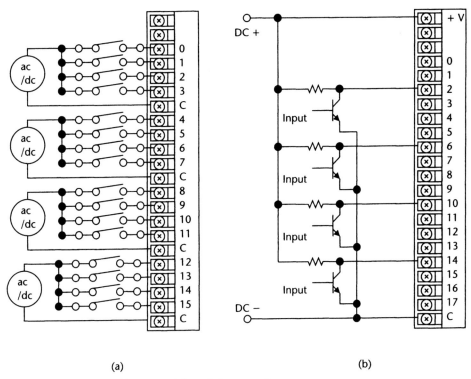

Figure 14.5 Examples of input module switch connections.

be segmented into groups of 4, 8, or 16. Some examples of how the wiring can be implemented in various modules are shown. In Figure 14.5(a), switches are connected in blocks of 4 with ac or dc power supplies. An open switch gives a "0" level input, and a closed switch gives a "1" level input. In Figure 14.5(b), the inputs are transistor logic levels, and the logic output transistor with load is shown. If the transistor is On, the input is a "0" level, and if Off, the input is a "1" level.

The input stage of a dc or ac module is used to detect presence or absence of a voltage, and to convert the input voltage to a logic 5V level. Figure 14.6 shows the block diagram of a discrete input module. The front ends of both the dc and ac modules are shown. With a high dc input voltage, the voltage is stepped down to a low voltage, which then goes through a de-bounce circuit with a noise filter, and threshold detector for "1" or "0" detection, followed by optical isolation, so that the signal can be referenced to the signal ground of the processor. The ac module input

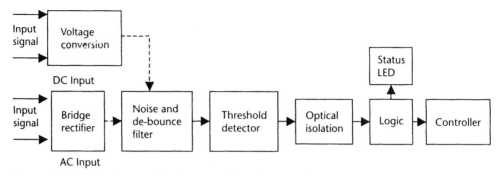

Figure 14.6 Block diagram of an input module with ac or dc input.

uses a bridge rectifier to convert the ac to dc, then uses the same circuit blocks as in the dc module. The LED is used to indicate the input logic level of the input signal. The input level LED indicators are normally located above the tag strip.

14.4.2 Analog Input Modules

Typical analog input modules are used to convert analog signals to digital values or words. Analog signals are derived from temperature, pressure, flow, position, or rate measurements. Analog signals can be single-ended or differential. Figure 14.7 shows the block diagram of a differential analog receiver. The signal is converted into a digital word. The resolution of the ADC is normally 305.176 μV when using voltage input, and 1.2207 μA when using a current input. The ADCs are 14 to 16 bits in length. The inputs are floating, so that the output from the ADC uses an optoisolator to reference the digital signal to the ground level of the controller.

Analog input applications are temperature, humidity, pressure, flow rate, load, thermocouple, RTD, magnetic, and acceleration sensors. The typical input voltage levels are from 50 to 500 mV, up to ±10V, and current ranges from 20 to 50 mA.

14.4.3 Special Function Input Modules

To satisfy some special cases, a variety of discrete modules are available to process or interface special signals. A list of these functions is given in Table 14.2.

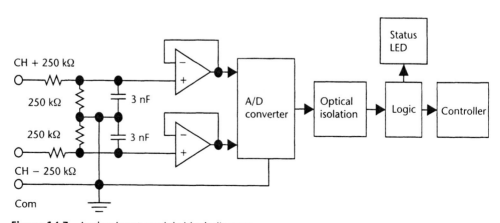

Figure 14.7 Analog input module block diagram.

14.4 Input/Output Modules

Table 14.2 Special Function Input/Output Modules

Module Type	Application
Interrupt input	Immediate response to signal changes
Voltage comparator input	Analog set-point comparison
Latching input	Detection of short duration signals
Fast input	Fast response to dc level changes
Rapid response I/O	Provides fast input/output response
Relay contact output	High current switching and signal multiplexing
Wire fault input	Wire break and short circuit detection

The interrupt function module is used to interrupt the processor's scan sequence, in order to perform a task that requires immediate attention.

The voltage comparator module is used to compare the amplitude of the input to an internally generated voltage or an externally derived voltage.

To detect fast transients of a few microseconds that would normally be missed by standard input modules, the latching input module is used to detect transients and set a latch.

The fast input performs a similar function to the latching module, but does not latch the transient. It only holds the information for a scan cycle, so that it can be detected and recorded.

The rapid response module is similar to the latching input module, but can immediately enable an output without having to wait for a scan cycle.

The relay output module has isolated relay contacts to handle high currents and to multiplex signals.

The fault input is used to interface wire fault detection circuits to the processor.

14.4.4 Discrete Output Modules

Discrete output modules are used to interface output information from the controller to peripheral units, to provide electrical isolation, and to provide the data in a suitable format for use by the external units. The output from the modules can be either discrete ac or dc outputs, or relay contacts. The output voltage can be from 12V to 230V, ac and dc, and TTL levels with multiple or isolated contacts. Table 14.3 shows a list of discrete output applications.

Figure 14.8 shows the block diagram of solid state discrete output drivers using TRIACs. Only two drivers are shown, but normally the drivers would be in groups of four or eight drivers in a module. The outputs have filtering and surge suppression to protect the drivers against transients and inductive spikes, and are fused for protection against overloads. The LED is located above the tag strip, and is used to indicate the logic state of the output.

Table 14.3 Discrete Output Applications

Type of Output	Application
Discrete outputs	Motor starters, solenoids, alarms, horns, buzzers, pilot lights, fans
TTL output levels	TTL and CMOS logic devices
ac, dc outputs	General purpose high-, medium-, or low-load, ac or dc
Parallel outputs	Seven segment displays, BCD controlled message displays, DAC, BCD/parallel data input devices

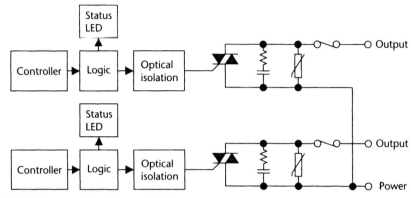

Figure 14.8 Discrete ac output module.

14.4.5 Analog Output Modules

Analog outputs from the PLC drive analog meters, chart recorders, proportional valves, and variable speed drive controllers. They also are used for current or voltage to pneumatic transducers. The voltage and current output ranges are the same as the input ranges. Figure 14.9 shows the block diagram of an analog output stage. The digital output from the controller is fed via an optical isolator to an ADC to reference the signal to the ground of the peripheral device. The analog output of the converter is amplified and fed to a voltage or current driver, which can have a single-ended output or a differential output. The output signal also will meet the standard voltage or current control ranges.

14.4.6 Smart Input/Output Modules

A number of specialized modules have been developed to interface to the controller. They normally contain their own processor and memory, and can be programmed to perform operations independent of the central processor. These modules are categorized in Table 14.4 [6].

Serial and network modules are used for data communication. The serial modules communicate between other PLCs, message displays, operator terminals, and intelligent devices. The network modules are used for LAN and WAN. The Manufacturing Automation Protocol (MAP) is used for communication to robotic devices and a variety of computers of differing manufacturers that can support MAP [7].

The coprocessor modules are basically used for housekeeping functions, such as math functions, algorithms, data manipulation, outputs of reports, outputs to

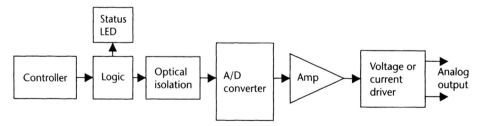

Figure 14.9 Analog output block diagram.

14.4 Input/Output Modules

Table 14.4 Intelligent Input/Output Module Categories

Intelligent Category	Intelligent I/O Module Type
Serial and network communications	ASCII communications module
	Serial communication module
	Loop controller interface module
	Proprietary LAN network module
	MAP network module
Computer coprocessor	PC/AT computer module
	Basic language module
	I/O logic processor module
Closed loop	PID control module
	Temperature control module
Position and motion	High-speed counter module
	Encoder input module
	Stepper positioning module
	Servo positioning module
Process specific modules	Parison control module
	Injection molding module
	Press controller module
Artificial intelligence module	Voice output module
	Vision input module

printer, displays, and mass data storage, using basic programming language functions that would be hard for the main processor to perform with ladder logic.

Intelligent modules performing closed-loop control algorithms are required for PID functions, such as maintaining temperature, pressure, flow, and level at set values. However, the introduction of smart sensors reduces the load on the processor and communication to the processor in this case is via the Fieldbus.

The internal control function of the PID module using an analog loop is shown in Figure 14.10. The analog temperature signal from the furnace is the input to the module, where it can be converted to a digital signal and recorded in the computer memory, fed to the PID controller via a gain control. It is then sent to an analog differencing circuit, where it is compared to the set point signal from the processor. The furnace signal and set point are subtracted, giving an analog error voltage. The error voltage is fed to the PID controller. The controller then produces an analog

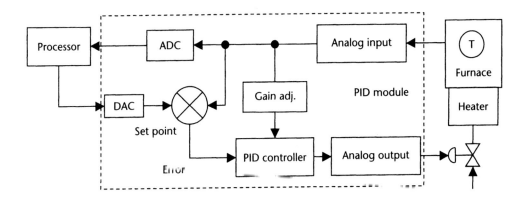

Figure 14.10 Supervisory PID control using an analog control loop.

control voltage to adjust the valve controlling the fuel flow to the furnace. The PID module could also use digital techniques, in which case, the analog input signal would be converted into a digital signal, and compared to the set point in a digital comparator to generate the error signal. The controller will use the error signal to generate a PWM signal to control the furnace fuel actuator [8].

The temperature control module normally controls from 8 to 16 temperature zones. The module is configured for two-position control (heat on/heat off), or three-position control (heat on/heat off/cool). The set points are stored in the processor. A typical application is large building HVAC, or controlling the zone temperatures required in plastic injection molding machines.

Position and motion modules enable PLCs to control stepper and servo motors in feedback loops, to measure and control rotation speeds and acceleration, and to control precision tools. This category of devices uses high-speed counting, rotational and linear position decoders, and open and closed loop control techniques, in order to measure axis rotation and linear speed and position. Typical applications of position and motion modules are given in Table 14.5.

Process specific modules are intelligent modules designed to perform a specific control function or a specific series of operations. Many machines built by different vendors perform similar functions and are similar in operation, using similar inputs and outputs. These modules were developed to interface with such machines. The operations they perform are normally repetitive, requiring precise measurements and complex numerical algorithms. Typical applications are profiling and controlling plastic molding and injection systems [9].

Artificial intelligence modules have a number of industrial applications in voice recognition, synthesized speech, and visual inspection. The sound module can give alarm announcements, voice recognition, and echo evaluation, when using sound waves for flaw detection. The video module can provide dimension gauging, visual inspection, flaw and defect detection, position analysis, and product sorting.

Table 14.5 Applications of Position and Motion Modules

Module	Application
High-speed counter module	Up/down counting
	Generate interrupt for set count
	Generate gating
	Generate delays
Encoder input module	Absolute position tracker
	Incremental position tracker
Stepper-positioning module	Open loop position
	Setting dwell times
	Define motion speed
	Motion acceleration
Servo-positioning module	Transfer and assembly lines
	Material handling
	Machine tool setting
	Table positioning
	Precision parts placement
	Automatic component insertion

14.5 Ladder Diagrams

The ladder diagram is universally used as a symbolic and schematic way to represent the interconnections between the circuit elements used in programming a PLC. The ladder network also is used as a tool for programming the operation of the PLC [10]. The elements are interconnected between the supply lines for each step in the control process, giving the appearance of the rungs in a ladder. A number of programming languages are in common use for controllers, as follows:

- Ladder;
- Instruction list;
- Boolean flowcharts;
- Functional blocks;
- Sequential function charts;
- High-level languages (ANSI, C, structured text).

14.5.1 Switch Symbols

Figure 14.11 shows the typical switch symbols with switch number used for the elements in a ladder diagram. A number of momentary action switches are shown. These are, from top to bottom: a push to close [normally open (NO)] and push to open [normally closed (NC)]. These switches are the normal momentary action panel mounted operator switches.

Position limit switches sense the position of an object, and close or open when a desired position is reached. Pressure, temperature, and level switches set limits, and can be designed to open or close when the set limits are reached. The level switch shown is the lower limit switch (LLS), The upper limit switch (ULS) or full switch (FS) are not shown. The output from the most common flow meter produces pulses that go to a counter. The symbol for the meter is shown.

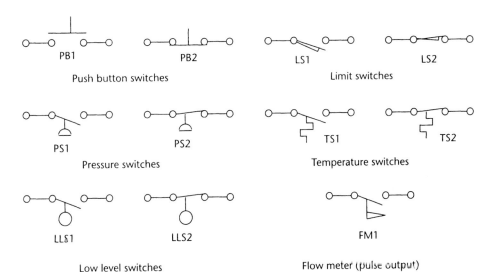

Figure 14.11 Switch symbols in use for ladder diagrams.

14.5.2 Relay and Timing Symbols

Figure 14.12 shows the symbol for relays, timers and counters. A control relay is a circle with the designation CR, followed by the number of the device to distinguish between the various relays used. This is shown with the symbols for its NO and NC contacts. These contacts will carry the same CR and number as the relay. A timer relay has the designation TR in a circle with a number, and its associated NO and NC contacts will be likewise named and numbered. A timer has two sets of contacts: (1) the On-delay contacts that operate after a time delay, and (2) the Off-delay contacts that operate as soon as the delay is initiated and reset at the end of the set time period. The delay time also will have the delay time specified. A counter has the designation CTR with its number. A third lead on the counter is used to activate the count, and to reset the counter after the count has finished. The counter will specify the number of counts.

14.5.3 Output Device Symbols

Symbols for output devices are shown in Figure 14.13. A circle with the letter M and an appropriate number represents a motor. A circle with radiating arms and a letter to indicate its color (e.g., R = red, B = blue, O = Orange, and G = green), plus its number represents an indicator. A solenoid that is used to operate a flow valve or move an object, or for a similar, unspecified, use, is designated SOL with number. Other output elements without specific symbols, such as alarm, heater, or fuse, are represented by boxes, as shown, with the name of the element and its number in the box to distinguish between similar types of elements used in different places.

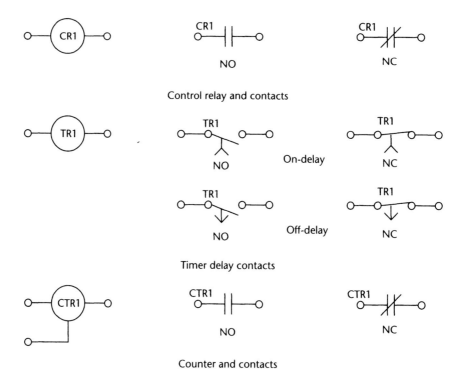

Figure 14.12 Symbols for control relays, timers, counter, and contacts.

14.5 Ladder Diagrams

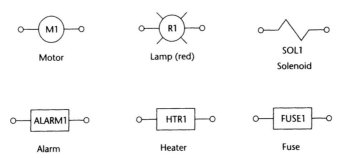

Figure 14.13 Symbols for output devices.

14.5.4 Ladder Logic

The verticals forming the sides of the ladder represent the supply lines. The elements are connected serially between the supply lines, as in a normal electrical schematic, to form the rungs of the ladder. Each ladder rung or step is numbered using the hexadecimal numbering system, with a note describing the function of the rung. The notes are required for debugging and for assisting in future faultfinding, upgrading, and modification. Figure 14.14 shows an electrical component wiring diagram and the equivalent ladder diagram. The components are represented using the open and closed contact symbols in the ladder diagram [11].

14.5.5 Ladder Gate Equivalent

Ladder diagrams can be made from logic diagrams, Boolean expressions, or component electrical wiring diagrams, as shown in Figure 14.14. When using more complex logic diagrams, the logic can be optimized for component count using Boolean

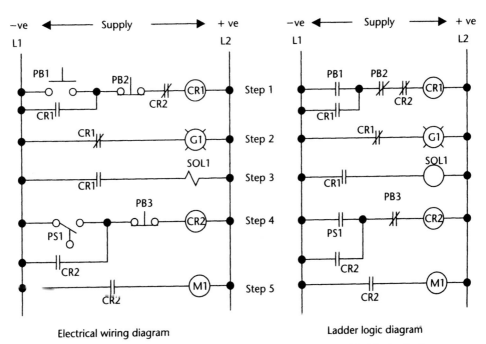

Figure 14.14 Comparison of component electrical wiring diagram and ladder logic diagram.

algebra to minimize the switch count and the number of operations required by the PLC. Figure 14.15 shows how the switch contacts in the ladder diagram are arranged to give the same function as the gate logic. The ladder equivalent of the inverter, AND, NAND, OR, and NOR gate logic are shown.

14.5.6 Ladder Diagram Example

The concept of making a ladder diagram using sequential logic can be best understood by an example, such as by using Figure 16.1 from Example 16.1. The switches and actuators have been assigned numbers, as shown in Figure 14.16.

Example 14.1

A jar-filling system is shown in Figure 14.16. The reservoir contains a mixture of two liquids that must be heated to a preset temperature before the jars can be filled. When the liquid is below a preset level, the heater must be Off. The incoming liquids must be turned Off when the reservoir is full, and not turned On until the liquid level reaches the low-level sensor. The filling of the jars cannot proceed until the liquid reaches a set temperature and above the lower set level in the reservoir. The conveyor belt moves the jars into a filling position that is sensed by a limit switch. When in position, the filling starts. When the jar is full, a level sensor senses the level, and the liquid to the jar turned off. The conveyor belt is then started, and the next jar is moved into position for filling. Design a ladder diagram for a PLC to perform the above control function.

Figure 14.17 shows a possible solution to Example 14.1. The input and output devices are shown down the left-hand side of the ladder. The ladder rungs are

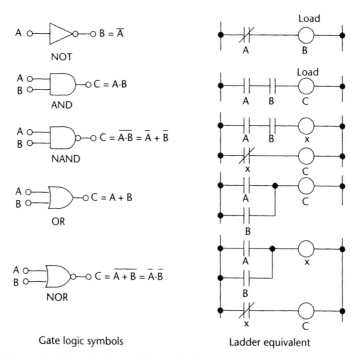

Figure 14.15 Logic ladder equivalent of electronic logic gates.

14.5 Ladder Diagrams

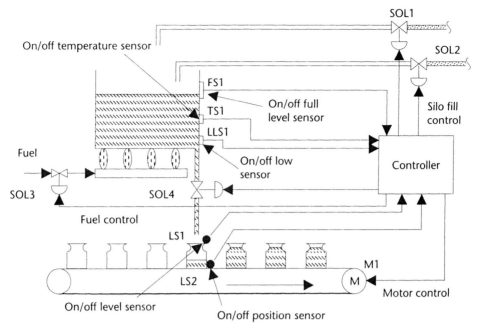

Figure 14.16 On/Off controls used in filling jars.

numbered using hexadecimal numbers, with the rung comments on the right-hand side of the ladder. The first five rungs control the filling of the reservoir, the seventh rung controls the fuel to the heater, and the remaining rungs control the filling and placement of the jars on the conveyor belt. The operation of the system is as follows:

1. Ladder rung S01, the NO contacts of the temperature sensor TS1, controls relay CR1 when the set temperature is reached the relay is energized.
2. Ladder rung S02, the NO contacts of the lower limit level sensor, will energize relay CR2 when the liquid level reaches the set minimum level.
3. Ladder rung S03, the NC relay contacts of CR1, are in series with the NC contacts of the full sensor FS1, whose contacts open when the reservoir is full. CR3 is energized, and opens valves SOL 1 and 2 in rungs S04 and S05 to fill the reservoir. A set of NO contacts of CR3 is in parallel with the NC set of contacts of CR1. CR3 contacts are closed as it is energized, so that when the liquid level reaches the lower limit and LLS1 contacts close, CR1 is energized and its NC contacts in rung S03 will open. CR3 will remain energized until the FS1 contacts open, which deenergizes CR3 and stops the filling. CR3 will not be reenergized until the liquid drops to the lower set limit.
4. Ladder rung S06, the NO contacts from CR1 and CR2, are connected in series, so that CR4 only will be energized when the temperature and liquid levels are above the set minimums, which will prevent filling of the jars until the contacts are energized.
5. Ladder rung S07 prevents fuel to the heater from being turned On via solenoid SOL3 when the liquid level is low, and turns Off the fuel when the set temperature is reached.

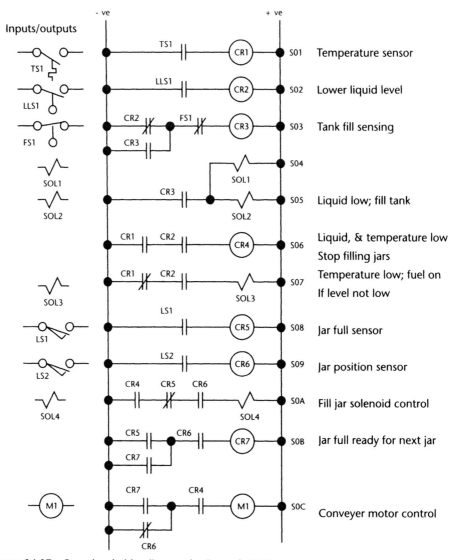

Figure 14.17 Complete ladder diagram for Example 14.1.

6. Ladder rung S08, the NO contacts of the jar full limit switch LS1, energizes relay CR5 when the jar is full.
7. Ladder rung S09, the NO contacts of the jar position limit switch LS2, energizes relay CR6 when the jar is in a filling position.
8. Ladder rung S0A, SOL4, is energized to fill the jar when CR4 is energized, the position limit switch is closed, and the jar is not full.
9. Ladder rung S0B, CR7, is energized when the jar is full. A set of contacts from CR7 are across the full contacts CR5 to keep CR7 energized until CR6 is deenergized, which prevents CR7 from deenergizing, due to the liquid level dropping due to motion.
10. Ladder rung S0C, the conveyor motor M1, is energized when CR4 and CR7 are energized. When CR7 is deenergized by CR6, a set of contacts on CR6 will supply power to the motor until the next jar is in position.

14.6 Summary

This chapter introduced the PLC, and described how it is used for sequential logic control and continuous control. The modular design of the PLC allows expansion from small to large systems as the need arises, and the use of only the required types of plug-in units. The PLC has the ability to interface to sensors, memory devices, printers, alarms, networks, and the Fieldbus. Interface modules, which are used for discrete and analog inputs and outputs, and intelligent modules, which are used for specific functions, are described. The intelligent modules are used for PID, network interface, coprocessors, position and motion control, injection molding, and artificial intelligence functions.

Programming of the PLC using ladder diagrams was discussed. Other forms of programming are instruction lists, Boolean logic, and sequential or high-level languages. The equivalent logic and conversion to ladder functions are shown, as well as the symbols used in ladder diagrams. A comparison between ladder symbols and ladder layout was shown, and an example of sequential control was given with the resulting ladder diagram.

References

[1] Sondermann, D., "Getting Control of the Process," *Sensors Magazine*, Vol. 15, No. 10, October 1998.

[2] Humphries, J. T., and L. P. Sheets, *Industrial Electronics*, 4th ed., Delmar, 1993, pp. 572–590.

[3] Caro, R. H., "Fieldbuses in Process Control," *Proceedings Sensor Expo*, September 1994, pp. 369–374.

[4] Cushing, M., "Redundancy and Self-Diagnostics Enhance Process Safety Systems," *Sensors Magazine*, Vol. 21, No. 10, October 2004.

[5] Dunning, G., *Introduction to Programmable Logic Controllers*, 2nd ed., Delmar, 2002, pp. 275–285.

[6] Jones, C. T., *Programmable Logic Controllers*, 1st ed., Patrick-Turner Publishing Co., 1996, pp. 174–200.

[7] Rinaldi, J. S., "Industrial Automation Networking 2004 and Beyond," *Sensors Magazine*, Vol. 21, No. 1, January 2004.

[8] Carrell, B., "Trends in Electronic Flow Computers," *Sensors Magazine*, Vol. 16, No. 10, October 1999.

[9] Moldoveanu, A., "Trends in Automation," *Sensors Magazine*, Vol. 18, No. 3, March 2001.

[10] Johnson, C. D., *Process Control Instrumentation Technology*, 7th ed., Prentice Hall, 2003, pp. 399–414.

[11] Battikha, N. E., *The Condensed Handbook of Measurement and Control*, 2nd ed., ISA, 2004, pp. 261–265.

Signal Conditioning and Transmission

15.1 Introduction

The amplitude of physical variables are converted into measurable parameters by sensors. The measurable parameters can be a visual indication, or an electrical signal which can be used as an actuator control signal or as a signal to a controller. Many sensors do not have a linear relationship between physical variable and output signal, and are temperature-sensitive. The output signals need to be corrected for the nonlinearity in their characteristics or conditioned for transmission, so that the necessary valves or actuators can accurately corrected for variations in the measured variable in a process control system. Signal conditioning refers to modifications or changes necessary to correct for variations in a sensor's input/output characteristics.

In the case of process control, the accuracy of transmission of the value of the variable is very important. Any errors introduced during transmission will be acted upon by the controller, and degrade the accuracy of the signal. Control signals can be transmitted pneumatically or electrically. Electrical or optical transmission are now the preferred methods, due to the following characteristics of pneumatic transmission: inflexible pluming, high cost, slow reaction time, limited range of transmission, lower reliability, lower accuracy, and increased requirements of the control systems. Electrical signals can be transmitted in the form of voltages or currents, as digital signals, and as wireless signals. Electrical signals can be converted to light signals and transmitted optically. Unfortunately, the terms transducer, converter, and transmitter are often confused and used interchangeably.

15.2 General Sensor Conditioning

The choice of a sensor for a specific application can be determined by its transfer characteristics. However, in many cases the choice of sensors is limited. Sensor transfer characteristics are normally nonlinear, temperature-sensitive, have high noise levels, requires span adjustment, and are offset from zero. The situation is aggravated when precise measurements and a fast correction time are required. A linear relationship is required between process variables and the output signal for actuator control. In analog circuits, linearization is very hard to achieve and requires the use of specialized networks.

15.2.1 Conditioning for Offset and Span

Offset and span are the adjustments of the output level of the sensor, corresponding to the minimum variable value to the zero signal value of the range used, and the system sensitivity. These adjustments assure that the sensor output corresponding to the maximum variable value gives the maximum signal value of the range being used.

Figure 15.1(a) shows the output of a sensor when measuring a variable, and Figure 15.1(b) shows the idealized current output obtained from a conditioning circuit after adjustment of the gain and bias (zero level), as required on most types of sensor outputs.

The accuracy of the sensor signal is not only dependent on the sensor characteristics, but is mainly dependent on the applied conditioning. Many processes require variables to be measured to an accuracy of less than 1% over the full range, which requires not only very accurate sensing, but also temperature compensation, linearization, zero set, and span adjustment. Op-amp offset and amplification are affected by supply voltages, so that these will have to be regulated, and care must be taken with the grounding of the system, in order to minimize ground noise and zero offset. Careful selection is needed in the choice of components. The use of close tolerance components and impedance matching devices is required to prevent the introduction of errors in conditioning networks.

Example 15.1

The output voltage from a sensor varies from 0.7V to 0.35V, as the process variable varies from low to high over its measurement range. However, the sensor output goes to equipment that requires a voltage from 0V to 10V for the range of the variable. Design a circuit to meet these requirements.

The circuit required for changing the output levels is shown in Figure 15.2. The +ve input to the reference buffer is set by the 2 kΩ potentiometer to 0.7V to offset the sensor signal level for the minimum level of the process variable. This will give 0V output when the input to the signal buffer is 0.7V. The gain of the amplifier then can be set to 28.6, giving 10V output with 0.35V input [i.e., 10/(0.7 − 0.35) = 28.6]. Note the use of impedance matching buffers that would be used in instrumentation, and the signal inversion to accommodate the negative signal. This circuit

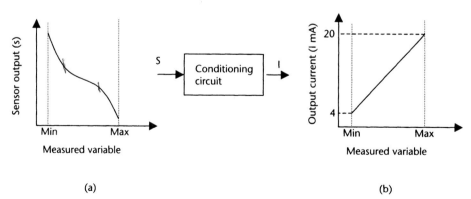

Figure 15.1 (a) Input and ideal output of an ideal linearization circuit, and (b) instrument circuit used for zero and span adjustments, as used in Example 15.1.

15.2 General Sensor Conditioning

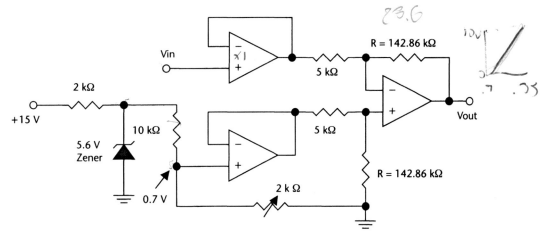

Figure 15.2 Instrument sensor compensation circuit.

does not correct for temperature sensitivity of the sensor, physical variable. The amplifier gain is given by:

$$R = 28.6 \times 5 \text{ k}\Omega = 142.86 \text{ k}\Omega$$

15.2.2 Linearization in Analog Circuits

Example 15.1 shows how to correct for offset and span. Another problem is the nonlinearity in the relation between the measured variable and the sensor output. The approach in analog systems and digital systems will be different. Linearization is difficult unless there is a relatively simple equation to describe the sensor's characteristics. In some applications, a much more expensive linear transducer may have to be used, due to the difficulties in analog circuits to linearize the signal conversion and the possibility of introducing a linearity circuit that is temperature sensitive. Figure 15.3(a) shows the circuit of a logarithmic amplifier. Figure 15.3(b) shows the variations in characteristics obtained with various resistor values that can be used in signal linearization. When $R_2 = \infty$ and $R_3 = 0$, the amplifier has a logarithmic relation between input and output, as shown. When the value of R_3 is larger than zero, the gain is higher at the upper end of the scale, and the curve is shown in the lower shaded portion. If R_2 is a high value resistor less than ∞, then the effect is to reduce the gain at the lower end of the scale, so that the curve is in the upper shaded portion, and its position will depend upon the value of R_2. Multiple feedback paths can be used with nonlinear elements and resistors to approximately match the amplifier characteristics to those of the sensor. Similarly, R_1 can be replaced by a nonlinear element to obtain an antilogarithmic function, or a mix of each can be used to obtain any number of complex transfer functions. The logarithmic and antilogarithmic circuits were discussed in Section 4.3.5.

15.2.3 Temperature Correction

Sensors are notoriously temperature sensitive. That is, their output zero as well as span will change with temperature, and in some cases, the change is nonlinear.

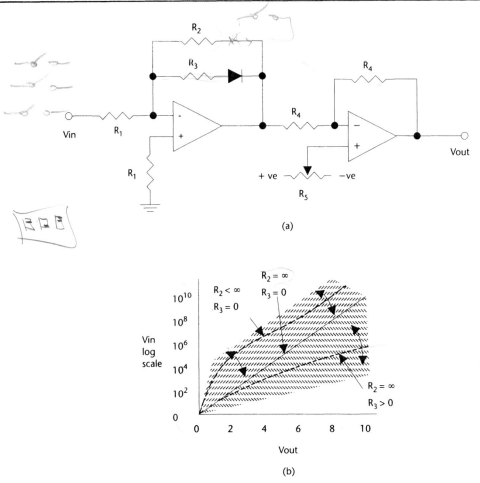

Figure 15.3 (a) Nonlinear amplifier circuit, and (b) characteristics of nonlinear circuit with different feedback values.

Physical variables are also temperature-sensitive and require correction. Correction of temperature effects requires a temperature-sensitive element to monitor the temperature of the variable and the sensor. Correction voltages then can be generated to correct the set zero and span. In Figure 15.2, the 10 kΩ resistor in the biasing network can be placed at the same temperature as the sensor, and can be designed to have the same temperature coefficient as the zero offset of the sensor, in order to compensate for zero drift. A temperature-sensitive resistor in the amplifier feedback (R) can compensate sensor span drift or gain drift with temperature. This feedback resistor also will need to be at the same temperature as the sensor, and track the changes in the sensitivity of the sensor. The temperature compensation in analog circuits will depend on the characteristics of the sensor used, as noted in Example 15.1. Because the characteristics of the sensors vary, the correction for each type of sensor may take a different form.

Temperature compensation is achieved in many sensors by using them in bridge circuits, as shown in Section 4.3.6. Further compensation may be needed to correct for changes in the physical variable due to temperature. Sensor temperature correction requirements can be obtained from the sensor manufacturer application notes and sensor datasheets.

15.2.4 Noise and Correction Time

Differencing amplifiers with high common mode rejection ratios are used to amplify low-level signals in high-noise environments, to obtain a high signal-to-noise ratio. These amplifiers were discussed in Section 4.3.1. The time elapsed from the detection of an error signal to the correction of the error is discussed in Chapter 16.

15.3 Conditioning Considerations for Specific Types of Devices

The method of signal conditioning can vary depending on the destination of the signal. For instance, a local signal for a visual display will not require the accuracy of a signal used for process control.

15.3.1 Direct Reading Sensors

Visual displays are not normally temperature-compensated or linearized. They often use mechanical linkages, which are subject to wear over time, resulting in a final accuracy from 5% to 10% of the reading, with little or no conditioning. However, with very nonlinear sensors, the scale of the indicator will be nonlinear, to give a more accurate indication. These displays are primarily used to give an indication that the system is either working within reasonable limits, or is within broadly set limits (e.g., tire pressure, air conditioning systems, and so forth).

A few sensors have outputs that are suitable for direct reading at the point of measurement, but cannot be used for control or transmission. Such devices include: sight glasses for level indication; liquid in glass, for temperature, rotameter, for flow; hydrometer, for density or specific gravity; and possibly, a liquid-filled U-tube manometer, for differential or gauge pressure measurements.

Visual indicators should be clear and the scale well-defined. Rotameters need to be selected for flow rates and fluid density, and their output values should be corrected for temperature variations from lookup tables. Care needs to be taken to ensure that thermometer bulbs are correctly placed in the fluid for temperature measurement, and do not touch the container walls, since this can effect the temperature reading. When measuring liquid levels, and liquid and gas pressures, the instrument should have conditioning baffles to minimize pressure and level fluctuations, which can introduce uncertainties into the readings.

The Bourdon tube, capsule, and bellows convert pressure into mechanical motion, which is well-suited for conversion to direct visual indication, as discussed in Section 7.3. These devices are cost-effective and in wide use, but are not temperature-compensated, and the cheaper instruments do not have zero or span adjustment. More expensive devices may have screw adjustments and a limited temperature range.

15.3.2 Capacitive Sensors

Capacitive sensing devices can use single-ended sensing or differential sensing. Single-ended sensing capacitance is measured between two capacitor plates, as shown in Chapter 8, Figure 8.7. Differential sensing can be used when there is a capacitor plate on either side of, and in close proximity to, a central plate or diaphragm, as shown in Chapter 7, Figure 7.5(a). In differential sensing, the two capacitors (A and B) can be used to form two arms of an ac bridge, or switch capacitor techniques can be used. For single-ended sensing, a fixed reference capacitor (B) can be used with a variable capacitor (A). Capacitive sensing can use ac analog or digital measuring techniques.

Figure 15.4 shows an ac bridge that can be used with capacitive sensing. Initially, the bridge is balanced for zero offset with potentiometer R_3, and the output from the bridge is amplified and buffered. The signal will be converted to a dc signal and further amplified for transmission.

Switch capacitor sensing techniques can use open loop or closed loop sensing techniques. Figure 15.5 shows an open loop switch circuit for sensing capacitance changes. The top capacitor is switched from V_{REF} to $0.5V_{REF}$, and the bottom plate is switched from $0.5V_{REF}$ to ground. Any difference in the capacitance of the upper and lower plates will appear as a charge on the input to the first amplifier. This amplifier is used as a charge amplifier and impedance matching circuit. The output of the first amplifier goes to a sample and hold circuit, where the charges are held in a capacitor and then become a voltage, which is amplified by the second amplifier to give a dc output voltage that is proportional to the capacitance difference. The second amplifier also modulates the $0.5V_{REF}$ voltage and feeds it back to the switches, so that the voltage across each capacitor is proportional to the distance between the capacitor plates. This prevents electrostatic forces due to the driving voltages from producing a deflection force on the diaphragm [1]. This can be a problem for micromachined devices where the capacitor spacing is less than 3 μm. This type of technique gives good linearity (better than 1%). In applications such as capacitive level sensors, the temperature of the liquid also must be measured, so that corrections can be made for the changes in the dielectric constant of the liquid due to temperature changes.

15.3.3 Magnetic Sensors

The resistance of MRE devices change in a fluctuating magnetic field, and MRE devices are also temperature-sensitive. Figure 15.6 shows the circuit used to condition the signal from an MRE into a digital signal in on/off applications. The MRE sensor contains four elements to form a bridge circuit. The four elements are

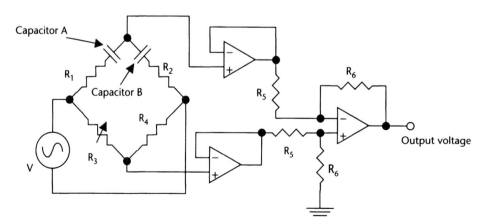

Figure 15.4 (a) Capacitive diaphragm pressure sensor, and (b) ac bridge for use with a capacitive sensor.

15.3 Conditioning Considerations for Specific Types of Devices 257

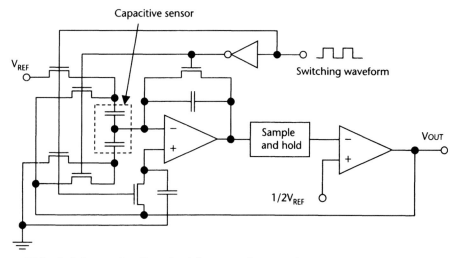

Figure 15.5 Switch capacitor filter circuit for measuring capacitance.

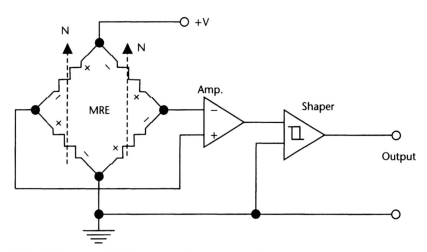

Figure 15.6 MRE magnetic field sensing device and circuit.

connected so that their resistance change is additive in a magnetic field, but that the temperature effects on resistance cancel. The output from the bridge is amplified and goes to a pulse-shaping circuit. When the Hall and MRE devices are being used as switches in a digital configuration, and they do not normally require temperature compensation for sensitivity changes. When used in turbine flow meters some conditioning may be required for the density changes in the liquid caused by temperature changes. For high and low flow rates, the conditioning will depend on the requirements of the application and manufacturers' specifications [2].

15.3.4 Resistance Temperature Devices

Sensor using resistance temperature devices (RTD) measure the change in electrical resistance of a wire-wound resistor with temperature. Typically, a platinum resistance element is used. RTD elements can be connected directly to the controller peripheral sensing circuits, using a two-, three-, or four-wire lead configuration, as shown in Figure 15.7. The resistance change can be measured in a bridge circuit, or the resistor can be driven from a constant current source, and the voltage developed across the resistor measured. The resistance of the element is low (100Ω) to minimize temperature changes due to internal heating of the resistor. If heating occurs, pulse techniques can be used to prevent the internal heating. In this case, the current is turned on for a few milliseconds, the voltage is measured, and then turned off for approximately 1 second. Figure 15.7(a) shows the simplest and cheapest connection to the RTD with just two leads, and the meter is connected to the current supply leads. The resistance of the leads between the detector and the resistor in the two lead wires can be significant, giving a relatively high degree of error. The meter is measuring the voltage drop across the current lead resistance and junctions as well as the RTD.

The three-wire connection Figure 15.7(b) is a compromise between cost and accuracy, and the four-wire connection Figure 15.7(c) is the most expensive but most accurate. The wires in all cases will be in screened cables [3].

The three-wire connection was discussed in Section 3.4.2, Figure 3.11. With the four-wire connection, the voltmeter is connected directly to the RTD, as shown in Figure 15.7(c). Since no current flows in the leads to the voltmeter, there is no voltage drop in the measuring leads due to the supply current, and a very accurate RTD voltage reading can be obtained. The accuracy of RTDs is typically <0.5% using the three-wire configuration, but can be improved to <0.1% of FSD by limiting the temperature range from −200° to +280°C in the four-wire configuration. The response time of the element is typically 0.5 seconds, but can be as high as 5 seconds with a stainless steel shell.

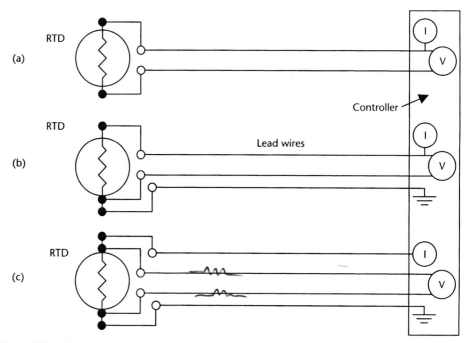

Figure 15.7 Alternative connection schemes between an RTD and a controller: (a) two-lead, (b) three-lead, and (c) four-lead.

15.3.5 Thermocouple Sensors

Thermocouples have several advantages over other methods of measuring temperature. They are very small in size; have a low time response (from 10 to 20 ms, compared to several seconds for some elements); are reliable; have good accuracy; operate over a wide temperature range; and can convert temperature directly into electrical units. The disadvantages are the need for a reference and the low signal amplitude. Thermocouples are compensated, as shown in Figure 10.6. The amount of conditioning required by a thermocouple will depend on its temperature-measuring range. An accuracy of ±1% over a limited temperature range without compensation can be obtained, as shown in Table 10.6, but conditioning (linearizing) is needed if used over its full operating range. Nonlinear amplifiers can be used to obtain an accuracy of ±0.5% of FSD. The set zero conditioning is a part of the reference temperature correction. Thermocouple voltages also can be sensed directly by the controller using an internal amplifier, and then conditioned internally. This is discussed in Section 10.3.4.

Controller peripheral modules are available for amplification of several thermocouple inputs with cold junction correction. Figure 15.8(a) shows a differential connection between the amplifier and the thermocouple as twisted pairs of wires that are screened to minimize noise. Other configurations of thermocouples are shown in Figures 15.8(b, c). In Figure 15.8(b), the thermocouples measure the average temperature at three points in a material, and in Figure 15.8(c), the thermocouples measure the temperature difference between two points in a material.

15.3.6 LVDTs

Sensors such as capsules and bellows normally use LVDTs as a motion to electrical transducer. The LVDT device is rugged and has excellent resolution with low

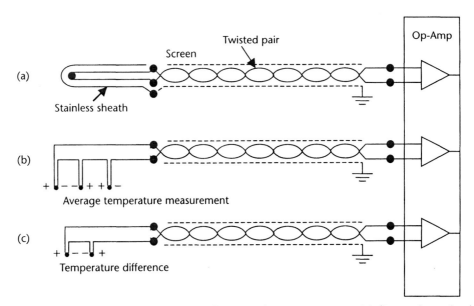

Figure 15.8 Different types of thermocouple connections to an op amp: (a) direct, using twisted pair to a reference and amplifier; and (b, c) for average temperature measurement and differential temperature measurement.

hysteresis, but is large, expensive, sensitive to stray magnetic fields, and has poor linearity, as shown in Section 11.2.1. LVDTs should be screened from magnetic fields, and should be used in closed-loop configurations, as shown in Chapter 7, Figure 7.6. The closed-loop configuration has several advantages over linear conversion. The feedback counteracts mechanical movement, and therefore linearizes the conversion, minimizes hysteresis effects, and reduces strain in the mechanical driving device.

15.3.7 Semiconductor Devices

Wide ranges of measurements are made using semiconductor devices. These devices are used to measure pressure, acceleration (MEMS), temperature, light intensity, strain, force, and so forth. These integrated electronics in these devices give high sensitivity and conditioning, do not suffer fatigue, do not need recalibration, and can handle large overloads, but have a limited temperature operating range from −40° to +150°C. Information on device characteristics and usage can be obtained from manufacturers' data sheets and application notes.

15.4 Digital Conditioning

Many analog signals are converted to digital signals for transmission. In many cases, the output from a sensor can be converted directly into a digital signal, and optoisolators can provide ground isolation. The value of capacitors and resistive-type devices can be accurately sensed using digital techniques, eliminating the need for analog amplification, but conditioning of the sensor signal may still be required. However, all of the conditioning in the digital domain can be performed by the processor in the controller, using software or lookup tables obtained from a knowledge of the temperature characteristics of the sensing device, and using physical variables.

15.4.1 Conditioning in Digital Circuits

Conditioning is performed for nonlinear devices by using equations or memory lookup tables [4]. If the relationship between the values of a measured variable and the output of a sensor can be expressed by an equation, then the processor can be programmed based on the equation to linearize the data received from the sensor. An example would be a transducer that outputs a current (I) related to flow rate (v) by:

$$I = Kv^2 \qquad (15.1)$$

where K is a constant.

The current numbers from the sensor are converted into binary signals, where the relationship still holds. In this case, a linear relationship is required between current and flow rate. This can be obtained by multiplying the I term by itself. The resulting number is proportional to v^2, and the generated number and flow now have a linear relationship. Span and offsets now may require further adjustment.

There are many instances in conversion where there is not an easily definable relationship between variable and transducer output, and it may be difficult or impossible to write a best fit equation that is adequate for linearization of the variable. In this case, lookup tables are used. The tables correlate transducer output to the true value of the variable, and these values are stored in memory. The processor can retrieve the true value of the variable from the transducer and temperature reading by consulting its lookup tables. This method is extensively used, for instance, with thermocouples.

15.5 Pneumatic Transmission

Pneumatic signals were used for signal transmission, and are still in use in older facilities, or in applications where electrical signals or sparks could ignite combustible materials. Pneumatic transmission of signals over long distances requires an excessively long settling time for modern processing needs, especially when compared to electrical signal transmissions. Pneumatic signal lines are also inflexible, bulky, and costly, compared to electrical signal lines, and are not microprocessor-compatible. They will not be used in new designs, except possibly in the special circumstances mentioned above. Pneumatic transmission pressures were standardized into two ranges—3 to 15 psi (20 to 100 kPa), and 6 to 30 psi (40 to 200 kPa). The 3 to 15 psi range is now the preferred range for signal transmission. Zero is not used for the minimum of the ranges, since low pressures do not transmit well. The zero level can then be used to detect system failure.

15.5.1 Signal Conversion

Pneumatic signals as well as electrical signals can be used to control actuators. Signal conversion is required between low-level signals and high-energy control signals for actuator and motor control. Electrical control signals can be either digital, analog voltage, or analog current. It is sometimes necessary to convert electrical signals to pneumatic signals for actuator control [5].

One of the many designs of a current to pressure converter is shown in Figure 15.9(a), in which the spring tends to hold the flapper closed, giving a high-pressure

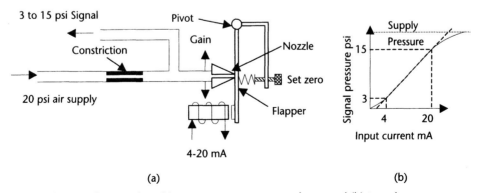

Figure 15.9 Signal conversion: (a) current to pressure transducer, and (b) transducer characteristic.

output (15 psi). When current is passed through the coil, the flapper moves towards the coil, closing the air gap at the nozzle and increasing the output air pressure. The output air pressure is set to the maximum of 3 psi by the set zero adjustment when the current through the coil is 3 mA. Moving the nozzle along the flapper sets the system gain and span. There is a linear relationship between current and pressure, as can be seen from the transducer characteristic shown in Figure 15.9(b).

A linear pneumatic amplifier or booster can be used to increase the pressure from a low-level pressure signal to a high-pressure signal for actuator control.

15.6 Analog Transmission

15.6.1 Noise Considerations

Analog voltage or current signals are hardwired between the transmitter and the receiver. These signals can be relatively slow to settle compared to digital signals, due to the time constant of the lead capacitance, inductance, and resistance, but the signals are still very fast in terms of the speed of mechanical systems. Analog signals can lose accuracy if signal lines are long with high resistance. The signals can be susceptible to ground offset, ground loops, radio frequency (RF) and EMI noise from transmitters and motors, and so forth.

To reduce these problems, the following precautions should be taken:

- The dc supply to the transmitter is generated from the ac line voltage via an isolation transformer and voltage regulators, to minimize noise from the power supply.
- The ground connection is used only for the signal return path.
- The signal and ground return leads are a screened twisted pair, with the screen grounded at one end only.

Other necessary compensations can include: filtering to remove unwanted frequencies, such as pickup from the 60 Hz line frequency, noise, or RF pickup; dampening out undulations or turbulence to give a steady average reading; and correcting for time constants and for impedance matching networks.

15.6.2 Voltage Signals

Voltage signals are normally standardized in the voltage ranges 0V to 5V, 0V to 10V, or 0V to 12V, with 0V to 5V being the most common. The requirements of the transmitter are: a low output impedance, to enable the amplifier to drive a wide variety of loads without a change in the output voltage; low temperature drift; low offset drift; and low noise. Improved voltage signal transmission can be obtained using a differential signal, as shown in Figure 15.10. In this case, the transmitter sends a differential signal via a screened twisted pair. Because any RF and EMI pickup will affect both signal lines by the same amount, any noise will cancel in the differential receiver in the controller. Ground noise and offsets do not normally affect differential signals.

15.6 Analog Transmission

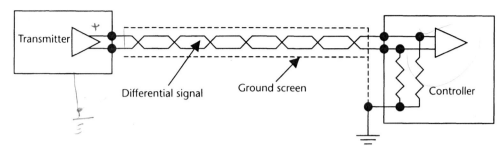

Figure 15.10 Screened differential signal connection between the controller and the transmitter.

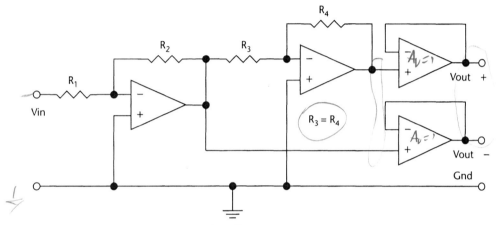

Figure 15.11 Differential amplifier with buffer outputs.

A differential output voltage signal can be generated using the circuit shown in Figure 15.11. The output stages used are buffers to give low output impedance, and are driven from a unity gain inverter to generate equal and opposite phase signals. Op-amps also are commercially available with differential outputs, which can be used to drive the output buffer stages.

Figure 15.12 shows a transmitter with a voltage output signal and line impedances. The low output impedance of the drivers enables them to charge up the line capacitance, achieving a quick settling time. However, the input voltage to the controller (V_{in}) can be less than the output voltage (V_{out}) from the transmitter, due to resistance losses in the cables (i.e., if the receiver is drawing any current).

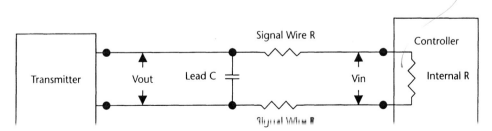

Figure 15.12 Line impedances.

$$V_{in} = \frac{V_{out} \times \text{Internal } R}{\text{Internal } R + 2\text{Wire } R} \tag{15.2}$$

Thus, the internal R of the controller must be very high compared to the resistance of the wire and connections to minimize signal loss, which is normally the case.

15.6.3 Current Signals

Current signals are standardized into two ranges—4 to 20 mA, and 10 to 50 mA, where 0 mA is a fault condition. The latter range formerly was the preferred standard, but has now been replaced by the 4 to 20 mA range as the accepted standard. The requirements of the transmitter are: high output impedance, so that the output current does not vary with load; low temperature; low offset drift; and low noise. Figure 15.13 shows an output current driver that gives low output impedance. The emitter follower transistor in the feedback loop is used to reduce the output impedance of the op-amp, to <20Ω. Because the transistor is within the feedback loop, it does not affect the overall gain of the circuit, which is set by the feedback resistors.

Figure 15.14 shows a transmitter with a current output. The main disadvantage of the current signal is the longer settling time due to the high output impedance of the driver, which limits the current available to charge up the line capacitance. The signal current at the controller after the line capacitance is charged is the same as the signal current from the transmitter, and is not affected by normal changes in lead resistance. The internal resistance of the controller is low for current signals (e.g., a few hundred ohms), which helps to lower the time constant of the transmission line.

15.7 Digital Transmission

15.7.1 Transmission Standards

Digital signals can be transmitted without loss of integrity, via a hardwired parallel or serial bus, radio transmitter, or fiber optics. Digital data transmission speeds are

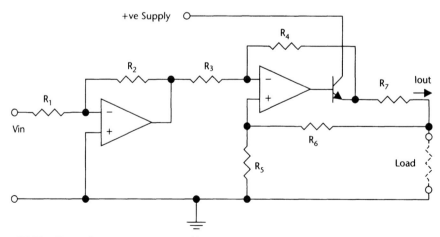

Figure 15.13 Transmitter current output driver.

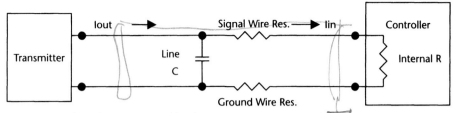

Figure 15.14 Effect of resistance and lead capacitance on current signals.

higher than with analog data. Another advantage is that digital transmitters and receivers require much less power than analog transmission devices.

The IEEE defines communication standards for digital transmission between computers and peripheral equipment. The standards are the IEEE-488 or RS-232 [6]. However, several other standards have been developed and are now in use. The IEEE-488 standard specifies that a digital "1" level will be represented by a voltage of 2V or greater, and a digital "0" level will be represented by a voltage of 0.8V or less. The signal format to be used is also specified. The RS-232 standard specifies that a digital "1" level will be represented by a voltage of between +3V and +25V, and a digital "0" level will be represented by a voltage of between −3V and −25V. The signal format to be used is also specified. Fiber optics are now being extensively used to give very high speed transmission over long distances, and are not affected by electromagnetic or RF pickup.

Digital signals can be transmitted without loss of accuracy, and can contain codes for limited automatic error correction or for automatic requests of data retransmission. These networks are known as LANs when used in a limited area, and as WANs when used as a global system. A typical LAN network is shown in Figure 15.15. Engineers, programmers, supervisors, and operators can communicate with the process controllers over the LAN, for functions such as process monitoring, cost information scheduling and entering or modifying programs.

15.7.2 Foundation Fieldbus and Profibus

The development of the smart sensor has caused a revolution in the method of communication between the sensing elements, actuators, and controllers. The controller now takes on a supervisory role of the smart sensor. The controller receives data

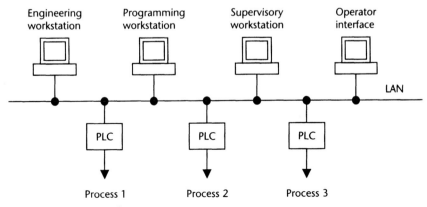

Figure 15.15 A LAN network.

from the smart sensor, and updates the smart sensor as required, using a digital serial bus [7].

The Foundation Fieldbus (FF) and Profibus are the two most universal serial data bus formats that have been developed for interfacing between a central processor and smart sensing devices in a process control system. The FF is primarily used in the United States and the Profibus format is primarily used in Europe. Efforts are being made for a universal acceptance of one bus system. Process control equipment is presently being manufactured for either one system or the other, but global acceptance of equipment standards would be preferred. A serial data bus is a single pair of twisted copper wires, which enables communication between a central processing computer and many monitoring points and actuators when smart sensors are used. This is shown in Figure 15.16. Although initially more expensive than direct lead connections, the advantages of the serial bus are minimal bus cost and installation labor. The system replaces the leads to all the monitoring points by one pair of leads. New units can be added to the bus with no extra wiring (i.e., a plug and play feature), giving faster control, and programming is the same for all systems. Accuracies are obtained that are higher than from using analog, and more powerful diagnostics are available. As the cost of integration and development decreases, the bus system with its features will become more cost-effective than the present systems. This was discussion in Section 6.5.1.

The bus system uses time division multiplexing, in which the serial data word from the central processor contains the address of the peripheral unit being addressed in a given time slot, and the data being sent. In the FF, current from a constant current supply is digitally modulated. Information on the FF is given in the ISA 50.02 standards.

One disadvantage of the FF is that a failure of the bus, such as a broken wire, can shut down the entire process, whereas with the direct connection method, only one sensor is disabled. This disadvantage can be overcome by the use of a redundant or a backup bus in parallel to the first bus, so that if one bus malfunctions, then the backup bus can be used [8].

A comparison of the characteristics of the serial data buses is given in Table 15.1. The original Foundation Fieldbus was designated H1. A new generation of the H1 is the HSE, which will use an Ethernet LAN bus to provide operation under the TCP/IP protocol used for the internet. The advantages are increased speed, unlimited addresses, and standardization [9].

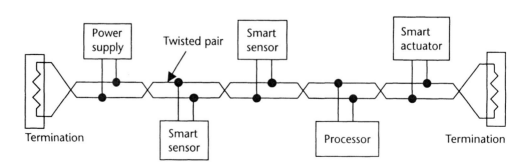

Figure 15.16 Foundation Fieldbus cable connection.

15.8 Wireless Transmission

Wireless signal transmission can be divided into short range networks, and telemetry for long range.

15.8.1 Short Range Protocols

A number of protocols are available for short range digital communication. The more common of these are Wi-Fi, Zigbee, and Bluetooth. These devices were designed for operation in high noise environments, using frequency hopping techniques. Zigbee and Bluetooth were both designed for battery operation, with low power consumption. The basic characteristics of these devices are:

- Wi-Fi IEEE 802.11(a, b, and g), range 1m to 100m;
- Zigbee IEEE 802.15.4, range 1m to 75m;
- Bluetooth IEEE 802.15.1 range 1m to 10m.

15.8.2 Telemetry Introduction

Telemetry is the wireless transmission of data from a remote location to a central location for processing and/or storage. This type of transmission is used for sending data over long distances (e.g., from weather stations), and data from rotating machinery where cabling is not feasible. More recently, wireless communication is being used to eliminate cabling or to give flexibility in moving the positioning of temporary monitoring equipment. Broadcasting information in a wireless transmission using amplitude modulation (AM) or frequency modulation (FM) techniques is not accurate enough for the transmission of instrumentation data, since reception quality varies and the original signal cannot be accurately reproduced. In telemetry, transmitters transmit signals over long distances using a form of FM or a variable width AM signal. When transmitting from battery-operated or solar cell–operated equipment, it is necessary to obtain the maximum transmitted power for the minimum power consumption. FM transmits signals at a constant power level, whereas AM transmits at varying amplitudes, using pulsing techniques. AM can transmit only the pulse information needed, which conserves battery power, so that for the transmission of telemetry data pulse, AM is preferred.

Table 15.1 Comparison of Bus Characteristics

	Fieldbus (H1)	Profibus	Fieldbus (HSE)
Bus type	Twisted pair copper	Twisted pair copper or fiber optic cable	Twisted pair copper and fiber optic cable
Number of devices	240 per segment; 65,000 segments	127 per segment; 65,000 segments	Unlimited
Length	1,900m	100m copper plus 24 km fiber	100m copper plus 24 km fiber
Maximum speed	31.25 Kbps	12 Mbps	100 Mbps
Cycle time (ms)	<600	<2	<5

15.8.3 Width Modulation

Width-coded signals or PWM are blocks of RF energy whose width is proportional to the amplitude of the instrumentation data. Upon reception, the width can be accurately measured and the amplitude of the instrumentation signal reconstituted. Figure 15.17(a) shows the relation between the voltage amplitude of the instrumentation signal and the width of the transmitted pulses, when transmitting a series of 1V signals and a series of voltages through 10V.

For further power saving, PWM can be modified to pulse position modulation (PPM). Figure 15.17(b) shows a typical PWM modulation and the equivalent PPM signal. The PWM signal shows an Off period for synchronization of transmitter and receiver. The receiver then synchronizes on the rising edge of the transmitted zero. The first three pulses of the transmission are calibration pulses, followed by a stream of width-modulated data pulses. In the case of the PPM, narrow synchronization pulses are sent, and then only a pulse corresponding to the lagging edge of the width-modulated data is sent. Once synchronized, the receiver knows the position of the rising edge of the data pulses, so that only the information on the lagging edge is required for the receiver to regenerate the data. This form of transmission has the advantage of greatly reducing power consumption and extending battery life [10].

15.8.4 Frequency Modulation

When using frequency modulation, the unmodulated carrier has fixed amplitude and frequency. When modulated, the frequency of the transmitted signal is varied in

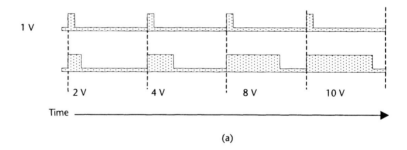

(a)

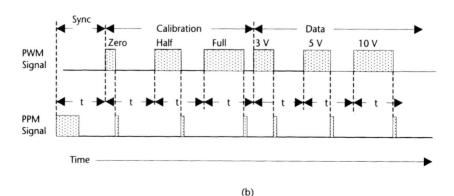

(b)

Figure 15.17 (a) Amplitude modulated waveform, in which the width of the modulations corresponds to voltage levels; and (b) comparison of PWM and PPM waveforms.

proportion to the amplitude of the variable signal, and the amplitude of the transmitted signal does not change. Upon reception, the base frequency of the transmission can be subtracted from the received signal, leaving the frequency of the modulating signal. This signal frequency then can be measured, and the data reconstituted to determine the amplitude of the variable.

15.9 Summary

Various methods of analog and digital signal conditioning to correct for nonlinear characteristics in sensors, temperature effects, and offset zero were discussed. Methods of correction include logarithmic amplifiers, impedance matching amplifiers, and bias adjustment in linear conditioning, and lookup tables and equations in digital conditioning. The conditioning of the various types of sensors was considered, together with their amplification and the signal transmission mode. Both pneumatic and electrical transmissions were considered, as well as the span adjustment of the signal. When using electrical transmission, voltage, current, or digital modes can be used, and their relative merits were discussed. Digital transmission is preferred, due to lower power requirements, greater integrity of transmission, higher speed, minimized noise, and increased direct interfacing with the controller. Digital signal transmission standards, as well as the use of Fieldbus, to reduce cost and the load on the controller, were discussed.

Wireless transmission falls into two categories: short range, such as Bluetooth and Zigbee; and long range, in which telemetry signals can be transmitted using pulse modulation techniques to minimize power requirement in battery operated equipment.

Definitions

Converters are devices that convert a signal format without changing the type of energy (e.g., an op-amp that converts a voltage signal into a current signal).

Offset refers to the low end of the operating range of a signal. When performing an offset adjustment, the output from the transducer is being set to give the minimum output (usually zero) when the input signal value is a minimum.

Sensors are devices that sense a variable, and give an output (e.g., mechanical, electrical) that is related to the amplitude of the variable.

Span references the range of the signal (i.e., from zero to full scale deflection). The span setting (or system gain) adjusts the upper limit of the transducer with maximum signal input. There is normally some interaction between offset and span. The offset should be adjusted first, and then the span.

Transducers are systems that change the output from a sensor into some other energy form, so that it can be amplified and transmitted with minimal loss of information.

Transmitters are devices that accept low-level electrical signals and format them, so that they can be transmitted to a distant receiver without loss of integrity.

References

[1] O'Grady, A., "Harness the Power of Sigma-Delta and Switched-Capacitor Technologies in Signal Processing Components," *Sensors Magazine*, Vol. 15, No. 11, November 1998.

[2] Caruso, M., et al., "A New Perspective on Magnetic Field Sensors," *Sensors Magazine*, Vol. 15, No. 12, December 1998.

[3] Jones, C. T., *Programmable Logic Controllers*, 1st ed., Patrick-Turner Publishing Co., 1996, pp. 150–164.

[4] Bryzek, J., "Signal Conditioning for Smart Sensors and Transducers," *Proceedings Sensors Expo*, October 1993.

[5] Johnson, C. D., *Process Control Instrumentation Technology*, 7th ed., Prentice Hall, 2003, pp. 322–330.

[6] Dunning, G., *Introduction to Programmable Logic Controllers*, 2nd ed., Delmar, 2002, pp. 287–295.

[7] Pullen, D., "Smart Sensors and Sensor Signal Processors: Foundation for Distributed Sensing Systems," *Proceedings Sensor Expo*, October 1993.

[8] Battikha, N. E., *The Condensed Handbook of Measurement and Control*, 2nd ed., ISA, 2004, pp. 171–173.

[9] Potter, D., "Plug and Play Sensors," *Sensors Magazine*, Vol. 19, No. 12, December 2002.

[10] Humphries, J. T., and L. P. Sheets, *Industrial Electronics*, 4th ed., Delmar, 1993, pp. 512–540.

CHAPTER 16
Process Control

16.1 Introduction

The two basic modes of process control are sequential control, or On/Off action, and continuous control action. This chapter deals with process control, discussing both sequential and continuous control. Sequential control is an event-based process, in which one event follows another until a process sequence is complete. Many process control functions are sequential.

Continuous control requires continuous monitoring of the process variables, so that they can be held at set levels. Process control systems vary extensively in complexity and industrial application. For example, controllers in the petrochemical industry, automotive industry, and soda processing industry have completely different types of control functions. The control loops can be very complex, requiring microprocessor supervision. Some of the functions need to be very tightly controlled, with tight tolerance on the variables and a quick response time, while in other areas, the tolerances and response times are not so critical. These systems are closed loop systems. The output variable level is monitored against a set reference level, and any difference detected between the two is amplified and used to control an input variable that will maintain the output at the set reference level.

16.2 Sequential Control

There are many processes in which a variable does not have to be controlled, but a sequence of events has to be controlled. The sequential process, or batch process, can be event-based, time-based, or a combination of both. An event-based system is one in which the occurrence of an event causes an action to take place. For example, a level sensor detects that a container is full, and switches its output from On to Off, which sends a signal to the controller to turn off the filling liquid. A time-based sequence is one in which an event is timed by the controller and is not waiting for an input signal. A batch process can be distinguished from a continuous process as follows:

- All actions can be defined by On/Off states.
- Discrete quantities are handled individually at each step, rather than as a continuous flow.
- Each quantity is kept as a separate unit and can be individually identified.

- Each unit, unlike in continuous processing, can see different process steps.
- Each unit will not start a new step until the previous step has been completed.
- Batch processes can have process variations for each batch of new material, depending on previous processing.

It should be noted that a continuous process could be part of, or a step in, a discrete process.

Example 16.1

The On/Off controller action has many applications in industry. An example of some of these uses is shown in Figure 16.1. In this case, jars on a conveyor belt are being filled from a silo. When a jar is full, it is sensed by the level sensor, which sends a signal to the controller to turn Off the hot liquid flowing from the silo, and to start the conveyer moving. As the next jar moves into the filling position, it is sensed by the position sensor, which sends a signal to the controller to stop the conveyor belt, and to start filling the jar. Once it is full, the cycle repeats itself [1].

Level sensors in the hopper senses when the hopper is full, and when it is almost empty. When empty, the sensor sends a signal to the controller to turn On the feed valve to the silo; when full, the level is detected and a signal is sent to the controller to turn the feed to the silo Off. When the silo is full, the heater is turned On to bring the liquid up to temperature before the jars can be filled. An On/Off temperature sensor senses the temperature, which informs the controller when to resume filling, and to stop the operation if the temperature in the silo falls below the minimum set temperature.

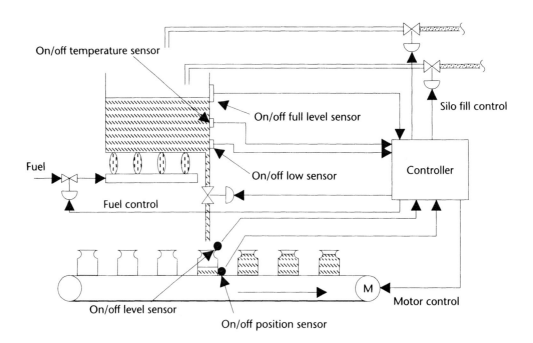

Figure 16.1 Example of the use of On/Off controls used for jar filling.

16.3 Discontinuous Control

Discontinuous control mode is frequently used in process control, and could be considered as a discrete On/Off mode. In reality, discontinuous control is the basis of continuous control, in its simplest and least expensive form. It is used where its disadvantages can be tolerated. Such systems have high inertia, which prevents the system from continuously cycling [2].

16.3.1 Discontinuous On/Off Action

The measured variable is compared to a set reference. When the variable is above the reference, the system is turned On, and when below the reference, the system is turned Off, or vice versa, depending upon the system design. This could make for rapid changes in switching between states. However, some systems, such as heating systems, normally have a great deal of inertia or momentum, which produces overswings and introduces long delays or lag times before the variable again reaches the reference level. Figure 16.2 shows an example of a simple room heating system. The top graph shows the room temperature or measured variable, and the lower graph shows the actuator signal. The room temperature reference is set at 70°F. When the air is being heated, the temperature in the center of the room has already reached above 73°F before the temperature at the sensor reaches the reference temperature of 70°F. Similarly, as the room cools, the temperature in the room will drop to below 67°F before the temperature at the sensor reaches 70°F. The room temperature will go approximately from 66°F to 74°F due to the inertia in the system.

16.3.2 Differential Closed Loop Action

Differential or delayed On/Off action is a mode of operation where the simple On/Off action has hysteresis or a dead-band built in. Figure 16.3 shows an example of the fluctuations in a room heating system with delayed action. This is similar to Figure 16.2, except that, instead of the thermostat turning On and Off at the set reference of 70°F, the switching points are delayed by ±3°F. As can be seen in the top

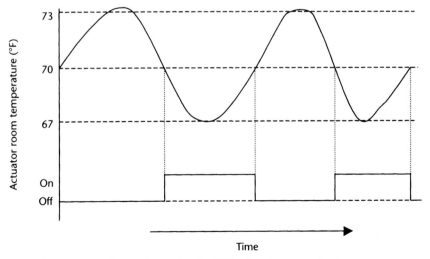

Figure 16.2 Temperature fluctuations using On/Off action in a room heating system.

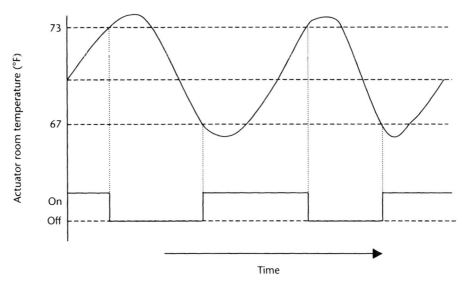

Figure 16.3 Temperature fluctuations using differential On/Off action in a room heating system.

graph, the room temperature reaches 73°F before the thermostat turns "Off the actuator, and the room temperature falls to 67°F before the actuator is turned On, giving a built-in hysteresis of 6°F. There is, of course, still some inertia. The room temperature will go approximately from 65°F to 75°F.

16.3.3 On/Off Action Controller

There are many ways of designing and implementing a two-position controller. An example of a typical bimetallic On/Off switch used in a room temperature controller is shown in Figure 16.4. The bimetallic element operates a mercury switch, as shown in Figure 16.4(a). At low temperatures, the bimetallic element tilts the mercury switch down, causing the mercury to flow to the end of the glass envelope, making an electrical short between the two contacts, which would operate a low voltage relay to turn On the blower motor and the heating element. When the room temperature rises to a predetermined set point, the bimetallic strip tilts the mercury switch back, as shown in Figure 16.4(b), causing the mercury to flow away from the contacts. The low voltage electrical circuit is turned "Off, the relay opens, and the

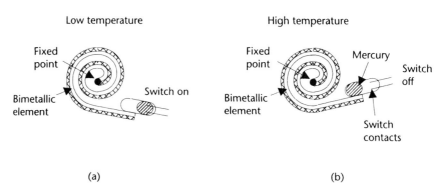

Figure 16.4 Simple On/Off room heating controller using a bimetallic switch.

power to the heater and the blower motor is disconnected. Bimetallic action is discussed in Section 10.3.1.

16.3.4 Electronic On/Off Controller

An electronic two-way On/Off controller is shown in Figure 16.5. The LM 34 can be used as a temperature sensor in a room controller application. The output of the LM 34 changes 10 mV/°F. If the nominal room temperature is 75°F, then the output of the LM 34 is 0.75V, and with a ratio for R_2/R_1 of 10, the output of the amplifier is −7.5V. If the set point voltage (V_{set}) is −7.5V, the output of the comparator will switch from "0" to "1" when the temperature increasing from 0°F reaches 75°F. The feedback via resistor R_{HIS} will give hysteresis, so that when the temperature is dropping, the comparator will switch Off a few degrees below the turn On point, depending on the R_1/R_{HIS} ratio.

Example 16.2

In Figure 16.5, the comparator is set to switch from "0" to "1" at 75°F (0.75V input). What is the value of R_{HIS} for the comparator to switch from "1" to "0" at 70°F? Assume $R_1 = 10$ kΩ and the output "1" level is 5V.

Turn On input V = 0.75V

Turn Off input V = 0.70V

Therefore 0.70V = 0.75V − (10 kΩ/R_{HIS}) × 5V

R_{HIS} = 10 × 5/0.05 kΩ = 1 MΩ

16.4 Continuous Control

In continuous control (modulating control) action, the feedback controller determines the error between a set point and a measured variable. The error signal is then used to produce an actuator control signal, which is used to control a process input variable. The change in input variable will reduce the change in the measured

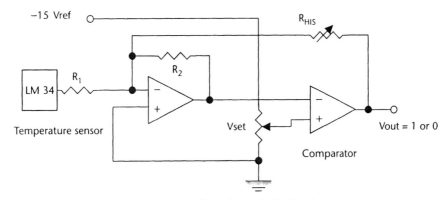

Figure 16.5 Electronic temperature controller switch with built-in hysteresis.

output variable, reducing the error signal. This type of control continuously monitors the measured variable, and has three modes of operation: proportional, integral, and derivative. Controllers normally use the proportional function on its own, or with one or both of the other functions, as required [3].

16.4.1 Proportional Action

Continuous industrial process control action uses proportional control action. The amplitude of the output variable from a process is measured and converted to an electrical signal. This signal is compared to a set reference point, and any difference in amplitude between the two (error signal) is amplified and fed to a control valve (actuator) as a correction signal. The control valve controls one of the inputs to the process. Changing this input will result in the output amplitude changing, until it is equal to the set reference, or the error signal is zero. The amplitude of the correction signal is transmitted to the actuator controlling the input variable, and is proportional to the percentage change in the output variable amplitude measured with respect to the set reference. In industrial processing, a situation exists that is different than a room heating system. The industrial system has low inertia. Overshoot and response times must be minimized for fast recovery and to keep processing tolerances within tight limits. In order to achieve these goals, fast reaction and settling times are needed. There also may be more than one variable to be controlled, and more than one output being measured in a process.

The change in output level may be a gradual change, a large on-demand change, or a change caused by a change in the reference level setting. An example of an on-demand change would be hot material flowing to a number of processing stations, using a heater to give a fixed temperature, as shown in Figure 16.6(a). At one point in time, the demand could be very low, with a low flow rate, as would be the case if only one process station were in use. If processing commenced at several of the other stations, then the demand could increase in steps, or there could be a sudden rise to a very high flow rate, and the increased flow rate would cause the material temperature to drop. The drop in material temperature would cause the temperature sensor to send a correction signal to the actuator controlling the fuel flow, in order to increase the fuel flow, raising the temperature of the material to bring it back to the set reference level, as shown in Figure 16.6(b). The rate of correction will depend on factors such as: inertia in the system, gain in the feedback loop, allowable amount of overshoot, and so forth.

Proportional action responds to a change in the measured variable, but does not fully correct the change in the measured variable, due to its limited gain. As an example, if the gain in the proportional amplifier is 100, then when a change in load occurs, 99% of the change is corrected. However, a 1% error signal is required for amplification to drive the actuator to change the manipulated variable. The 1% error signal is effectively an offset in the variable with respect to the reference. There is also a delay between the change in variable, the correction signal to the actuator, and the correction in the measured variable. The time from when the change in the variable is detected to when the corrective action is started is called the "Dead Time," and the time to complete the corrective action is the "Lag Time."

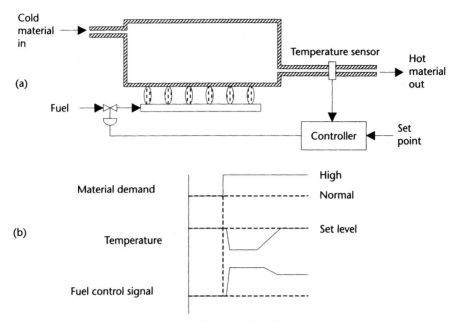

Figure 16.6 (a) Material heater showing a feedback loop for constant temperature output, and (b) effect of load changes on the temperature of the material from the heater.

In the proportional mode, there is a linear relationship between error and controller output p [4]. The range of error to cover the controller output from 0% to 100% is known as the proportional band (PB), and can be expressed by:

$$p = K_p e_p + p_o \tag{16.1}$$

where K_p is the proportional gain between error and controller output, expressed as a percentage per percentage, and p_o is the controller output with zero error, expressed as a percentage.

When there is zero error, the output equals p_o. When there is an error, a correction of K_p% is added to, or subtracted from, p_o for every 1% of error, providing the output is not saturated. The term gain (K_p) or *proportional band* can be used to describe the transfer function, and the relation between the two is given by:

$$PB = \frac{100}{K_P} \tag{16.2}$$

Example 16.3

In Figure 16.6, the hot material demand changes from a flow of 1.3 to 1.8 m/min. If the controller output is normally 50%, with a constant of $K_p = 12$% per percentage for the set temperature, then calculate the new controller output and offset error. Assume a temperature/flow scale factor of 0.028% controller output.

$$\text{New controller output} = (0.028 \text{ m/min/\%})(p\%) = 1.8 \text{ m/min}$$

$$p = 63.2\%$$

From (16.1),

$$e_p = (p - p_o)/K_p = (63.2 - 50)\%/12$$

$$e_p = 1.1\%$$

16.4.2 Derivative Action

Proportional plus derivative (PD) action was developed in an attempt to reduce the correction time that would have occurred using proportional action alone. Derivative action senses the rate of change of the measured variable, and applies a correction signal that is only proportional to the rate of change (this is also called rate action or anticipatory action). Figure 16.7 shows some examples of derivative action. As can be seen in these examples, a derivative output is obtained only when the load is changing. The derivative of a positive slope is a positive signal, and the derivative of a negative slope is a negative signal, while zero slopes give zero signals [5].

Figure 16.8 shows a changing measured variable and the resulting proportional and derivative waveforms. To obtain the control signal, the proportional and derivative waveforms are added, as shown in the composite waveform.

Figure 16.9 shows the effect of PD action on the correction time. When a change in loading is sensed, both the P and D signals are generated and added, as shown. The effect of combining these two signals is to produce a signal that speeds up the actuator's control signal. The faster reaction time of the control signal reduces the time to implement corrective action, thus reducing the excursion of the measured variable and its settling time. The amplitudes of these signals must be adjusted for optimum operation, otherwise, overshoot or undershoot can still occur.

In derivative controller action, the derivative of the error is the rate at which the error is changing, and is approximately given by:

$$p_t = K_D \frac{de_p}{dt} \qquad (16.3)$$

where the gain K_D is the percentage change in controller output for every percentage per second rate of change of error.

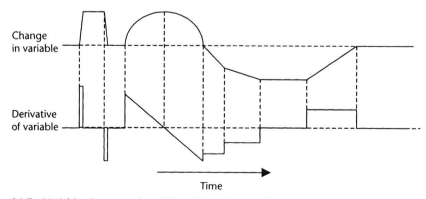

Figure 16.7 Variable change and resulting derivative waveform.

16.4 Continuous Control

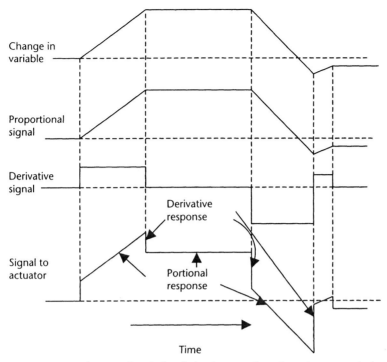

Figure 16.8 Generation of proportional plus derivative waveform for actuator control.

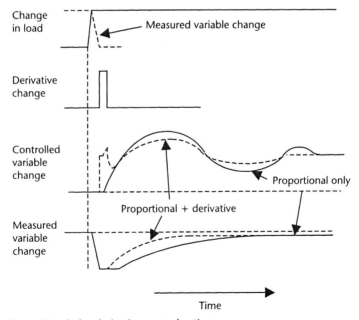

Figure 16.9 Proportional plus derivative control action.

Note that an output is obtained only when the error is changing.

In PD action, the analytic expression is given by the combination of (16.1) and (16.3), and is given by:

$$p = K_p e_p + K_p K_D \frac{de_p}{dt} + p_o \qquad (16.4)$$

16.4.3 Integral Action

Proportional plus integral (PI) action, also known as reset action, was developed to correct for long-term loads, and applies a correction that is proportional to the area under the change in the variable curve. Figure 16.10 gives some examples of the integration of a curve or the area under a curve. In the top example, the area under the square wave increases rapidly but remains constant when the square wave drops back to zero. The area increases more rapidly when the sine wave is at its maximum, and more slowly as it approaches the zero level. During the triangular section, the area decreases rapidly at the higher slopes, but increases slowly as the triangle slope reduces, as shown. Integral action gives a slower response to changes in the measured variable to avoid overshoot, but has a high gain, so that with long term load changes, it takes over control of the manipulated variable, and applies the correction signal to the actuator. Because of the higher gain, the measured variable error or offset is reduced to close to zero. This also returns the proportional amplifier to its normal operating point, so that it can correct other fluctuations in the measured variable. Note that these corrections are done at relatively high speeds, while the older pneumatic systems are much slower, and can take several seconds to make such a correction. When a change in loading occurs, the P signal responds to take corrective action to restore the measured variable to its set point. Simultaneously, the integral signal starts to change linearly to supply the long-term correction, thus allowing the proportional signal to return to its normal operating point. The integral signal can become complex. The expression for the integral process is given by:

$$p(t) = K_I \int_0^t e_p \, dt + p(0) \qquad (16.5)$$

where $p(0)$ is the controller output when the integral action starts, and the gain K_I is the controller output in percentage needed for every percentage time accumulation of error.

Alternatively the integral action can be found by taking the derivative of (16.5), which gives:

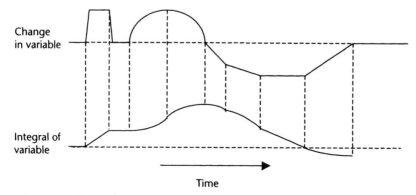

Figure 16.10 Integral variable change in response to a change in a variable, and in the area under the change in a variable.

$$\frac{dp}{dt} = K_I e_p \quad (16.6)$$

Equation 16.4 shows that if the error is not zero, then the output will change at a rate of $K_I\%$ per second for every 1% of error. The integral of a function determines the area under the function, so (16.4) is providing a controller output equal to the error under the error-time curve, multiplied by K_I. For every 1% second of accumulated error-time area, the output will be $K_I\%$.

When proportional mode is combined with integral mode, the equation for the control process is given by combining (16.1) and (16.5), to give:

$$p = K_p e_p + K_p K_I \int_0^t e_p \, dt + p_I(0) \quad (16.7)$$

where $p_I(0)$ is the integral term value at $t = 0$ (initial value).

16.4.4 PID Action

A combination of all three of the actions described above is more commonly referred to as PID action. The waveforms of PID action are illustrated in Figure 16.11. The proportional, derivative, and integral signals are generated from the change in measured variable. The P and D signals are summed, and combined with the I signal, to generate the PID signal for the controlled variable. The integral signal

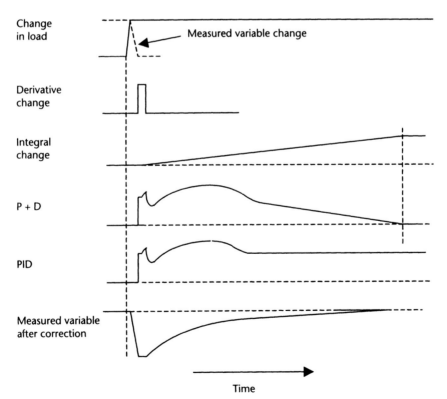

Figure 16.11 (a) Waveforms for proportional plus integral action, and (b) waveforms for proportional plus derivative and integral action.

is shown taking over from the proportional signal and reducing the offset to zero. PID is the most complex corrective action used for process control. However, there are many other types of control actions based upon PID action. Understanding the fundamentals of PID action gives a good foundation for understanding other types of controllers. The waveforms have been idealized for ease of explanation, and are only an example of what may be encountered in practice. Loading is a function of demand, and is not affected by the control functions or actions. The control function is to ensure that the variables are within their specified limits.

The effects of combining P, I, and D, can be found by combining (16.4) and (16.7) the typical PID equation is as follows:

$$p = K_p e_p + K_p K_D \frac{de_p}{dt} + k_p K_I \int_0^t e_p \, dt + p_I(0) \tag{16.8}$$

Where p is the controller output in percentage of full scale, e_p is the process error in percentage of the maximum, K_p is the proportional gain, K_I is the integral gain, K_D is the derivative gain, and $p_I(0)$ is the internal controller integral output.

This equation can be implemented using op-amps. An alternative expression when combining the effects of P, I, and D is give by:

$$\text{Output} \; \alpha \; \text{gain} \left(e + \frac{1}{T_1} \int_0^t e \, dt + T_d \frac{de}{dt} \right) \tag{16.9}$$

where, Output is the controller output, e is the error (variable − set point), T_i is integral time in minutes, t = time (minutes), and T_d is the derivative time in minutes.

To give an approximate indication of the use of PID controllers for different types of loops, the following are general rules:

- Pressure control requires P and I, but D is not normally required.
- Level control uses P and sometimes I, but D is not normally required.
- Flow control requires P and I, but D is not normally required.
- Temperature control uses P, I, and D, usually with I set for a long time period.

However, these are general rules, and each application has its own requirements.

Typical feedback loops have been discussed. However, the reader should be aware that there are other kinds of control loops used in process control, such as cascade, ratio, and feed-forward controls.

Cascade control is two feedback loops operating together in a series mode. The interaction that occurs between two control systems can sometimes be used to achieve an overall improvement in performance. This could be achieved in a feedback loop by measuring a second variable to give enhanced control. An example of cascade control is shown in Figure 16.12. The primary controller measures the temperature of the fluid and adjusts the fuel flow into the heater to control the temperature. A secondary control loop is introduced to measure the rate of flow of fuel. The primary controller now adjusts the set point of the secondary loop to give improved fuel flow and better temperature control.

16.4 Continuous Control

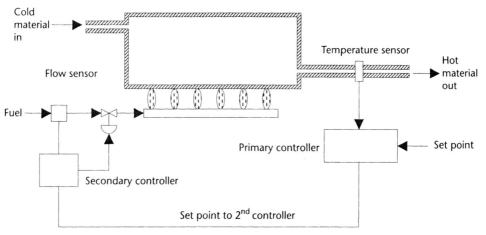

Figure 16.12 Cascade feedback loop.

Ratio control is used to reference one variable to another variable. An example of this is shown in Figure 16.13, where a controlled flow rate is set to a primary flow rate by a fixed ratio. The primary flow rate is measured and the ratio is used as the set point to the controller, setting the secondary flow rate. The primary flow rate (wild flow) is uncontrolled, and can be used as a reference to control several other variables.

Feed-forward control is used to anticipate a change and on its own is an open loop control, but is normally used in conjunction with a feedback control system. In Figure 16.14, the flow sensor is used to detect the increase (or decrease) in the rate of material flow, which will eventually lead to a temperature drop in the measured variable. The flow sensor will increase the fuel to the heater in anticipation that the temperature of the material is going to drop, because of the increase in demand. The change in flow rate can be sensed before the actual temperature drops, because of

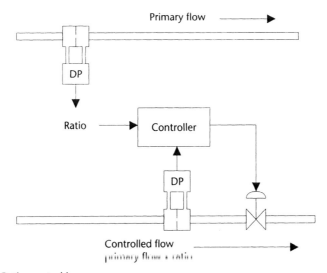

Figure 16.13 Ratio control loop.

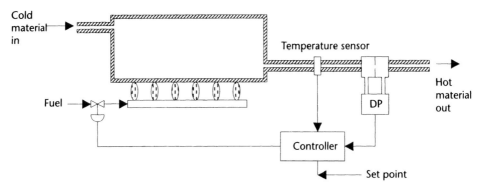

Figure 16.14 Feed-forward feedback loop.

the reservoir of hot material in the heater. The feedback loop can then fine-tune the amount of fuel required to maintain constant temperature.

16.4.5 Stability

In a closed loop feedback system, settings are critical. If the system has too much gain and/or is underdamped, then the amplitude of the correction signal is too great and/or changing too rapidly. This can cause the controlled variable to overcorrect for the error, which can cause the system to become unstable and oscillate, or can cause an excessively long settling or lag time. If the gain in the system is too low or overdamped, then the correction signal is too small or slow, and the correction never will be fully completed, or an excessive amount of time will be taken for the output to reach the set reference level.

This effect, as shown in Figure 16.15, can be seen in comparing the overcorrected (excessive gain) or underdamped, and the undercorrected (too little gain) or overdamped to the optimum gain case (with just a little overshoot) and correct damping. It takes a much longer time for the variable to implement the

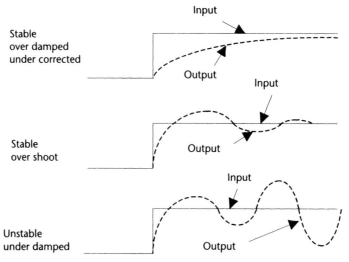

Figure 16.15 Effect of loop gain and damping on correction time using proportional action, with overcorrection and undercorrection.

correction than in the optimum case. In many processes, this long delay or lag time is unacceptable.

A criterion sometimes used for the optimum stable case for a damped cyclic response to a disturbance is that the overshoot decreases by one-quarter of its amplitude every successive cycle, as shown in Figure 16.16. This setting is a good compromise between minimum deviation and minimum duration. Further fine-tuning may be required to optimize a specific process.

16.5 Process Control Tuning

The performance of a control loop depends on system design, loop implementation, and setup of the control functions. The factors affecting the quality of control are:

1. The accuracy and conditioning of the measuring and control devices;
2. Stability in the control loops;
3. The ability of the system to return a measured variable to its set point after a disturbance.

Tuning is the adjustment of P, I, and D in the correct combination to give the optimum performance of the process control loop. Idealized settings may not give optimum performance, since settings are application-dependent. There are several methods of tuning a control loop, but it is generally done automatically, manually, or through adjustments based on experience.

Loops can be tuned for:

1. Minimum area, which produces a longer lasting deviation from the set point, and is used where high peak values can be detrimental to the process;
2. Minimum cycling, which minimizes the number of disturbances and time duration, and gives better stability;

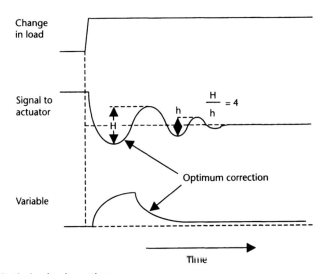

Figure 16.16 Optimized gain settings.

3. Minimum deviation, which minimizes the deviation amplitude, but has cycling around the set point, and is the most commonly used process.

16.5.1 Automatic Tuning

Programs are available from software and hardware suppliers that will include features to automatically set up and tune the control loops, although some final fine-tuning for optimum performance may still be required [6].

16.5.2 Manual Tuning

Manual tuning requires a good understanding of the control loop being tuned, good engineering skills, and much patience. There are two basic methods of manual tuning—open loop tuning and closed loop tuning. Open loop tuning is typically used in applications that have long delays, such as in analysis and temperature loops. Closed loop tuning is used in applications with short time delays, such as in pressure, flow, and level loops.

Open loop requires the opening of the feedback loop at the output of the controller and applying a step change S (5% to 10%) to the actuator or valve. The following measurements from the feedback element are then made and recorded, as shown in Figure 16.17.

1. The reaction rate $R = T/A$.
2. Find the unit reaction rate $R_U = R/S$.
3. Measure the lag L.
4. Set the controller PID values:
    a. Gain $K_P = 1.2/(R_U \times L)$
    b. Integral $T_I = 2L$ (min)
    c. Derivative $T_D = 0.5L$ (min)

Testing and fine-tuning may still be required.

Closed loop, also known as the Ziegler-Nichols closed loop method or ultimate cycle method, uses the following steps:

1. Use P only with set point at 50%, setting I and D to minimum.
2. Move the controller set point 10%, and hold a few minutes.

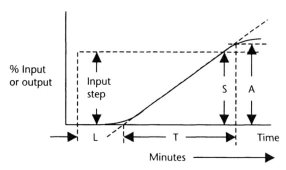

Figure 16.17 Reaction curve.

3. Return the set point to its 50% point.
4. Adjust the gain Gc until a stable continuous oscillation is obtained.
5. Measure period of cycle T_c.
6. Set the controller values for three-mode control, as follows:
   a. Gain $K_p = 0.6G_c$;
   b. Integral $T_I = T_c/2$ (min);
   c. Derivative $T_D = 0.125T_c$ (min).

The system can now be tested and fine-tuned.
To give a quarter amplitude response, we set:

a. Integral $T_I = T_c/1.5$;
b. Derivative $T_D = T_c/6$;
c. Adjust the proportional gain for quarter amplitude response.

Test and fine-tune as required.

Rules of thumb have been developed by experienced operators for values of P, I, and D. The controller is set up with these values, but since they are only approximations, the system will still require fine-tuning to obtain acceptable PID settings for a specific process. Typical rules of thumb are described in Table 16.1.

16.6 Implementation of Control Loops

Implementation of the control loops can be achieved using pneumatic controllers, or analog or digital electronic controllers. The first process controllers were pneumatic. However, these are largely obsolete and in most cases have been replaced by electronic systems. Electronic systems have improved reliability, less maintenance, easier installation, easier adjustment, higher accuracy, lower cost, higher operating speeds, and they can be used with multiple variables.

16.6.1 On/Off Action Pneumatic Controller

Figure 16.18 shows a pneumatic control system using a pneumatic On/Off controller. In this case, the transducer moves a flapper that controls the airflow from a nozzle. When the spring force on the flapper reaches the reference point, the transducer moves the flapper away from the nozzle, letting air flow from the nozzle, which allows the pressure to the actuator to drop. This reduction in pressure opens the valve. When the force from the transducer drops below the set level, the flapper

Table 16.1 Typical Rules of Thumb for PID Settings

Type of Loop	Gain (K_p)	T_c (min)	T_D (min)
Line pressure	0.5	0.2	0
Tank pressure	2.0	2.0	0.5
Temperature	2.0	2.0	2.0
Flow	0.7	0.05	0
Level	1.7	5.0	0

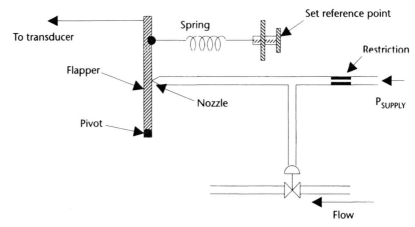

Figure 16.18 On/Off pneumatic controller.

moves, stopping the air flow from the nozzle increasing the air pressure to the valve, which in turn opens the actuator and closes the valve [7].

16.6.2 Pneumatic Linear Controller

A force to pressure converter is shown in Figure 16.19. When the force F is increasing, the flapper pivots to close the nozzle. This in turn increases the pressure in the feedback bellows, which tries to rotate the flapper to open the nozzle and balance the force of the input. A balance condition occurs when the torque exerted by the force equals the torque exerted by the feedback bellows, giving:

$$(P_{out} - P_o)A \times t = F \times d$$

solving for the output pressure

$$P_{out} = \frac{F \times d}{A \times t} + P_o \tag{16.10}$$

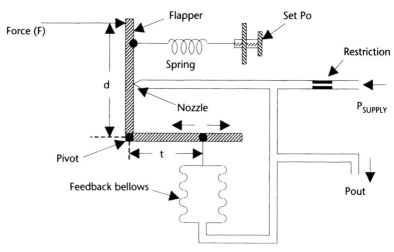

Figure 16.19 Linear pneumatic controller.

where, P_{out} is the output pressure, P_o is the zero error pressure, A is the area of the bellows, t is the distance between the pivot and bellows, F is the applied force, and d is the distance from the force to the pivot.

For pressure transducers P_o is normally 3 psi (20 kPa). Since d, A, and t are fixed, there is a linear relation between P_{out} and F. The gain of the system is set by the ratio d/t, which is chosen so that when F is at its maximum, $P_{out} = 15$ psi (100 kPa).

16.6.3 Pneumatic Proportional Mode Controller

Proportional mode control can be achieved with the setup shown in Figure 16.20. The operation is basically the same as in the force to pressure converter. The torque is produced by the error signal. The torque produced by the output balances the difference between the reference and input pressure (P_{in}), giving the following:

$$(P_{out} - P_o)A \times t = (P_{in} - P_{ref})A_{in} \times d$$

solving for the output pressure

$$P_{out} = \frac{A_{in} \times d}{A \times t}(P_{in} - P_{ref}) + P_o \qquad (16.11)$$

where A_{in} is the area of the input and reference bellows, and P_{ref} is the reference pressure.

This gives a linear relation between P_{in} and P_{out}.

16.6.4 PID Action Pneumatic Controller

Many configurations for PID pneumatic controllers have been developed over the years, have performed well, and are still in use in some older processing plants. With the advent of the requirements of modern processing and the development of electronic controllers, pneumatic controllers have achieved the distinction of becoming

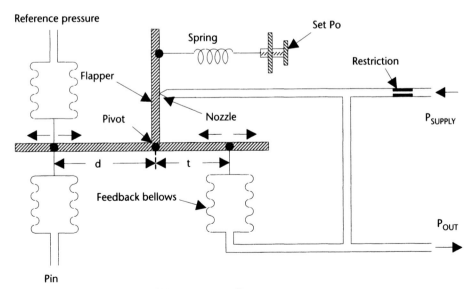

Figure 16.20 Pneumatic On/Off furnace controller.

museum pieces. Figure 16.21 shows an example of a pneumatic PID controller. The pressure from the sensing device (P_{in}) is compared to a set or reference pressure (P_{ref}) to generate a differential force (error signal) on the flapper, which moves the flapper in relation to the nozzle, giving an output pressure proportional to the difference between P_{in} and P_{ref}. If the derivative restriction is removed, then the output pressure is fed back to the flapper via the proportional bellows, to oppose the error signal and to give proportional action. System gain is adjusted by moving the position of the bellows along the flapper arm (i.e., the closer the bellows is positioned to the pivot, the greater the movement of the flapper arm).

By putting a variable restriction between the pressure supply and the proportional bellows, a change in P_{in} causes a large change in P_{out}, since the feedback from the proportional bellows is delayed by the derivative restriction. This gives a pressure transient on P_{out} before the proportional bellows can react, thus giving derivative action. The size of the bellows and the setting of the restriction set the duration of the transient.

Integral action is achieved by the addition of the integral bellows and restriction, as shown. An increase in P_{in} moves the flapper towards the nozzle, causing an increase in output pressure. The increase in output pressure is fed to the integral bellows via the restriction, until the pressure in the integral bellows is sufficient to hold the flapper in the position set by the increase in P_{in}, creating integral action.

16.6.5 PID Action Control Circuits

PID action can be performed using either analog or digital electronic circuits. This section discusses the use of analog circuits for proportional, derivative, and integral action. A large number of circuit variations can be used to perform these analog functions, but only the basic circuits used in control circuits are discussed.

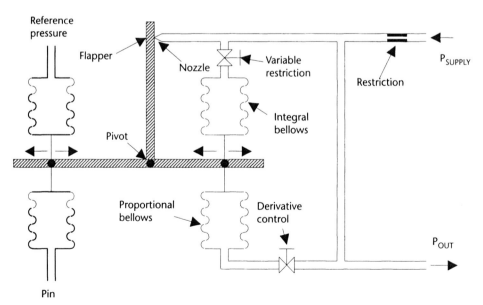

Figure 16.21 Pneumatic PID controller.

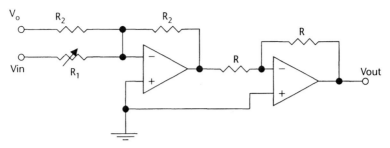

Figure 16.22 Circuits used in PID action: (a) error generating circuit, and (b) proportional circuit.

The *proportional mode*, shown in Figure 16.22, is a circuit that can be used to generate an amplified voltage that is proportional to the error signal voltage. The proportional signal is given in (16.1). This is the equation of a summing amplifier, which, when used in a voltage amplifying circuit, can be modified to:

$$V_{out} = A \times V_{in} + V_o \quad (16.12)$$

where V_{out} is p (16.1), A is gain or K_p, V_{in} is e_p, and V_o is p_o, or output voltage with zero error input; that is, the measured variable and the reference voltage are equal. Proportional action is achieved by amplifying the error signal (V_{in}). The stage gain is the ratio of R_2/R_1, and the gain is adjusted by potentiometer (R_1). In (16.1), p_o is the controller output with zero error, because the R_2 input has unity gain. This input can be used to set the zero error output voltage P_o, by making $V_o = P_o$. The output is inverted to keep the signal positive.

The *derivative mode* is never used alone because it cannot provide a dc input to control the level of the output. Implementation of the action is shown so that it can be used to work with other modes. The circuit for derivative action is shown in Figure 16.23. The feedback resistor R_2 can be replaced with a potentiometer to adjust the differentiation amplitude. The output signal of the differentiator is inverted, but is changed to a noninverted signal with an inverting amplifier stage. When using voltage amplifiers, (16.3) can be modified to:

$$V_{out} = R_2 C_1 (dV_{in}/dt) \quad (16.13)$$

In practice, this circuit tends to be unstable due to high gain at high frequencies. Using ac analysis and impedances, the amplifier equation becomes:

$$\frac{V_{in}}{-jX_c} = \frac{V_{out}}{R_2}$$

where $Xc = 1/2\pi f C_1$, from which we get:

$$V_{out} = -2j\pi f C_1 R_2 V_{in}$$

Because only magnitudes are of interest, the equation can be written:

$$|V_{out}| = 2\pi f R_2 C_1 |V_{in}| \quad (16.14)$$

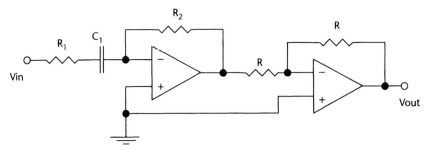

Figure 16.23 Derivative amplifier: (a) circuit, and (b) waveforms.

This equation shows the gain increasing linearly with frequency, and the possibility of instability at high frequencies. Resistor R_1 is put in series with C_1 to limit the gain of the amplifier at high frequencies, ensuring high frequency stability. The maximum frequency response f_m of the system is estimated, for stability:

$$2\pi f_m R_2 C_1 = 0.1 \tag{16.15}$$

where C_1 is found from the mode derivative gain requirements, and R_2 then can be calculated.

The *PD mode* can be obtained by combining the circuits shown in Figures 16.22 and 16.23, as shown in Figure 16.24. Derivative action is obtained by the input capacitor C_1 with R_3, and proportional action is obtained by the ratio of the resistors R_2 to R_1 in series with R_3. From analysis of the circuit and using (16.4), we get:

$$V_{out} = \left(\frac{R_2}{R_1 + R_3}\right) V_{in} + \left(\frac{R_2}{R_1 + R_3}\right) R_1 C_1 \frac{dV_{in}}{dt} + V_0 \tag{16.16}$$

where the proportional gain is $R_2/(R_1 + R_3)$, and the derivative gain is $R_1 C_1$.

The *integral mode* can be obtained using the circuit shown in Figure 16.25. Capacitive feedback around the amplifier prevents the output from the amplifier from following the input change, giving integral action. The output changes slowly and linearly when there is a change in the measured variable (V_{in}). The feedback C_1 and the input resistance R_1, which can be adjusted, set the required integration time. Integral action is given by (16.5), which, when applied to the voltage amplifier in Figure 16.25, results in:

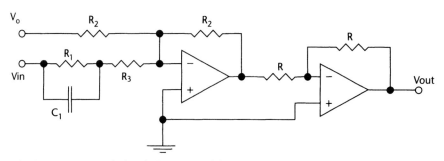

Figure 16.24 Proportional plus derivative amplifier: (a) circuit, and (b) waveforms.

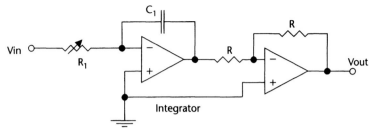

Figure 16.25 Integrating amplifier: (a) circuit, and (b) waveforms.

$$V_{out} = \frac{1}{R_1 C_1} \int_0^1 V_{in} dt + V_0 \qquad (16.17)$$

The *PI* mode is a combination of the proportional mode and the integral mode. The resulting circuit is shown in Figure 10.26. Using circuit analysis and (16.7), results in:

$$V_{out} = \frac{R_3}{R_1} V_{in} + \frac{R_3}{R_1} \frac{1}{R_3 C_1} \int_0^1 V_{in} dt + V_0 \qquad (16.18)$$

16.6.6 PID Electronic Controller

Figure 16.27 shows the block diagram of an analog PID controller. The measured variable from the sensor is compared to the set point in the summing circuit, and its output is the difference between the two signals. This is the error signal. This signal is fed to the integrator [8].

When there is a change in the measured variable, the error signal is seen by the proportional amplifier, the differentiator, and the integrator. The output from the amplifiers is combined in a summing circuit, amplified, and fed to the actuator motor to change the input variable. Although the integrator sees the error signal, it is slow to react, so its output does not change immediately, but starts to integrate the error signal. If the error signal is present for an extended period of time, then the integrator will supply the correction signal via the summing circuit to the actuator.

The circuit implementation of the PID controller is shown in Figure 16.28. Each of the amplifiers is shown doing a single function, and is only used as an example. In practice, there are a large number of circuit component combinations that can be used to produce PID action.

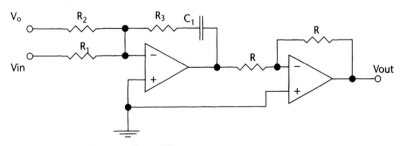

Figure 16.26 Proportional integral amplifier.

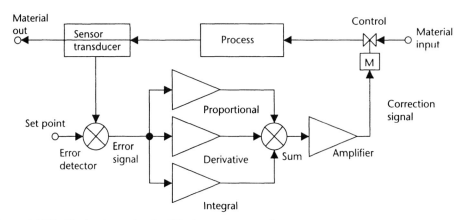

Figure 16.27 Block schematic of a PID electronic controller.

A single amplifier also can be used to perform several functions, which would greatly reduce the circuit complexity. Such a circuit is shown in Figure 16.29, where the PD action circuit is combined with the integral circuit. This is only one of many alternatives [9].

In new designs, the PID action can be performed in the PLC processor using digital techniques, and by the processor in a smart sensor [10].

16.7 Summary

This chapter discussed process control, the terminology used, and the various methods of implementation of the controller functions. The differences between sequential, On/Off feedback control, and continuous feedback control were given. Continuous feedback control can be broken down into proportional, derivative, and integral action. Other types of feedback control used in process control are cascade, ratio, and feed-forward. The stability of feedback control systems is discussed, and the various methods of tuning the system are given.

Pneumatic hardware for performing On/Off and PID control is described, and the operation of a typical controller is given. The use of digital electronics for

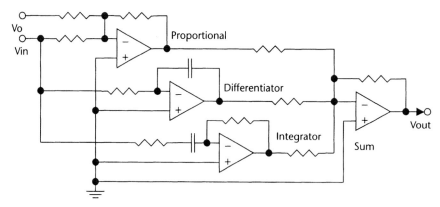

Figure 16.28 Circuit of a PID action electronic controller.

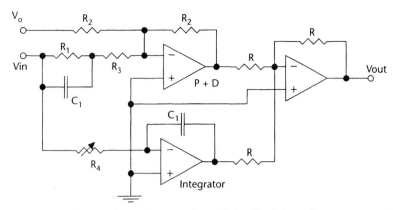

Figure 16.29 Circuit of a PID electronic controller with feedback from the actuator position.

sequential control functions was discussed, and analog circuits for proportional, derivative, and integral action were given. The combination of these circuits in PID feedback loops is shown, and their operations were discussed. Together with methods of tuning the feed back loops for stability. An understanding of these circuits enables the reader to extend these principles to other circuits and methods of control.

Definitions

Controlled variable is an input variable to a process that is varied by a valve to keep the output variable (measured variable) within its set limits.

Control parameter range is the range of the controller output required to control the input variable to keep the measured variable within its acceptable range.

Dead-band is a set hysteresis between detection points of the measured variable when it is going in a positive or a negative direction. This band is the separation between the turn On set point and the turn "Off set point of the controller, and is sometimes used to prevent rapid switching between the turn On and turn Off points.

Dead time is the elapsed time between the instant an error occurs and when the corrective action first occurs.

Error signal is the difference between a set reference point and the amplitude of the measured variable.

Lag time is the time required for a control system to return a measured variable to its set point after there is a change in the measured variable, which could be the result of a loading change or set point change.

Measured variable is an output process variable that must be held within given limits.

Offset is the difference between the measured variable and the set point after a new controlled variable level has been reached. It is this portion of the error signal that is amplified to produce the new correction signal, and produces an "Offset" in the measured variable.

Set point   is the desired amplitude of an outpoint variable from a process.

Transient   is a temporary variation of a load parameter, after which the parameter returns to its nominal level.

Variable range   is the acceptable limits within which the measured variable must be held, and can be expressed as a minimum and a maximum value, or as a nominal value (set point) with a plus-or-minus spread, expressed as a percentage.

References

[1] Dunning, G., *Introduction to Programmable Logic Controllers*, 2nd ed., Delmar, 2002, pp. 177–200.

[2] Pummer, A., "Controlling Your HVAC System," *Sensors Magazine*, Vol. 19, No. 9, September 2002.

[3] Jones, C. T., *Programmable Logic Controllers*, 1st ed., Patrick-Turner Publishing Co., 1996, pp. 183–197, 294–296.

[4] Johnson, C. D., *Process Control Instrumentation Technology*, 7th ed., Prentice Hall, 2003, pp. 472–485.

[5] Humphries, J. T., and L. P. Sheets, *Industrial Electronics*, 4th ed., Delmar, 1993, pp. 450–460.

[6] Battikha, N. E., *The Condensed Handbook of Measurement and Control*, 2nd ed., ISA, 2004, pp. 147–158.

[7] Johnson, C. D., *Process Control Instrumentation Technology*, 7th ed., Prentice Hall, 2003, pp. 485–490.

[8] Cheng, D., "Applying PID Control Algorithms in Sensor Circuits," *Sensors Magazine*, Vol. 21, No. 5, May 2004.

[9] Harrold, S., "Designing Sensor Signal Conditioning with Programmable Analog ICs," *Sensors Magazine*, Vol. 20, No. 4, April 2003.

[10] Nachtigal, C. L., "Closed-Loop Control of Flow Rate for Dry Bulk Solids," *Proceedings Sensors Expo*, September 1994, pp. 49–55.

CHAPTER 17
Documentation and P&ID

17.1 Introduction

A vast amount of documentation is required for the design and construction of a process facility, which are front-end and detailed engineering drawings. The main engineering documents used on a regular basis by the engineering staff for smooth and efficient running, maintenance, and upgrading of the facility are Alarm and Trip Systems, PLC documentation, and Pipe and Instrumentation Diagrams (P&ID). As in all engineering disciplines, the initial accuracy of these documents, and the regular updating of them when changes are made, is critical, and one of the most important aspects of engineering. For this reason, documentation is discussed in this chapter. Documentation standards and symbols for all aspects of process control have been set up and standardized by the ISA, in conjunction with the ANSI [1].

17.2 Alarm and Trip Systems

The purpose of an alarm system is to bring a malfunction to the attention of operators and maintenance personnel, whereas the purpose of a trip system is to shut down a system in an orderly fashion when a malfunction occurs, or to switch failed units over to standby units. The elements used in the process control system are the first warnings of a failure. This could show up as an inconsistency in a process parameter, or as a parameter going out of its set limits. The sensors and instruments used in the alarm and trip system are the second line of defense, and must be totally separate from those used in the process control system. Alarm and trip system information and its implementation are given in ANSI/ISA-84.01-1996—Application of Safety Instrumented Systems for the Process Control Industry.

17.2.1 Safety Instrumented Systems

The alarm and trip system, or Safety Instrumented System (SIS), has its own sensors, logic, and control elements, so that under failure conditions, it will take the process to a safe state to protect the personnel, facility, and environment. To ensure full functionality of the SIS, it must be regularly tested. In an extreme situation, such as with deadly chemicals, a second or third SIS system with redundancy can be used in conjunction with the first SIS system, to ensure as close to 100% protection as possible. The sensors in the SIS usually will be of a different type than those used for process control. The control devices are used to accurately sense varying levels in the measured variable, whereas the SIS sensor is used to sense a trip point, and will be a much more reliable, rugged, and high-reliability device. The use of redundancy in a system cannot be used as a justification for low reliability and inexpensive components. The most commonly used high performance SIS system is the dual redundancy system, which consists of the main SIS with two redundant systems. In this case, a two-out-of-three logic monitoring system determines if a single monitor or the entire system has failed. If a single failure is detected, then the probability is that a sensor, its associated wiring, or logic has failed. If more than one failure is detected, then the indication is a system failure. A two-out-of-three logic circuit is shown in Figure 17.1(a), and the truth table is shown in Figure 17.1(b). With correct operation, the inputs are normally low (0). If one input goes high (1), it would indicate a sensor failure, and the sensor failure output would go from 0 to 1 to give warning of a sensor failure, but the system failure output would remain at 0. If two or more inputs go high, it would indicate a system failure, and the system failure output would go from 0 to 1, as shown.

In SIS systems failure analysis, the rate of component failure is as follows:

- Logic, 8%;
- Sensors, 42%;
- Control devices, 50%.

17.2.2 Safe Failure of Alarm and Trip

No system is infallible, and failures are going to occur. A good philosophy is the fail-safe approach, where each valve will trip to a predetermined fail position when they are deenergized. Even with an uninterruptible power system, power wires can get cut, fuses can blow, or cables can break, cutting off power. In some cases, this approach is not feasible, and extra safeguards are necessary to maintain safety when the SIS fails.

There are typically three levels of safety, and the systems normally associated with the safety levels are:

- Level 1—Single sensor with a one-out-of-one logic detection and single final control.

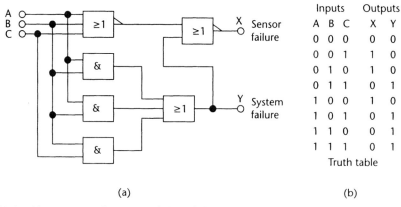

Figure 17.1 (a) Monitor and two-out-of-three failure indicator, and (b) truth table.

- Level 2—More diagnostics than Level 1, plus redundancy for each stage.
- Level 3—Minimum of two systems with redundancy, or a two-out-of-three sensing system.

Components in an SIS system should be high-grade, with a high mean time between failures (MTBF). Relays were the preferred choice due to the capability of multiple contacts and isolation. However, semiconductor devices have an excellent MTBF, and they are replacing relay logic. A good design will take into account the integrity of all the components in an alarm system, as well as interactions between the components.

Testing of the alarm system is required on a regular basis to uncover faults or potential failures, which require corrective action. Testing is of prime importance in SIS applications. An SIS is designed to detect hazardous conditions, so it must be able to sense a malfunction of the logic, measuring device, and final alarms during testing. The requirements and testability of the SIS must be factored in at the system design stage.

17.2.3 Alarm and Trip Documentation

Good, up-to-date documentation is a prerequisite in alarm and trip systems, and must be initiated at the design stage. Hazard analysis must be performed on the facility to determine all areas that require alarms or trips. The SIS devices should be clearly marked and numbered. System drawings must show all SIS devices using standard symbols, their locations, functions, and set limits. Drawings must include lock and logic diagrams [2].

The types of information required in Alarm and Trip documentation are:

1. Safety requirement specifications;
2. Logic diagram with functional description;
3. Functional test procedures and required maintenance;
4. Process monitoring points and trip levels;
5. Description of SIS action if tripped;
6. Action to be taken if SIS power is lost;
7. Manual shutdown procedures;
8. Time requirements to reach safe status;
9. Restarting procedures after SIS shutdown.

Test procedures are needed to verify operation of the total SIS. These procedures must not pose any hazards or cause spurious trips, and must have the ability to detect wear, slow operation, leaking shutoffs, and sticking devices. A test procedure is necessary for an SIS, and should be available for all alarm and trip devices. The test procedure should contain the following information:

1. Frequency of testing;
2. Hazards that may be encountered;
3. Drawing and specification information;
4. Test equipment;

5. Performance limits;
6. Test procedure.

The results of the system testing must record any problem areas found, and the corrective action taken. Typical SIS test results will have the following information:

1. Time and date of test;
2. Test personnel;
3. System identification;
4. Test procedure;
5. Results of test;
6. Corrective action taken;
7. Follow-up required;
8. SIS operational.

17.3 PLC Documentation

The PLC documentation is a very important engineering record of the process control steps, and, as with all technical descriptions, accurate detailed engineering records are essential. Without accurate drawings, changes and modifications needed for upgrading and diagnostics are extremely difficult or impossible. Every wire from the PLC to the monitoring and control equipment must be clearly marked and numbered at both ends, and recorded on the wiring diagram. The PLC must have complete up-to-date ladder diagrams (or other approved language), and every rung must be labeled with a complete description of its function [3].

The essential documents in a PLC package are:

1. System overview and complete description of control operation;
2. Block diagram of the units in the system;
3. Complete list of every input and output, destination, and number;
4. Wiring diagram of I/O modules, address identification for each I/O point, and rack locations;
5. Ladder diagram with rung description, number, and function.

It is also necessary to have the ability to simulate the ladder program off-line on a personal computer, or in a background mode in the PLC, so that changes, upgrades, and fault simulations can be performed without interrupting the normal operation of the PLC, and the effects of changes and upgrades can be evaluated before they are incorporated [4].

17.4 Pipe and Instrumentation Symbols

The electronics industry has developed standard symbols to represent circuit components for use in circuit schematics, and, similarly, the processing industry has developed standard symbols to represent the elements in a process control system.

Instead of a circuit schematic, the processing industrial drawings are known as P&ID (not to be confused with PID), which represent how the components and elements in the processing plant are interconnected. Symbols have been developed to represent all of the components used in industrial processing, and have been standardized by ANSI and ISA. The P&ID document is the ANSI/ISA 5.1—1984 (R 1992)—Instrumentation Symbols and Identification Standards. An overview of the symbols used is given in this chapter, but the list is not complete. The ISA should be contacted for a complete list of standard symbols.

17.4.1 Interconnect Symbols

The standard on interconnections specifies the type of symbols to be used to represent the various types of connections in a processing plant. The list of assigned symbols for instrument line connections is given in Figure 17.2. Interconnect lines can be solid bold lines, which are used to represent the primary lines used for process product flow, and solid narrow lines, which are used to represent secondary flows, such as steam for heating or electrical supplies. One signal line symbol is undefined, and can be assigned at the user's discretion for a special connection not covered by any of the assigned interconnection symbols. The binary signal lines can be used for digital signals or pulse signals. The pneumatic signal lines can represent any gas used for signal transmission, and the gas may be specified next to the line. Electromagnetic lines can be any EM waves, such as light, nuclear, or radio frequencies [5].

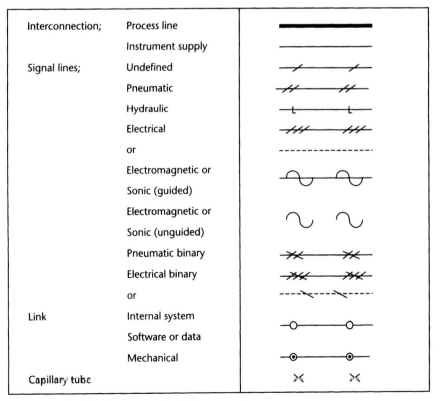

Figure 17.2 Symbols for instrument line interconnection.

Abbreviations to define the type of interconnect secondary flow lines are given in Table 17.1. The abbreviations are placed adjacent to the lines. Descriptive information can be added to signal lines to show on the P&ID, such as the signal's content and range. Electrical signals can be either current or voltage signals, and would be marked as such. Examples of signal lines with the signal's content and range marking are shown in Figure 17.3. The first two lines are supply lines. The first line is a 48V ac line, and the second a 60-psi nitrogen supply line. Lines 3 and 4 are signal lines. The third line carries an analog signal that ranges from 1V to 10V, and the fourth line carries a binary signal with 0 = 0V and 1 = 5V.

17.4.2 Instrument Symbols

Figure 17.4 shows the symbols designated for instruments. Discrete instruments are represented by circles, shared instruments by a circle in a rectangle, computer functions by hexagons, and PLC functions by a diamond in a rectangle. A single horizontal line, no line, dashed line, or double line through the display is used to differentiate between location and accessibility to an operator. A line through an instrument may indicate that the instrument is in a panel in the control room giving full access; no line could mean that the instrument is in the process area and inaccessible to the operator; and a double line could indicate that the instrument is in a remote location, but accessible to the operator. An instrument symbol with a dashed horizontal line means that it is not available, by virtue of being located in a totally inaccessible location [6].

17.4.3 Functional Identification

All instruments and elements will be identified according to function, and should contain the loop numbers. The letters are a shorthand way of indicating the type of instrument and its function in the system. Typically, two or three letters are used. The first letter identifies the measured or initiating variable, the second letter is a modifier, and the remaining letters identify the function. Table 17.2 defines some of the meanings of the assigned instrument letters.

Table 17.1 Abbreviations for Secondary Flow Lines

AS	Air Supply	HS	Hydraulic Supply
IA	Instrument Air	NS	Nitrogen Supply
PA	Plant Air	SS	Steam Supply
ES	Electric Supply	WS	Water Supply
GS	Gas Supply		

Figure 17.3 Method of indicating the content of a line.

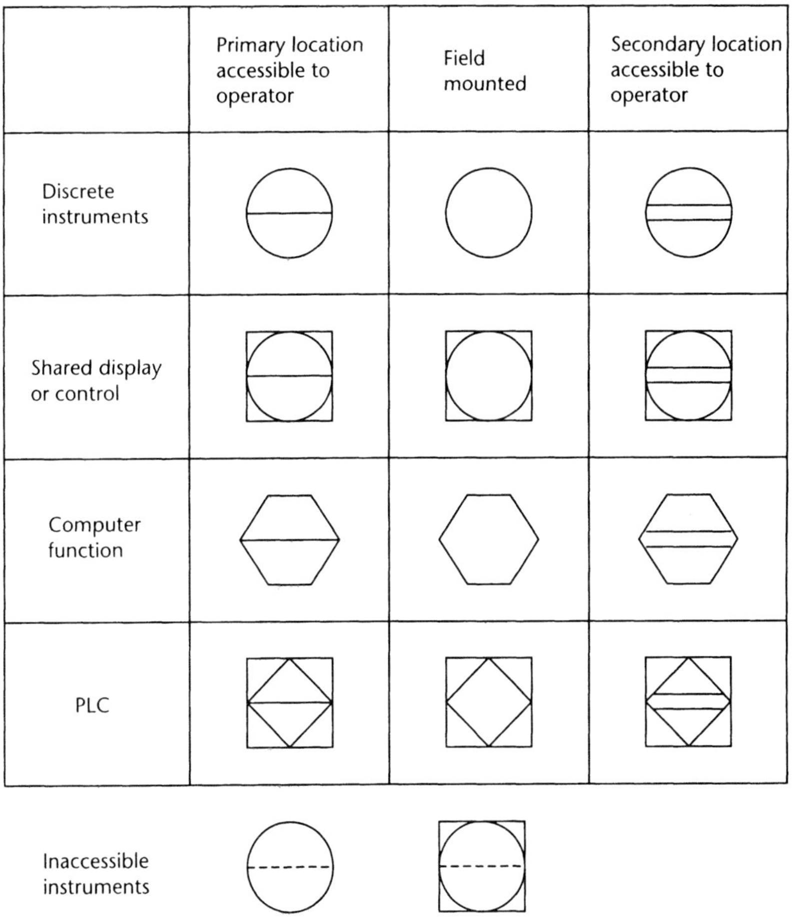

17.4 Pipe and Instrumentation Symbols

	Primary location accessible to operator	Field mounted	Secondary location accessible to operator
Discrete instruments	⊖	○	⊜
Shared display or control	⊟	⊡	⊟
Computer function	⬡	⬡	⬡
PLC	◈	◈	◈
Inaccessible instruments	⊖ (dashed)	⊡ (dashed)	

Figure 17.4 Standardized instrument symbols.

Examples of the use of instrument identification letters and numbers are shown in Figure 17.5. The instrument identification can be determined as follows:

(a) The flow control loop number 14 is shown. an orifice plate that has an electrical transmitter (FT14) measures the flow. The first letter, F, denotes that the function is flow, the second letter, T, denotes transmitter, and the dashed line is an electrical signal ranging from 0V to 10V. The output goes to a PLC (FC14) denoting flow control. The output is a current signal ranging from 4 to 20 mA, and this signal goes to a signal converter FY14, which converts the signal into a pressure signal ranging from 3 to 15 psi to drive the control valve FV14.

(b) The tank has a direct reading level indicator LI17, a high-level detector LSH17, and a low-level detector LSL17, where the first L denotes level, S denotes switch, H denotes high, and the subsequent L denotes low. The output from the level switch goes to an alarm (note the shared instrument symbol) LAHL 17, where A denotes alarm, H is high and L is low, showing that the alarm will be activated if the fluid level is above the set high level or below the low set level.

Table 17.2 Instrument Identification Letters

	First Letter + Modifier		Succeeding Letters		
	Initiating or Measured Variable	Modifier	Readout or Passive Function	Output Function	Modifier
A	Analysis		Alarm		
B	Burner, combustion		User's choice	User's choice	User's choice
C	User's choice			Control	
D	User's choice	Differential			
E	Voltage		Sensor		
F	Flow rate	Ratio			
G	User's choice		Glass, viewing device		
H	Hand				High
I	Current		Indicate		
J	Power		Scan		
K	Time	Time rate of change		Control station	
L	Level		Light		Low
M	User's choice	Momentary			Middle
N	User's choice		User's choice	User's choice	User's choice
O	User's choice		Orifice		
P	Pressure		Test point		
Q	Quantity		Integrate, totalize		
R	Radiation		Record		
S	Speed, frequency	Safety		Switch	
T	Temperature			Transmit	
U	Multivariable		Multifunction	Multifunction	Multifunction
V	Vibration, mechanical analysis			Valve, damper, louver	
W	Weight, force		Well		
X	Unclassified	x-axis	Unclassified	Unclassified	Unclassified
Y	Event, state, or presence	y-axis		Ready, compute, convert	
Z	Position, dimension	z-axis		Driver, actuator	

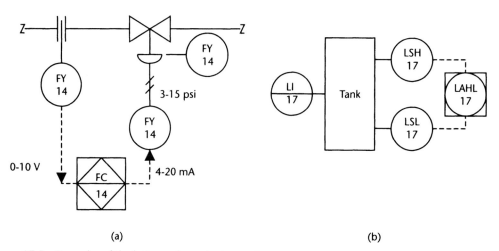

Figure 17.5 Examples of the letter and numbering codes.

17.4 Pipe and Instrumentation Symbols

17.4.4 Functional Symbols

A number of functional symbols or pictorial drawings are available for most P&ID elements. A few examples are given here to acquaint the reader with these elements. They have been divided into valves, actuators, temperatures, pressures, flows, levels, math functions, and others. The list is by no means complete, and a complete list of symbols can be obtained from the ISA—ISA-5.1-1984 (R1992).

Valve symbol examples are shown in Figure 17.6. Each type of valve has its own symbol. The first row shows a control valve, an angle valve, a three-way valve, and a four-way valve. The three-way valve has an arrow indicating that if power is lost, the fail-safe position is an open path between A and C ports. The second row of valves shows the fail-safe indication used for control valves, a globe valve, and a butterfly valve symbol. The last row shows other types of valves. In practice, each valve will have a balloon with functional information and loop numbers [7].

Actuator symbols are shown in Figure 17.7. Examples of eight types of valve actuators are shown. These actuators control the valves directly. The first row shows hand and electrical actuators, and the second row shows examples of pneumatic and hydraulic actuators.

Temperature symbol examples are shown in Figure 17.8, with six temperature functions shown: basic thermometer, thermometer in a well, capillary symbol, transmitter, radiation device, and high-level switch. Note the changes in symbols for different types of thermometer, letters for device functions, and loop numbers.

Pressure symbol examples are given in Figure 17.9, with six pressure sensors and regulators shown: basic pressure symbol, diaphragm isolated pressure symbol, pressure transmitter, two regulators, and pressure release rupture disk. Note the use of function indicators and loop numbers.

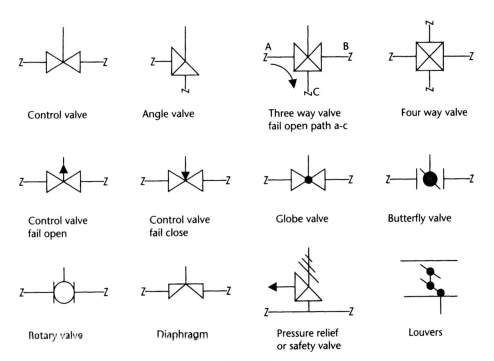

Figure 17.6 Examples of valve symbols used in P&ID.

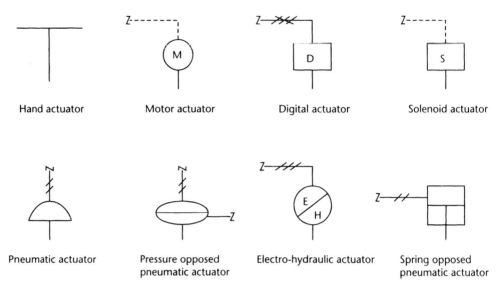

Figure 17.7 Examples of basic actuator symbols.

Flow symbol examples are given in Figure 17.10, with six flow measuring devices shown: orifice, internal flow instrument, venture tube, turbine, variable area, and magnetic instrument with transmitter. Functional letters and loop numbers are shown.

Level symbol examples are given in Figure 17.11, with three level measuring devices shown: basic two-connection level instrument with electrical output, single-connect instrument with electrical output, and float instrument. Letters are used for function and numbers for the loop.

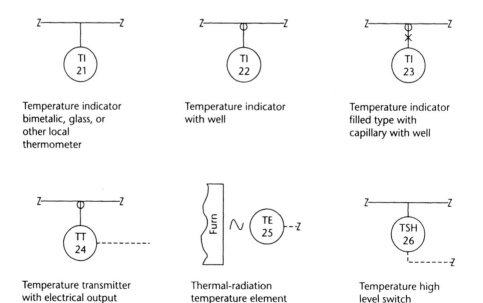

Figure 17.8 Examples of temperature symbols.

17.4 Pipe and Instrumentation Symbols

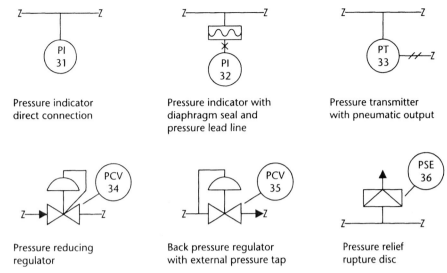

Figure 17.9 Examples of pressure symbols used in P&ID.

Other symbols are given in Figure 17.12, with six instruments shown: counting devices using a light source and detector, conveyer thickness measuring instrument, weight measurement, vibration, heat exchanger, and speed sensor. In loop 54, the QQS represents quantity, totalize, switch, or totalize or count number of switch operations.

Math functions can be performed digitally in PLCs using software. However, these functions were performed using hardware or analog devices (e.g., use of a square root to convert a pressure measurement to flow data). These functions have been symbolized. Some examples of these math symbols are shown in Figure 17.13: root, multiplication, division, derivative function, and subtraction.

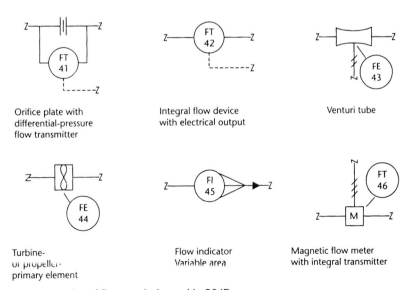

Figure 17.10 Examples of flow symbols used in P&ID.

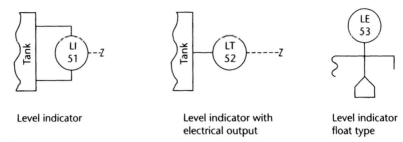

Figure 17.11 Examples of level symbols used in P&ID.

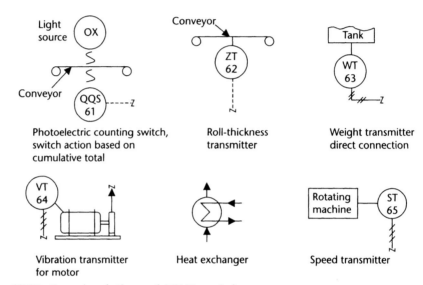

Figure 17.12 Examples of other useful P&ID symbols.

17.5 P&ID Drawings

All processing facilities will have a complete set of drawings using the standardized ISA symbols. These are the P&IDs or engineering flow diagrams that were developed for the detailed design of the processing plant. The diagrams show complete details and locations of all the required plumbing, instruments, signal lines, control loops, control systems, and equipment in the facility. The drawings normally consist of one or more main drawings depicting the facility on a function basis, along with support drawings showing details of the individual functions. In a large processing plant, these could run into many tens of drawings. Each drawing should be numbered, have a parts list, and have an area for revisions, notes, and approval signatures [8]. The process flow diagrams and plant control requirements are generated by a team from process engineering and control engineering. Process engineering normally has the responsibility for approving changes to the P&ID. These engineering drawings must be correct, current, and rigorously maintained. A few minutes taken to update a drawing can save many hours at a later date, trying to figure out a problem on equipment that has been modified, but whose drawings have not been updated. Every P&ID change must be approved and recorded. If not, time

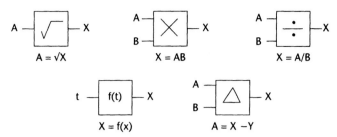

Figure 17.13 Examples of Math symbols used in P&ID.

is lost in maintenance, repair, and modifications. Using obsolete drawings can result in catastrophic errors.

P&IDs typically show the following types of information:

1. All plant equipment and vessels, showing location, capacity, pressure, liquid level operating range, usage, and so forth;
2. All interconnection signal lines, distinguishing between the types of interconnection (e.g., gas or electrical), and the operating range of the signal in the line;
3. All motors, giving voltage, power, and other relevant information;
4. All instrumentation, showing location of instrument, its major function, process control loop number, and range;
5. All control valves, giving type of control, type of valve, type of valve action, fail-save features, and flow and pressure information;
6. All safety valves and pressure regulators, giving temperature and operating ranges;
7. All sensing devices, recorders, and transmitters, with control loop numbers.

Figure 17.14 shows an example of a function block. The interconnection lines and instruments are clearly marked, and control loops are numbered.

Figure 17.15 shows the typical information that appears on each sheet of the P&ID. The information should contain a parts list with an area for notes, a sign-off sheet for revision changes, and the diagram name and original drafter, with approval signatures.

17.6 Summary

This chapter introduced the documentation for alarm and trip systems, PLCs, and P&IDs, and the standards developed for the symbols used in PID drawings. Alarm and trip systems were discussed. Alarm systems bring malfunctions to the attention of operators and maintenance personnel, whereas trip systems shutdown a system in an orderly fashion, if necessary. Such systems trip to a safe mode with loss of power, and are designed for high reliability using reliable components, redundancy, and regular testing. Alarm and trip documentation covers safety requirement specifications, a full system description, actions to be taken if the SIS shuts down, test equipment, test procedures, recordings of failures, and test results.

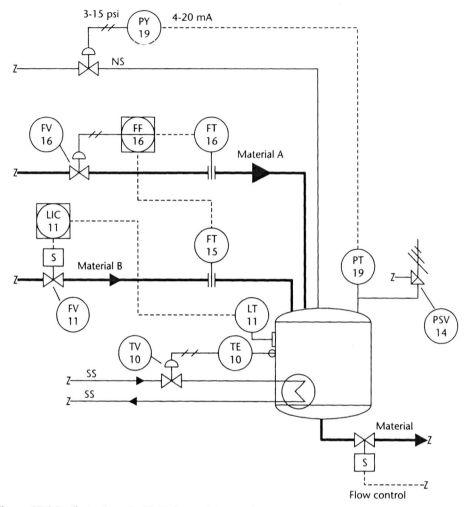

Figure 17.14 Illustration of a P&ID for a mixing station.

PLC documentation needs to be up-to-date, in order to have an accurate record of the programming used in the control of the process. Updates and changes are virtually impossible without accurate records.

The plumbing and signal lines in a process facility are shown in the P&ID. Standard symbols developed by the ISA for all of the various instruments, types of interconnections, actuator valves, and instrument functions are shown, together with an example of a facility P&ID.

References

[1] Mulley, R. M., *Control System Documentation*, Research Triangle Park, NC: ISA, 1998.
[2] Battikha, N. E., *The Condensed Handbook of Measurement and Control*, 2nd ed., ISA, 2004, pp. 241–263.
[3] Jones, C. T., *Programmable Logic Controllers*, 1st ed., Patrick-Turner Publishing Co., 1996, pp. 344–351.

17.6 Summary

Materials list			
Tag #	Manufacturer	Model	Part #
XV-213	Galant IND.	425 CV	66742-07

Notes

Revisions					
Rev #	Date	Description	By	Chk	Aprv

Industrial fuel processing Mixing station			
Dwn #		Sheet #	
Dwn by		Date	
Chq by		Date	
Aprv by		Date	

Figure 17.15 Typical information that appears on each sheet of the P&ID, such as drawing parts lists, notes, revisions, and sign-offs.

[4] Dunning, G., *Introduction to Programmable Logic Controllers*, 2nd ed., Delmar, 2002, pp. 387–397.

[5] Battikha, N. E., *The Condensed Handbook of Measurement and Control*, 2nd ed., ISA, 2004, pp. 9–18.

[6] Johnson, C. D., *Process Control Instrumentation Technology*, 7th ed., Prentice Hall, 2003, pp. 619–625.

[7] Stuko, A., and J. D. Faulk, *Industrial Instrumentation*, 1st ed., Delmar Publishers, 1996, pp. 305–308.

[8] Johnson, C. D., *Process Control Instrumentation Technology*, 7th ed., Prentice Hall, 2003, pp. 33–36.

Glossary

Absolute position measurement   Position measured from a fixed point.

Absolute pressure   Pressure measured with reference to a perfect vacuum.

Accelerometer   A sensor for measuring acceleration or the rate of change of velocity.

Accuracy   A measure of the difference between the indicated value and the true value.

Actuator   A device that performs an action on one of the input variables of a process according to a signal received from the controller.

Analog to digital converter   A device that converts an analog voltage or current into a digital signal.

Aneroid barometer   A barometer that uses an evacuated capsule as a sensing element.

Anticipatory action   See derivative action.

Bellows   A pressure sensor that converts pressure into linear displacement.

Bernoulli equation   A flow equation based on the conservation of energy, which includes velocity, pressure, and elevation terms.

Beta ratio   The ratio of the diameter of a restriction to the diameter of the pipe containing the restriction.

Bimetallic   A thermometer with a sensing element made of two dissimilar metals with different thermal coefficients of expansion.

Bode plot   A plot of a transfer function (gain in decibels and phase in degrees) versus frequency (log scale).

Bourdon tube   A pressure sensor that converts pressure to movement. The device is a coiled metallic tube that straightens when pressure is applied.

British Thermal Unit   A measure of heat energy, which is defined as the amount of heat required to raise 1 lb of water 1°F at 68°F and 1 atm.

Buffer amplifier   A circuit for matching the output impedance of one circuit to the input impedance of another.

Buoyancy   The upward force on an object floating or immersed in a fluid caused by the difference in pressure above and below the object.

Cascade control   A control system composed of two loops, where the set point of the inner loop is the output of the outer loop.

Calorie   A measure of heat energy, which is defined as the amount of heat required to raise the temperature of 1g of water 1°C.

Coefficient of heat transfer   A term used in the calculation of heat transfer by convection.

Coefficient of thermal expansion   A term used to determine the amount of linear expansion due to heating or cooling.

Comparator   A device that compares two signals and outputs the difference.

Concentric plate   A plate with a hole located at its center (orifice plate) that is used to measure flow by measuring the differential pressures on either side of the plate.

Continuity equation   A flow equation which states that, if the overall flow rate is not changing with time, then the flow rate past any section of the system must be constant.

Continuous level measurement   A level measurement that is continuously updated.

Controlled variable   The variable measured to indicate the condition of the process output.

Controller   The element in a process control loop that evaluates any error of the measured variable and initiates corrective action by changing the manipulated variable.

Converter   A device that changes the format of a signal (e.g., voltage to current) but not the type of energy used as the signal carrier.

Correction signal   The signal to the manipulated variable.

Digital to analog converter   A device that converts a digital signal into an analog voltage or current.

Data acquisition system   The Interface between many log signals and the computer.

Direct digital control   Action in which a computer performs all the functions of error detection and controller action.

Dead weight tester   A device for calibrating pressure-measuring devices that uses weights to provide the forces.

Derivative action   Action that is proportional to the rate at which the measured variable is changing.

Dew point   The temperature at which the water vapor in a mixture of water vapor and gas becomes saturated and condensation starts.

Dielectric constant   The factor by which the capacitance between two plates changes when a material fills the space between the plates.

DP cell   A differential pressure sensor.

Dry-bulb temperature   The temperature indicated by a thermometer whose sensing element is dry.

Dynamic pressure   The part of the total pressure in a moving fluid caused by the fluid motion.

Dynamometer   An instrument used for measuring torque or power.

Eccentric plate   An orifice plate with a hole located below its center to allow for the passage of suspended solids.

Electromagnetic flow meter   A flow-measuring device that senses a voltage between two electrodes induced by a magnetic field when a conductive fluid is flowing.

Error signal   The difference in value between a measured signal and a set point.

Feedback   (1) The voltage fed from the output of an amplifier to the input, in order to control the characteristics of the amplifier. (2) The measured variable signal fed to the controller in a closed-loop system, so that the controller can adjust the manipulated variable to keep the measured variable within set limits.

Flow nozzle   A device placed in a flow line to provide a pressure drop that can be related to flow rate.

Flume   An open-channel flow-measuring device.

Form drag   The force acting on an object due to the impact of fluid.

Foundation Fieldbus   Process control bus used in the United States.

Free surface   The surface of the liquid in an open-channel flow that is in contact with the atmosphere.

Gauge pressure   The measured pressure above atmospheric pressure.

Gas thermometer   A temperature sensor that converts temperature to pressure in a constant volume system.

Hall effect sensor   A transducer that converts a changing magnetic field into a proportional voltage.

Head   Sometimes used to indicate pressure (e.g., 1 ft of "head" for water is the pressure under a column of water 1 ft high).

Hot-wire anemometry   A velocity-measuring device for gas or liquid flow that senses temperature changes due to the cooling effect of gas or liquid moving over a hot element.

Humidity ratio   The mass of water vapor in a gas divided by the mass of dry gas in the mixture.

Hygroscopic   A material that absorbs water, and whose conductivity changes with moisture content.

Hysteresis   The nonreproducibility in an instrument caused by approaching a measurement from opposite directions (e.g., going from low up to the value, or high down to the value).

Impact pressure   The sum of the static and dynamic pressure in a moving fluid.

Impedance   An opposition to ac current or electron flow caused by inductance, and/or capacitance.

Incremental position measurement   An incremental position measurement from one point to another, in which absolute position is not recorded and the position is lost if the power fails.

Indirect level-measuring device   A device that extrapolates the level from the measurement of another variable (e.g., liquid level from a pressure measurement).

Integral action   The action designed to correct for long term loads.

Ionization gauge   A low pressure sensor ($<10^{-3}$ atm) based on the conduction of electric current through ionized gas whose pressure is being measured.

Kirchoff's Current Law   Law that states that the sum of the currents flowing at a node is zero.

Kirchhoff's Voltage Law   Law that states that the algebraic sum of voltages around a closed path is zero.

Ladder logic   The programmable logic used in PLCs to control automated industrial processes.

Lag time   The time required for a control system to return a measured variable to its set point.

Laser   Light amplification by simulated emission of radiation, which is characterized by monochromaticity and high collimation.

Linearity   A measure of the direct proportionality between the actual value of the variable being measured and the value of the output of the instrument to a straight line.

Load   The process load is a term used to denote the nominal values of all variables in a process that affect the controlled variable.

Load cell   A device for measuring force.

Linear variable differential transformer   A device that measures displacement by conversion to a linearly proportional voltage.

Magnetorestrictive element   A magnetic field sensor that converts a changing magnetic field into a proportional resistance.

Manipulated variable   The variable controlled by an actuator to correct for changes in the measured variable.

Multiplexer   An analog or digital device to select one of many input signals.

Newtonian fluid   A fluid in which the velocity varies linearly across the flow section between parallel plates.

Nutating disk meter   A flow-measuring device using a disk that rotates and wobbles in response to the flow.

Offset   The nonzero output of a circuit when the input is zero.

Open-channel flow   The flow in an open conduit (e.g., as in a ditch).

Orifice plate   A plate containing a hole that, when placed in a pipe, causes a pressure drop, which can be related to flow rate.

Overpressure   The maximum amount of pressure a gauge can withstand without damage or loss of accuracy.

Overshoot   The overcorrection of the measured variable in a control loop.

Parabolic velocity distribution   Occurs in laminar flow when the velocity across the cross section takes on the shape of a parabola.

pH   A term used to indicate the activity of the hydrogen ions in a solution, it helps to describe the acidity or alkalinity of the solution.

Phons   A unit for describing the difference in loudness levels.

Photodiode   A sensor used to measure light intensity by measuring the leakage across a pn junction.

PID   Proportional control with integral and derivative action, where the feedback signal is directly proportional to the error signal, but the error is further reduced and has a faster response time due to integral, and derivative action.

P&ID   Stands for piping and instrument diagrams, representing process and material flow, and signal flow in process control.

Piezoelectric effect   The electrical voltage developed across certain crystalline materials when a force or pressure is applied to the material.

Piezoresistive effect   is a change in the resistance of a semiconductor element when it is under strain.

Pirani gauge   A gauge used primarily for low pressures (< 1 atm) based on the resistance of a heated wire whose temperature is a function of the pressure.

Pitot-static tube   A device used to measure the flow rate using the difference between dynamic and static pressures.

PLC   Programmable logic controller. A microprocessor based system that is easily programmed for the control of industrial processes

Pneumatic system   is a system that uses gas for control and signal transmission.

Poise   The measurement unit of dynamic or absolute viscosity.

Process   A sequence of operations carried out to achieve a desired end result.

Process control   The automatic control of certain process variables to hold them within given limits.

Processor   A digital electronic computing system that can be used as a control system.

Profibus   Process control bus used in Europe.

Proportional action   A controller action in which the controller output is directly proportional to the measured variable error.

Psychrometric chart   A chart dealing with moisture content in the atmosphere.

Pyrometer   An instrument for measuring temperature by sensing the radiant energy from a hot body.

Quarter amplitude   A process control tuning criterion for stability, where, after a transient change, the oscillations of the control signal decrease by one-quarter of its amplitude with each cycle.

Radiation   The emission of energy from a body in the form of electromagnetic waves.

Rate action   See Derivative action.

Relative humidity   The amount of water vapor present in a given volume of a gas, expressed as a percentage of the amount that would be present in the same volume of gas under saturated conditions at the same pressure and temperature.

Reset action   See Integral action.

Resistance thermometer   A sensor that provides temperature readings by measuring the resistance of a metal wire (usually platinum).

Resolution   The minimum detectable change of a variable in a measurement.

Reynolds number   A dimensionless number indicating whether the flow is laminar or turbulent.

Rotameter   A flow-measuring device in which a float moves in a vertical tapered tube.

Saturated   The condition when the maximum amount of a material is dissolved in another material at the given pressure and temperature conditions (e.g., water vapor in a gas).

Sealing fluid   An inert fluid used in a manometer that is used to separate the fluid whose pressure is being measured from the manometer fluid.

Segmented plate   An orifice plate with a hole that allows suspended solids to pass through.

Sensitivity   The ratio of the change in output to input magnitudes.

Serial transmission   A sequential transmission of digital bits.

Set point   The reference value for a controlled variable in a process control loop.

Signal conditioning   The linearization of a signal, temperature correction, and the setting of the signal's reference and sensitivity to set levels.

Sling psychrometer   A device for measuring relative humidity.

Smart sensor   The integration of a processor directly into the sensor assembly, which gives direct control of the actuator and digital communication to a central controller.

Sone   A unit for measuring loudness.

Sound pressure level   The difference between the maximum air pressure at a point and the average air pressure at that point.

Span   The difference between the lowest and highest reading for an instrument.

Specific gravity   The ratio of the specific weight of a solid or liquid material and the specific weight of water; or for a gas, the ratio of the specific weight of the gas and the specific weight of air under the same conditions.

Specific heat   The amount of heat required to raise a definite amount of a substance by one degree (e.g., 1 lb by 1°F, or 1g by 1°C).

Specific humidity   The mass of water vapor in a mixture divided by the mass of dry air or gas in the mixture.

Specific weight   The weight of a unit volume of a material.

Stoke   The measurement unit of kinematic viscosity.

Strain gauge   A sensor that converts information about the deformation of solid objects when the objects are acted upon by a force into a change of resistance.

Sublimation   The process of passing directly from solid to vapor, or from vapor to solid.

Telemetry   The electrical transmission of information over long distances, usually by radio frequencies.

Thermal time constant   The time required for a body to heat or cool by 63.2% of the difference between the initial temperature and the aiming temperature.

Thermocouple   A temperature-sensing device that uses dissimilar metal junctions to generate a voltage that is proportional to the differential temperature between the metal junctions.

Thermopile   A number of thermocouples connected in series.

Time constant (electrical)   The amount of time needed for a capacitance C, to discharge or charge through a resistance R, to 62.3% of the difference between the initial voltage and the aiming voltage. The product of RC gives the time constant in seconds.

Torque   The force moment that tends to create a twisting action.

Torr   The pressure caused by the weight of a column of mercury 1 mm high.

Transducer   A device that changes energy from one form to another.

Transfer function   An equation that describes the relationship between the input and output of the function.

Transmission   The transferring of information from one point to another.

Transmitter   A device that conditions the signal received from a transducer so that it is suitable for sending to another location with minimal loss of information.

Turbine flow meter   A flow-measuring device utilizing a turbine wheel.

Ultrasonic probe   An instrument using high frequency sound waves to measure fluid levels.

Vapor pressure thermometer   A temperature sensor, in which the pressure of a vapor in a closed system is a function of the temperature.

Vena contracta   The narrowing down of the fluid flow stream as it passes through an obstruction.

Venturi tube   A specially shaped restriction in a section of pipe that provides a pressure drop, which can be related to flow rate.

Viscosity   The resistance to flow of a fluid.

Vortex   The swirling or rotating motion of fluids when in motion.

Weir   An open-channel flow-measuring device.

Wet-bulb temperature   The temperature indicated by a thermometer whose sensing element is kept moist.

Wheatstone Bridge   The most common electrical bridge circuit used to measure small changes in the value of an element.

Zeigler-Nichols method   A method of determining the optimum controller settings when tuning a process control loop for stability.

About the Author

William Dunn has a B.Sc. in physics from the University of London, graduating with honors; he also has a B.S.E.E. He has over 40 years of industrial experience in management, marketing support, customer interfacing, and advanced product development in systems, microelectronics, and micromachined sensor development. Most recently he taught industrial instrumentation and digital logic at Ouachita Technical College as an adjunct professor. Previously he was with Motorola Semiconductor Product Sector working in advanced product development, designing micromachined sensors and transducers. He holds 30 patents in electronics and sensor design and has presented 20 technical papers in sensor design and application.

Index

A

Absolute accuracy, 6, 13
Absolute humidity, 194
Absolute position, 171, 190
Absolute pressure, 100, 114
Absolute zero, 149, 169
Acidity, 206–207
Acceleration, 175, 191
Accelerometer, 92–94, 175
Accuracy, 6–7, 12
   absolute, 6, 13
   percentage full-scale, 6
Actuator, 13, 220–221
Air Supply, 11
Alarm and trip documentation, 299–300
Alarm and trip systems, 297–300
Alkalinity, 206
American National Standards Institute (ANSI), 5, 11, 297
Amplification, 42–44
Amplifier, 44–57
   applications, 57
   bandwidth, 44
   buffer, 48
   common mode rejection, 49
   converters, 50–52
   current, 52–53
   differential, 53–54
   instrument, 55–56
   integral, 53
   nonlinear, 54–55
   operational, 41–45
   protection, 57
   summing, 46
   voltage, 45–50
Amplitude modulation, 267
Analog circuits, 41–57
Analog data, 10, 262–264
Analog-to-digital converter (ADC), 68–72
Anemometers, 144
Angular motion, 177, 191

Anticipatory action. See Derivative action, 278
Atmospheric pressure (atm), 102, 114
Automation, 1, 12

B

Ball valve, 218
Barometer, 109
Base units, 16–18
Bellows, 108
Bernoulli equation, 132–134, 148
Beta ratio, 138–139
Bimetalic, 157–158
Bourdon tube, 108–109
Bridge circuits, 34–39
   ac, 38–39
   current balanced, 37
   lead compensation, 36
   Wheatstone, 35–39
British Thermal Unit (Btu), 151, 169
Bubbler systems, 122
Buoyancy, 103
Butterfly valve, 217–218

C

Calibration flow-measuring devices, 147
   pH-measuring devices, 207
   pressure-sensing devices, 112–113
   temperature-measuring, 168
Calorie, 151, 169
Capacitive devices, 172–173
Capillary tube, 204
Capsule, 107
Celsius scale, 149, 169
Centimeter-gram-second (CGS), 15
Chemical sensors, 208–209
Closed-loop feedback, 4–5, 276, 282–283
Coefficient of heat transfer, 153
Coefficient of thermal expansion, 155–156
Common mode rejection, 49

Comparator, 62–63
Concentric plate, 137
Condensation, 197
Conductivity thermal (k), 153–154
Contactors, 221–222, 227
Continuous process control, 2, 275–287
   cascade control, 282
   derivative action, 278–280
   feed-forward control, 283–284
   integral action, 280–281
   PID action, 281–282
   proportional control, 276–278
   ratio control, 283
   stability, 284–285
   tuning, 285–287
Continuity equation, 131–132
Controlled variable, see Manipulated Variable,
Controller, 3–5, 273–275, 281–284, 287–290
   control loop, 4–5, 282–284
   on/off action, 3, 273–275, 287–288
   PID action, 281–282
   pneumatic, 289–290
   programmable logic controller (PLC), 233–248
Control loop implementation, 287–294
   derivative mode, 291–292
   integral mode, 292
   pneumatic, 287–290
   proportional mode, 291
   proportional derivative mode, 292
   proportional integral mode, 293
   PID mode, 293–294
Control parameter range, 296
Control System, 9
   evaluation, 9
   regulation, 9
   stability, 9, 284–285
   transient response, 9
Convection, 154
   forced, 154
   natural (free), 154
Conversion between units, 18–20
Converter, 13, 50–52, 61, 269
   analog, 50–52
   analog-to-digital, (ADC), 68–72
   digital-to-analog (DAC), 64–68
   signal, 13, 269
   voltage-to-frequency, 72–73
Coulomb (C), 17
Couple, 178, 191
Crystals, 85

D

Dall tube, 138–139
Data acquisition devices, 74–75
   demultiplexers, 74
   multiplexers, 74
   programmable logic arrays, 75
Dead band, 295
Dead time, 276, 295
Dead-weight tester, 112
Delayed on/off action 273–274
Density (ñ), See also Specific gravity, 113, 198–202
Density measuring devices, 199–201
   differential bubblers, 200
   gas density, 201
   hydrometers, 199
   induction hydrometer, 199
   pressure, 199
   radiation density sensors, 201
   vibration sensors, 199
Derivative action, 278–280, 291–292
Dew point, 197
Diaphragm, 106
Dielectric, 121, 172–173
Dielectric constant, 121, 172–173
Differential pressure, 3–4, 102, 114
Digital building blocks, 59–60
Digital communication, 264–266
   standards for, 264–265
   foundation fieldbus, 265–266
Digital data, 10, 264–266
Digital-to-analog converter (DAC), 64–68
Direct reading level sensors, 115–118
Discontinuous on/off action, 273
Displacers, 119–120, 127
Distance measuring devices, 171–175
Distributed systems, 95–96
Documentation, 297–310
Doppler effect, 174
Drag viscometer, 203
Drift, 6, 8, 13
Dry particulate flow rate, 144–145
Dual slope converters, 70–71
Dynamic pressure, 113, 131
Dynamometers, 185–186

E

Eccentric plate, 137
Elbow, 139
Electrical supply, 11
Electrical units, 17

Electromagnetic interference (EMI), 11, 262
Electromagnetic radiation, 186–187
Energy, 133
   kinetic, 133
   potential, 133
English system of units, 15–20
Error signal, 4, 14, 276–282, 295
Execution mode, 236
Expansion, 155–156
   linear, 155–156
   volumetric, 156–157

F

Facility requirements, 11–12
Fahrenheit scale, 149, 169
Falling-cylinder viscometer, 7
Feedback, 2–5, 9, 13, 275, 282–284
   cascade control, 282
   feed-forward control, 283–284
   ratio control, 283
Filters, 33–33, 94
Fitting losses, 135
Flash converters, 68–69
Floats, 116–117, 126
Flow, 129–147, 148
   Bernoulli equation, 132–134, 148
   Beta ratio, 138–139
   continuity equation, 131
   dry particulate, 144–145
   laminar, 129–130, 148
   losses, 134–136
   mass, 144, 148
   open-channel, 145
   Reynolds number, 129–130, 148
   total flow, 142–143, 148
   turbulent, 129–130, 148
   velocity, 130, 148
Flow control valves, see valves, 215–220
Flow sensors, 136–145
   anemometer, 144
   coriolis, 144
   Dall, 136, 138–139
   elbow, 139
   electromagnetic flow meter, 140–141
   flow nozzle, 136, 138
   moving vane, 140
   nutating disc, 143
   open-channel, 145
   orifice plate, 136–137
   paddle wheel, 145
   pilot static tube, 139
   piston, 143
   pressure meter, 140
   rotameter, 139–140
   turbine, 140
   ultrasonic, 141–142
   velocity meters, 144
   Venturi tube, 138, 139–140
   vortex flow meter, 140
Flow patterns, 129–130
Flow rate, 130–131, 148
Force, 177–186, 191
Force sensors, 81–82, 87–88, 181–185
Foundation fieldbus, 265–266
Frequency, 186–187
Frequency modulation (FM), 268–269
Frictional losses in liquid flow, 134–136

G

Gas thermometers, 159–160
Gauge factor, 180–181
Gauge pressure, 102–103, 114
Gaseous phase, 151
Globe valve, 215–217
Grounding, 11, 262

H

Hair hygrometer, 195
Hall effect, 82–83, 174
Head, 100
Heat, 151–155, 169
   British thermal unit, 151, 169
   calorie, 151, 169
   joule, 151, 169
   phase change, 151
   specific heat, 152–153
   thermal energy, 151–152
Heating ventilating and air conditioning (HVAC), 1, 3, 273–274
Heat transfer, 153–155
   conduction, 153–154
   convection, 154–155
   radiation, 155
Hexadecimal, 247
Hot-wire anemometry, 144
Hooke's law, 179–180
Humidity, 193–198, 209–210
   absolute, 194
   dew point, 197, 210
   hygrometers, 195–197
   moisture content, 197
   psychrometric chart, 194, 210
   relative, 193, 209

specific, 194, 210
Humidity ratio, 194
Humidity sensors, 194–197
    capacitive hygrometer, 196
    dew point, 197
    electrolytic hygrometer, 196
    hair hygrometer, 195
    infrared absorption, 197
    laminate hygrometer, 195
    microwave absorption, 197
    piezoelectric/sorption hygrometer, 196–197
    psychrometers, 194–195
    resistance capacitance hygrometer, 196
    resistive hygrometer, 195–196
    sling psychrometer, 195
Hydrometer, 199
Hydrostatic paradox, 100
Hydrostatic pressure, 99–100
Hygrometer, 195–197
Hygroscopic, 195
Hysteresis, 6, 7–8, 13, 63

I

Impact pressure, 113, 131, 139
Impedance (Z), 29
Indirect level-measuring devices, 118–124
Incremental position, 171, 190
Infrared waves, 186, 197
Input-output scan mode, 236
Installation and maintenance, 12
Institute of Electrical and Electronic Engineers (IEEE), 265
Institutions, 22
Instrument amplifiers, 55–56
Instrument Systems and Automation Society (ISA), 5, 11, 230, 297
Instruments, 5, 302
Integral action, 280–281, 292
Integrator, 53, 292
Ionization gauge, 110

J

Joules, 151, 169

K

Kelvin scale, 149, 169

L

Ladder diagrams, 243–248
Ladder logic, 245–248

Lag time, 276, 295
Laminar flow, 129–130, 148
Lasers, 188
Law of intermediate metals, 163
Lead compensation, 36
Level, 115–127
    direct measuring devices, 115–118
    indirect measuring devices, 118–124
    single point measurement, 124–125
Level regulators, 214
Level sensors, 115–126
    beam breaking, 125
    bubblers, 122, 127
    capacitive probes, 121–122, 127
    conductive probes, 124
    displacer, 119–120, 127
    float, 116–117, 126
    load cells, 123–124
    paddle wheel, 125
    pressure devices, 118–119, 127
    radiation, 125, 127
    resistive tape, 123
    sight glass, 115–116
    thermal probes, 125
    ultrasonic devices, 117–118, 127
Light intensity, 187–188
Light intensity sensors, 80–81, 188
Light Interference lasers, 173
Light emitting diodes, 189
Light measuring devices, 80–81, 188
Light sources, 188–189
Light to frequency converters, 80–81
Linear variable differential transformer (LVDT), 107, 171–172, 176
Linearity, 6, 13, 253–254
Linearization, 253
Liquid filled thermometers, 157
Liquid phase, 151
Litmus paper, 207
Load cells, 123, 183–184
Local area network (LAN), 233, 265
Logarithmic amplifiers, 54–55, 253
Logic gates, 59–60, 246
Loudness, 205

M

Magnetic control devices, 227
Magnetic field sensors, 82–84, 174, 256–257
    Hall effect device, 82–83
    Magneto-resistive element (MRE), 83
    Magneto-transistor, 84
Manipulated variables, 3–5, 13, 275, 295

Manometer, 105
Mass, 12, 191
Mass flow, 144, 148
Math function, 307
McLeod gage, 111
Measured variables, 3–7, 275, 295
Mercury thermometers, 157
Metric units, 20–21,
Micromechanical devices, 88–94
   bulk micromachining, 89–91
   surface micromachining, 91–94
   accelerometer, 92–94
   filters, 94
   gas flow sensors, 90–91
   pressure sensors, 89–90
Microphones, 205
Microwaves, 173–174, 186, 197
Moisture content, 197
Modulus of elasticity, 180
Motion, 171–177, 190–191
   acceleration, 175, 191
   angular, 171
   linear, 190
   measuring devices, 171–176
   velocity, 171, 174, 191
   vibration, 175, 191
Motors, 227–230
   position feedback, 228
   servo, 228
   stepper, 228–229
   synchronous, 229–230
Multiplexers, 74

N

National Institute of Standards and
   Technology (NIST), 6, 113
Neutron reflection, 197
Newtonian fluids, 202
Noise, 10, 255, 262
Nonlinear amplifiers, 54–55, 253
Nutating disk meter, 143

O

Offset, 6, 8, 13, 42–43, 252, 269, 296
Offset control, 43
On/off control action, 2–3, 273–275, 287–288
Operational amplifier (op-amp), 41–45
Optical devices, 174–175
Opto-couplers, 222
Orifice plate, 136–137
   concentric, 137

   eccentric, 137
Oscillation, 284
Oscillator, 86–87
Over pressure, 112
Overshoot, 276, 284

P

Paddle wheel, 145
Parallel transmission, 264–265
Parshall flume, 145–146
Pascal (Pa), 99
Pascal's Law, 104
Passive components, 25
   step input, 25
   sine wave input, 28–32
   time constants, 27–28
Peltier effect, 163
Percent reading, 6, 12
Percent span, 6
Percentage full-scale accuracy, 6
pH, 206–208
pH meters, 207
Phase change, 25–31
Phons, 204
Photocells, 188
Photoconductive devices, 188
Photodiodes, 189–190
Photoemissive materials, 188
Photosensors, 188
Physical variables, 13,
Piezoelectric devices, 84–85, 183
Piezoelectric sensors, 87
Piezoresistors, 90
Pilot-static tube, 139
Pipe and instrumentation diagrams (P&ID),
   300–310
Pipe and instrumentation drawings, 308–310
Pipe and instrumentation documentation,
   308–309
Pipe and instrumentation symbols, 300–308
Pirani gage, 111
Piston flow meter, 142–143
Pneumatic data, 10, 261–262
Pneumatic transmission, 261–262
Poise, 130, 202
Position, 171–174
   absolute measurement, 177, 190
   angular, 177, 191
   incremental measurement, 177 191
   measuring devices, 171–176
Positive displacement meters, 142–143
Potentiometers (pots), 171

Power control, 221–227
Power dissipation, 225
Power switches (electrical), 222–227, 231
  BJT, 226
  DIAC, 224
  IGBT, 226
  MCT, 226
  MOSFET, 226
  SCR, 222–224, 226
  TRIAC, 224–225
Precision, 8, 13
Prefix standards, 21–22
Pressure (P), 99–113, 114
  absolute, 102, 114
  atmospheric, 102, 114
  differential, 102, 114
  dynamic, 113, 130–131
  gauge, 102, 114
  head, 100
  hydrostatic, 99
  impact, 113, 130–131, 139
  static, 113, 130–131, 139
  vacuum, 101–102, 114
Pressure regulators, 211–213
  pilot operated, 212
  pressure, 212
  spring, 211
  weight, 212
Pressure sensors, 105–111
  barometers, 109
  bellows, 108
  bourdon tube, 108–109
  capacitive, 106
  capsules, 107–108
  diaphragm, 106–107
  ionization gauge, 110
  manometer, 105–106
  McLeod, 111
  piezoresistive, 109–110
  Pirani, 111
  selection, 111–112
  solid state, 89–90, 107
Probe capacitance, 121
  conductivity, 124
  thermal, 125
  ultrasonic, 125
Process control, 1–5, 271–294
  automatic tuning, 286
  elements, 4,
  implementation, 287–294
  manual tuning, 286–287
Process facility, 11–12

air supply, 11
electrical supply, 11
grounding, 11, 262
installation and maintenance, 12
water supply, 11
Processor, 75–76
Profibus, 265–266
Programmable logic arrays (PLA), 75
Programmable logic controller (PLC), 233–248
  documentation, 300
  input modules, 236–239
  ladder diagrams, 243–248
  operation, 235–236
  output modules, 239–240
  smart modules, 240–242
Proportional action, 276–278, 290–291
Proportional and derivative (PD) action,
    278–279, 282
Proportional and integral (PI) action, 280–281,
    293
Proportional with derivative and integral (PID)
    action, 281–282, 289–290, 293–294
Psychrometer, 194–195
  sling, 195
Psychrometric chart, 194
Pulse position modulation, 268
Pulse width modulation, 68, 268
Pyrometer, 164–165
PZT, 85, 88
  actuators, 88

Q
Quartz, 86

R
Radiation devices, 125, 201
Range, 8, 12
Rankine scale, 149, 169
Ramp converters, 69
Rate action. See Derivative action
Refractive index, 187
Regulation, 9, 211–213
Regulators, 221–214
Relative humidity, 193
Relays, 221–222, 227
Repeatability, 8, 13
Reproducibility, 7, 13
Reset action. See Integral action, 280
Resistance–capacitor filters, 32–33
Resistance temperature devices (RTD),
    160–161, 257–258

Index

Resistive hygrometer, 195–196
Resistor ladder network, 64
Resolution, 7, 13
Resonant frequency, 30–31
Reynolds number (R), 129–130, 148
Rotameters, 139–140
Rotary plug valve, 219
Rotating disc viscometers, 203–204
RS-232, 265

S

Safe failure of alarm and trip, 298–299
Safety, 11–12, 297–299
Safety Instrumented systems (SIS), 297–299
Safety valves, 213
Sample and hold, 72
Saybolt instrument, 204
Scan mode, 236
Scan time, 236
Seebeck effect, 162
Semiconductor micromachining, 89–94
  bulk, 89–91
  surface, 91–94
Sensitivity, 6, 7, 13
Sensors, 4, 5, 13, 80–84, 86–88, 105–111, 115–126, 136–145, 157–166, 171–176, 181–185, 194–197, 199–201, 203–204, 205–206, 207, 208–209, 269
  chemical, 208–209
  density, 199–201
  flow, 136–145
  force, 81–82, 87–88, 181–185
  humidity, 194–197
  level, 115–126
  light, 80–81, 188
  magnetic field, 82–84
  pH, 207
  position/motion, 171–176
  pressure, 105–111
  sound, 205–206
  temperature, 80, 157–166
  time measurements, 86–87
  vibration, 175–176
  viscosity, 203–204
Sequential process control, 2, 16, 271–272
Serial bus, 95, 266
Servo motors, 228
Set point, 3–5, 9, 13, 276, 295
Shear stress ($\hat{o}$), 179–180
Sight glass, 115–116
Signal, inversion, 3, 291

Signal conditioning, analog, 251–260
  capacitive sensors, 255–256
  direct reading sensors, 255
  linearization, 253
  LVDT, 259–260
  magnetic sensors, 256–257
  noise, 255
  offset and span, 252
  resistance temperature devices, 257–258
  semiconductor devices, 260
  temperature, 253–254
  thermocouple sensors, 259
Signal conditioning, digital, 260–261
Signal transmission, 261–269
  current, 264
  digital, 264–266
  telemetry, 267–269
  pneumatic, 261–262
  voltage, 262–264
  wireless, 267–269
Silicon diaphragms, 90, 107
S I units, 15–21
Sling psychrometer, 195
Slug, 15
Smart sensors, 10, 94–96, 265–266
Smoke detectors, 208
Solid phase, 151
Sound, 204–206
Sound measuring devices, 205–206
Sound pressure level (SPL), 204–205
Span, 8, 12, 269
Specific gravity (SG), 100, 113, 198–201
Specific heat, 152–153
Specific humidity. See Humidity ratio, 194
Specific weight (γ), 100, 113, 198–199
Spectrometer, 209
Spring transducer, 182
Stability, 9, 284–285
Standards, 22
  institutions, 22–23
Static pressure, 113, 131, 139
Stepper motors, 13
Still well, 146
Stoke, 130
Strain, 178–179
Strain gage, 81–82
Strain gauge sensors, 183–185
Stress, 178–179
Sublimation, 151, 169
Successive approximation, 69–70
Synchronous motors, 229–230

Systéme International D'Unités (SI) units, 15–21

T
Taguchi sensors, 208
Telemetry, 267–269
Temperature, 149–168, 169
    absolute, 149–150, 169
    ambient, 149–150
    Celcius, 149–150, 169
    Fahrenheit, 149–150, 169
    Kelvin, 149–150, 169
    phase change, 151
    protection, 168
    Rankine, 149–150, 169
Temperature sensors, 80, 157–166
    bimetallic, 157–159
    integrated, 80
    liquid expansion, 157
    mercury in glass, 157
    pressure spring, 159–160
    pyrometer, 164–165
    resistance temperature devices (RTD), 160–161
    solid state, 165–166
    thermistors, 161–162
    thermocouples, 162–164
Thermal
    conductivity, 153–154
    convection, 154–155
    energy, 151–152
    expansion, 155–157
    radiation, 155
    time constant, 167–168
Thermistors, 161–162
Thermocouple, 162–165, 259
    Peltier effect, 163
    Seebeck effect, 162
    Thompson effect, 163
Thermohydrometer, 199
Thermometer, 157, 159–160
Thermopile, 165
Thomson effect, 163
Time constant, 8, 27–28, 167–168
RC, 27–28
    sensor, 8
thermal, 167–168
Time measurements, 86–87
Time lag, 276, 295
Torque (moment), 177–178, 191
Torr, 101
Total flow, 142–144, 148

Transducers, 5, 7, 10, 13, 269
Transient, 290, 295
Transmitters, 13, 269
Transmission standards, digital, 264–265
Turbine flow meter, 140
Turbulent flow, 130–131, 148
Two-way globe valve, 13

U
Ultrasonic devices, 118, 125, 141–142, 173–174
Uninterruptible power supply (UPS), 11
Units and standards, 15–23
U-tube manometer, 105–106

V
Vacuum, 7, 114
Vacuum instruments, 7
Valves, 215–200, 230–231
    ball, 218
    butterfly, 217–218
    globe, 215–217
    plugs, 215–216
    rotary plug, 219
    fail safe, 219–220
    sizing, 219
    weir/diaphragm, 218
Vapor-pressure thermometer, 159
Variable signal, 3–7, 276
    manipulated, 3–5, 276
    measured, 3–7, 276
    range, 295
Velocity, 148, 171, 174, 191
Vena contracta, 137
Venturi tube, 137–138
Vibration, 175, 191
Vibration sensors, 175–176, 199
Viscosimeter (Viscometer), 203–204
Viscosity, 130, 148, 202–204
Viscosity measuring instruments, 203–204
    falling cylinder viscometer, 203
    rotating disc or drag viscometer, 203–204
    Saybolt viscometer, 204
Voltage dividers, 34–35
Voltage signals, 262–264
Voltage vectors, 29
Voltage to current converters, 4
Voltage to frequency converters, 72–73
Volume flow rate, 130
Vortex flow meter, 140

W

Water supply, 11
Wavelengths, 186–187
Weight, 15–16, 191
Weir valve, 218
Wheatstone bridge, 35–39
Wide area network (WAN), 233, 265
Width modulation, 268
Wireless transmission, 267–269

Y

Youngs modulus, 180

Z

Zero-voltages switches (ZVS), 225–226

Related Titles from Artech House

Chemical and Biochemical Sensing with Optical Fibers and Waveguides,
  Gilbert Boisdé and Alan Harmer

Fundamentals and Applications of Microfluidics, Nam-Trung Nguyen
  and Steve Werely

Introduction to Microelectromechanical Microwave Systems, Second Edition,
  Héctor J. De Los Santos

Introduction to Microelectromechanical Systems Engineering, Second Edition,
  Nadim Maluf and Kirt Williams

Measurement Systems and Sensors, Waldemar Nawrocki

MEMS Mechanical Sensors, Steve P. Beeby, Graham Ensel, and Neil M. White

Semiconductor Nanostructures for Optoelectronic Applications,
  Todd Steiner, editor

Sensor Technologies and Data Requirements for ITS, Lawrence A. Klein

Optical Fiber Sensors, Volume III: Components and Subsystems, Brian Culshaw
  and John Dakin, editors

Optical Fiber Sensors, Volume IV: Applications, Analysis, and Future Trends,
  Brian Culshaw and John Dakin, editors

*Signal Processing Fundamentals and Applications for Communications and Sensing
  Systems*, John Minkoff

Understanding Smart Sensors, Second Edition, Randy Frank

Wireless Sensor Networks: A Systems Perspective, Nirupama Bulusu and Sanjay Jha

For further information on these and other Artech House titles, including previously considered out-of-print books now available through our In-Print-Forever® (IPF®) program, contact:

Artech House
685 Canton Street
Norwood, MA 02062
Phone: 781-769-9750
Fax: 781-769-6334
e-mail: artech@artechhouse.com

Artech House
46 Gillingham Street
London SW1V 1AH UK
Phone: +44 (0)20 7596-8750
Fax: +44 (0)20 7630-0166
e-mail: artech-uk@artechhouse.com

Find us on the World Wide Web at: www.artechhouse.com